Khaled Zennir, Svetlin G. Georgiev

Distributional Nonlinear Wave Equations

De Gruyter Series in Nonlinear Analysis and Applications

Volume 45

Khaled Zennir, Svetlin G. Georgiev

Distributional Nonlinear Wave Equations

Well-Posedness and Stabilizability

DE GRUYTER

Mathematics Subject Classification 2020
Primary: 35L05, 35L20, 35L70; Secondary: 58G16, 93B05

Authors
Dr. Khaled Zennir
Department of Mathematics
Faculty of sciences
Qassim University
Buraydah 51921
Saudi Arabia
khaledzennir4@gmail.com

Dr. Svetlin G. Georgiev
Sorbonne University
1 Rue Victor Hugo
70005 Paris
France
svetlingeorgiev1@gmail.com

ISBN 978-3-11-163368-8
e-ISBN (PDF) 978-3-11-163378-7
e-ISBN (EPUB) 978-3-11-163403-6
ISSN 0941-813X

Library of Congress Control Number: 2024950855

Bibliographic information published by the Deutsche Nationalbibliothek
The Deutsche Nationalbibliothek lists this publication in the Deutsche Nationalbibliografie; detailed bibliographic data are available on the Internet at http://dnb.dnb.de.

Typesetting: VTeX UAB, Lithuania
Printing and binding: CPI books GmbH, Leck

www.degruyter.com
Questions about General Product Safety Regulation:
productsafety@degruyterbrill.com

Preface

In recent decades, a multilevel approach to describing the behavior of a structurally nonlinear wave system in fields of external/ internal influences has been intensively developed in bounded domain. However, for the unbounded, there have only been a few methods in literature until now. Partial differential equations in hyperbolic type are ubiquitous in mathematically oriented scientific fields such as physics and engineering. The nonlinear wave systems are foundational in the modern scientific understanding of sound, electrostatics, electrodynamics, thermodynamics, fluid dynamic elasticity, general relativity and quantum mechanics. They also arise from many purely mathematical considerations such as differential geometry and the calculus of variations in distributional spaces. This book is devoted on some recent investigations of some methods to be applied on classes of nonlinear wave models in Sobolev and functional spaces involving distributional structural damping terms.

In the present book, we discuss and develop some mainly different methods (arranged chronologically) to be used to analyze problems in $\mathbb{R}^n$ for some physical and mechanical systems governed by nonlinear wave equations. Among these properties, we mention such problems with compactly supported initial data, density and weighted functions, Fourier transform and some recent techniques.

Section snippets: The book contains eleven chapters introduced by an introductory description. The introduction is entirely devoted to the presentation of definitions and results necessary with description of some physical phenomena related with simple wave models. In particular, we introduce the different properties of the proposed methods of distributional solutions to certain times PDEs. Qualitative properties for the semilinear dissipative wave equations are discussed in Chapter 2. Chapter 3 is devoted to the study of the impact of viscoelastic term combined with sources on the stability of system. These two chapters are based on the solutions with compactly supported initial data. Chapter 4 aims to present results on the well-posedness, and behavior of solutions to a nonlinear viscoelastic wave equations in weighted spaces. Elements of theory of the Kirchhoff problem is introduced in Chapter 5. It is introduced at the same decay rate of second-order evolution equations with density. Chapter 6 is devoted to the original method for well-posedness and general decay for a wave equation with logarithmic nonlinearities. In Chapter 7, the uniform stabilization of the Petrovsky-Wave nonlinear coupled system with strong damping is investigated. The well-posedness in polynomial type and the lack of exponential stability are studied. The question of well-posedness and general energy decay of solutions for a system of three wave equations with a nonlinear strong dissipation are investigated in Chapter 8. From Chapter 4 through Chapter 8, we use the weighed function to compensate for the lack of Poincaré's inequality. Insofar as Chapter 9 and Chapter 10, we are concerned with damped nonlinear wave problems in Fourier spaces. The last Chapter 11 analyzes the existence/nonexistence of solutions for structural damped wave equations with nonlinear memory terms in $\mathbb{R}^n$.

https://doi.org/10.1515/9783111633787-201

This book addresses a wide audience of specialists such as mathematicians, physicists, engineers and biologists. It is prepared mainly for mathematics students interested in applications to quantum physics and electrical engineering. It will appeal to researchers, graduate math students and academic libraries interested in applications of mathematics to problems arising in physics, engineers, biology and natural sciences.

December 2024

Khaled Zennir (Riyadh)
Svetlin G. Georgiev (Paris)

Contents

1 Introduction

By $\mathbb{R}^+$, we mean the set

$$\{\tau \mid \tau \geq 0\}.$$

In the sequel, we omit the space variable x of $u(x,t)$, t is the time. For simplicity reasons, denote

$$u(x,t) = u, \quad \frac{\partial u}{\partial t}(x,t) = \frac{\partial u}{\partial t}, \quad \frac{\partial^2 u}{\partial t^2}(x,t) = \frac{\partial^2 u}{\partial t^2}$$

when no confusion arises. Also, the functions considered are all real valued. For

$$x = (x_1, x_2, \ldots, x_n) \in \mathbb{R}^n, \quad n \geq 1,$$

we denote by

$$|\nabla_x u|^2 = \sum_{i=1}^{n} \left(\frac{\partial u}{\partial x_i} \right)^2,$$

and

$$\Delta_x u = \sum_{i=1}^{n} \frac{\partial^2 u}{\partial x_i^2}.$$

The constants c and sometimes C used throughout this book is a positive generic constant, which may be different in various occurrences. Recall that the inner product (resp., norm) of H is denoted by $(\cdot,\cdot)_H$ (resp., $\|\cdot\|_H$). By

$$a \lesssim b,$$

we mean that there exists a constant $C > 0$ independent of a, b, such that

$$a \leq Cb.$$

We have

$$a \equiv b$$

means that both $a \lesssim b$ and $b \lesssim a$ hold. We will write throughout this book $f < g$ when there exists a nonnegative constant C such that $f \leq Cg$ and it is always supposed to be independent of $T > 0$. It is not the same in every step. For $1 \leq p \leq \infty$, we denote by $L^p(\mathbb{R}^n)$ the usual Lebesgue space on $\mathbb{R}^n$ with the norm $\|\cdot\|_{L^p}$. For a nonnegative integer m, H^m, $(\mathbb{R}^n)$ denotes the Sobolev space of $L^2(\mathbb{R}^n)$ functions on $\mathbb{R}^n$, equipped with the norm $\|\cdot\|_{H^m}$.

https://doi.org/10.1515/9783111633787-001

1.1 Wave equation models for some physical phenomena

Many wave motion problems in physics can be modeled by the standard linear wave equation, or a similar formulation with a system of first-order equations. There are some exceptions. Perhaps the most important are water waves. They are modeled by the Laplace equation with time-dependent boundary conditions at the water surface. Quantum mechanical waves constitute another example where the waves are governed by the Schrödinger equation, i. e., not by a standard wave equation. Many wave phenomena also need to take nonlinear effects into account when the wave amplitude is significant. Shock waves in the air is a primary example. This comprehensive section is prepared with meticulously organized content to provide a solid foundation for linear wave equation models while offering numerous examples that span the interdisciplinary of applied mathematics. With its wide applicability, this section aims to facilitate a deeper understanding of certain concepts of physical phenomena, practically in the following.

1.1.1 Waves on a string with fixed ends

The string can be modeled as a set of discrete point masses with elastic strings in between, and it has a large constant tension T. The displacement of mass m at mesh point $x \in \mathbb{R}$ is denoted by $u(t)$ at time t. With the constant mesh spacing $\Delta u = u_i - u_{i-1}$, we define the density $\rho = m/\Delta u$. The continuous model for waves on a string with a large constant tension T is then given by

$$\varrho \frac{\partial^2 u}{\partial t^2} = T \frac{\partial^2 u}{\partial x^2}. \tag{1.1}$$

The density ρ may change along the string, while the tension T is a constant. With variable wave velocity $c(x)\sqrt{T/\rho(x)}$, we can write the wave equation in the more standard form

$$\frac{\partial^2 u}{\partial t^2} = c^2(x) \frac{\partial^2 u}{\partial x^2}. \tag{1.2}$$

Because of the way ρ enters the equations, the variable wave velocity does not appear inside the derivatives as in many other versions of the wave equation. However, most strings of interest have constant ρ. The end points of a string are fixed so that the displacement u is zero. The boundary conditions are therefore $u = 0$.

Damping: The resistance (e. g., the air) and nonelastic effects in the string will contribute to reduce the amplitudes of the waves so that the motion dies out after some time. This damping effect can be modeled by a term $b\frac{\partial u}{\partial t}$ (named frictional term) on the left-hand side of the equation:

$$\varrho \frac{\partial^2 u}{\partial t^2} + b \frac{\partial u}{\partial t} = T \frac{\partial^2 u}{\partial x^2}. \tag{1.3}$$

The parameter $b \leq 0$ is small for most wave phenomena, but the damping effect may become significant in long time simulations. We assumed, in the derivation above, that it is linear, and it is easy to be considered nonlinear in a different type. These types of damping terms help to ensure the existence of the displacement u and their role lies in the qualitative properties of solutions, and their effect varies according to the type and nature of each term.

External forcing: It is easy to include an external force acting on the string (named source term). Say we have a vertical force f acting on mass m, modeling the effect of, for example, gravity on a string. This force affects the vertical component of Newton's law and gives rise to an extra term $f(x,t)$ on the right-hand side of (1.1). In model (1.2), we would add a term

$$f(x,t) = f(x,t)/\rho(x).$$

Equation (1.3) will be written as

$$\varrho\frac{\partial^2 u}{\partial t^2} + b\frac{\partial u}{\partial t} = T\frac{\partial^2 u}{\partial x^2} + f(x,t). \tag{1.4}$$

The additional term $f(x,t)$ situated in the right-hand side usually hinders and obstructs the solutions.

1.1.2 Elastic waves in a rod

When we consider an elastic rod subjected to a hammer impact at the end, it will give rise to an elastic deformation pulse that travels through the rod. A mathematical model for longitudinal waves along an elastic rod starts with the general equation with the density ρ and the displacement field u and external body forces f,

$$\varrho\frac{\partial^2 u}{\partial t^2} = \frac{\partial}{\partial x}\left(E\frac{\partial u}{\partial x}\right) + \rho f \tag{1.5}$$

where E is known as Young's modulus with boundary conditions.

1.1.3 Waves on a membrane

We think of a thin, elastic membrane with a shape-bounded dome on $\mathbb{R}^2$. This membrane can be brought into oscillatory motion and will develop elastic waves. We can model this phenomenon somewhat similar to waves in a rod: waves in a membrane are simply the two-dimensional counterpart. We assume that the material is deformed in the z direction only and write the elastic displacement field on the form $u(x,y,t)$. The

z coordinate is omitted since the membrane is thin and all properties are taken as constant throughout the thickness. Inserting this displacement field in Newton's second law of motion (1.5) results in

$$\varrho\frac{\partial^2 u}{\partial t^2} = \frac{\partial}{\partial x}\left(\mu\frac{\partial u}{\partial x}\right) + \frac{\partial}{\partial y}\left(\mu\frac{\partial u}{\partial y}\right). \tag{1.6}$$

1.1.4 The acoustic model for seismic waves

Seismic waves are used to infer properties of subsurface geological structures. The physical model is a heterogeneous elastic medium where sound is propagated by small elastic vibrations. The general mathematical model for deformations in an elastic medium is based on Newton's second law,

$$\rho\frac{\partial^2 u}{\partial t^2} = \nabla.\sigma + \rho f, \tag{1.7}$$

and a constitutive law relating σ to u, often Hooke's generalized law,

$$\boldsymbol{\sigma} = K\nabla\cdot\boldsymbol{u}\boldsymbol{I} + G\left(\nabla\boldsymbol{u} + (\nabla\boldsymbol{u})^T - \frac{2}{3}\nabla\cdot\boldsymbol{u}\boldsymbol{I}\right). \tag{1.8}$$

Here, u is the displacement field, σ is the stress tensor, I is the identity tensor, ρ is the medium's density, f are external body forces, K is the medium's bulk modulus and μ is the shear modulus. All these quantities may vary in space, while u and σ will also show significant variation in time during wave motion.

1.1.5 Sound waves in liquids and gases

Sound waves arise from pressure and density variations in fluids. The starting point of modeling sound waves is the basic equation for a compressible fluid where we omit viscous (frictional) forces, body forces (gravity, for instance) and temperature effects:

$$\begin{cases} \frac{\partial\rho}{\partial t} + \nabla.(\rho u) = 0, \\ \rho\frac{\partial u}{\partial t} + \rho u.\nabla u = -\nabla p, \\ \rho = \rho(p). \end{cases} \tag{1.9}$$

These equations are often referred to as the Euler equations for the motion of a fluid. The parameters involved are density ρ, velocity u, and pressure p. Equation reflects $(1.9)_1$ mass balance, $(1.9)_2$ is Newton's second law for a fluid, with frictional and body forces omitted and $(1.9)_3$ is a constitutive law relating density to pressure by thermodynamic considerations. A typical model for $(1.9)_3$ is the so-called isentropic relation, valid for adiabatic processes where there is no heat transfer:

$$\rho = \rho_0 \left(\frac{p}{p_0} \right)^{1/\gamma}. \tag{1.10}$$

Here, p_0 and ρ_0 are reference values for p and ρ when the fluid is at rest, and γ is the ratio of specific heat at constant pressure and constant volume ($\gamma = 5/3$ for air). The key approximation in a mathematical model for sound waves is to assume that these waves are small perturbations to the density, pressure and velocity. We therefore write

$$p = p_0 + \hat{p}, \tag{1.11}$$

$$\varrho = \varrho_0 + \hat{\varrho}, \tag{1.12}$$

$$u = \hat{u}, \tag{1.13}$$

where we have decomposed the fields in a constant equilibrium value, corresponding to $u = 0$, and a small perturbation marked with a hat symbol. By inserting these decompositions in $(1.9)_1$ and $(1.9)_2$, neglecting all product terms of small perturbations and/or their derivatives, and dropping the hat symbols, one gets the following linearized PDE system for the small perturbations in density, pressure and velocity:

$$\frac{\partial^2 p}{\partial t^2} = c^2 \nabla^2 p, \tag{1.14}$$

where $c = \sqrt{\gamma p_0/\rho_0}$ is the speed of sound in the fluid.

1.1.6 Waves in blood vessels

Blood flow in bodies is essentially the flow of fluids in a network of tubes. Unlike rigid tubes, the walls of blood vessels are elastic and will increase in diameter when pressure rises. The elastic forces will then push the wall back and accelerate the fluid. This interaction between blood flow and vessel wall deformation results in waves that travel through the blood vessels. Let x be the coordinate along the blood vessel and assume that all cross-sections are circular, even though the radius $R(x, t)$ is different. The quantity

$$A(x, t) = 2\pi \int_0^{R(x,t)} \tau d\tau,$$

is the cross-section area and $P(x, t)$ is the averaged pressure whereas

$$Q(x, t) = A(x, t) \frac{\partial^2 v(x, t)}{\partial x}$$

is the total volume flux, where $v_x(x,t)$ is the velocity of blood averaged over the cross-section at point x. Mass balance and Newton's second law lead to the PDEs

$$\frac{\partial A}{\partial t} + \frac{\partial Q}{\partial x} = 0, \tag{1.15}$$

and

$$\frac{\partial Q}{\partial x} + \frac{\lambda+2}{\lambda+1}\frac{\partial}{\partial x}\left(\frac{Q^2}{A}\right) + \frac{A}{\rho}\frac{\partial P}{\partial x} = -2\pi(\lambda+2)\frac{\mu}{\rho}\frac{Q}{A}, \tag{1.16}$$

$$\frac{\partial P}{\partial t} + \frac{1}{C}\frac{\partial Q}{\partial x} = 0. \tag{1.17}$$

Here, λ is a parameter related to the velocity profile, ρ is the density of blood, μ is the dynamic viscosity of blood and C is the compliance of the wall, given by

$$C = \frac{\partial A}{\partial P} + \frac{\partial A}{\partial t}. \tag{1.18}$$

Equation (1.17) is needed to relate the flow to the deformations of the wall. The equation for one-dimensional waves along a blood vessel can be derived from calculating the average fluid flow across the cross-section of the blood vessel. It is given as

$$\frac{\partial^2 Q}{\partial t^2} - c^2\frac{\partial^2 Q}{\partial x^2} = 0, \quad c = \sqrt{\frac{A}{\rho C}}. \tag{1.19}$$

1.1.7 Electromagnetic waves

Light and radio waves are governed by standard wave equations arising from Maxwell's general equations. When there are no charges and no currents, as in a vacuum, Maxwell's equations take the form

$$\begin{aligned} \nabla . E &= 0, \\ \nabla . B &= 0, \\ \nabla \times E &= -\frac{\partial B}{\partial t}, \\ \nabla \times B &= \mu_0 \epsilon_0 \frac{\partial E}{\partial t}, \end{aligned} \tag{1.20}$$

where $\epsilon_0 = 8.854187817620.10^{-12}$ (f/m) is the permittivity of free space, also known as the electric constant, and $\mu_0 = 1.2566370614.10^{-6}$ (H/m) is the permeability of free space, also known as the magnetic constant. Taking the curl of the two last equations and using the mathematical identity

$$\nabla \times (\nabla \times E) = \nabla(\nabla.E) - \nabla^2 E$$
$$= -\nabla^2 E, \quad \nabla.E = 0,$$

give the wave equation governing the electric and magnetic field:

$$\frac{\partial^2 E}{\partial t^2} = c^2 \nabla^2 E, \tag{1.21}$$

$$\frac{\partial^2 B}{\partial t^2} = c^2 \nabla^2 B. \tag{1.22}$$

Here, $c = \frac{1}{\sqrt{\mu_0 \epsilon_0}}$ as the velocity of light. Each component of E and B fulfills a wave equation and can hence be solved independently.

1.2 Basic helpful knowledge

1.2.1 Necessary contractions in C_0-semigroup approach

Consider the following abstract Cauchy initial value problem:

$$\begin{cases} Z'(t) = AZ(t), & t \geq 0, \\ Z(0) = Z_0, \end{cases} \tag{1.23}$$

where $(A, D(A))$ is an unbounded operator on a such Hilbert space $\mathcal{H}$ and $(T(t))_{t\geq 0}$ its associated semigroups.

Definition 1.1. A C_0-semigroup, $T(t)$, on a Hilbert space $\mathcal{H}$ is exponentially stable if there exist positive constants M and α such that

$$\|T(t)\| \leq Me^{-\alpha t} \quad \text{for } t \geq 0. \tag{1.24}$$

The real α is called the decay rate.

One can remark that if $T(t)$ is exponentially stable, then the mild solution to the abstract Cauchy problem (1.23) given by

$$Z(t) = T(t)Z_0,$$

and tends to zero exponentially fast as $t \to \infty$. Moreover, as an important criteria for exponential stability, we cite the following.

Lemma 1.1. *The C_0-semigroup $T(t)$ on $\mathcal{H}$ is exponentially stable if and only if for every $z \in \mathcal{H}$ there exists a positive constant $\gamma_z < \infty$ such that*

$$\int_0^\infty \|T(t)z\|^2 dt \leq \gamma_z. \tag{1.25}$$

This last lemma can be used to prove a Lyapunov-type result for establishing stability.

Theorem 1.1. [1] *Let $S(t) = e^{\mathcal{A}t}$ be a C_0-semigroup of contractions on Hilbert space. Then $S(t)$ is exponentially stable if and only if*

$$\rho(\mathcal{A}) \supseteq \{i\beta : \beta \in \mathbb{R}\} \equiv i\mathbb{R}$$

and

$$\overline{\lim_{|\beta|\to\infty}} \|(i\beta I - \mathcal{A})^{-1}\|_{\mathcal{L}(\mathcal{H})} < \infty.$$

Theorem 1.2. *Suppose that A is the infinitesimal generator of the C_0-semigroup $T(t)$ on the Hilbert space $\mathcal{H}$. Then $T(t)$ is exponentially stable if and only if there exists positive operator $P \in \mathcal{L}(\mathcal{H})$ such that*

$$\langle Az, Pz\rangle + \langle Pz, Az\rangle = -\langle z, z\rangle \quad \text{for all } z \in \mathbf{D}(A). \tag{1.26}$$

Equation (1.26) *is called a Lyapunov equation. If $T(t)$ is exponentially stable, then the unique self-adjoint solution of* (1.26) *is given by*

$$Pz = \int_0^\infty T(s)^* T(s) z ds \quad \text{for } z \in \mathcal{H}.$$

Definition 1.2. [2] We call the semigroup $T(t)$ polynomially stable if there exist the constants $M, \alpha, \beta > 0$, such that

$$\|T(t)x\| \leq Mt^{-\beta}\|x\|_{D(A^\alpha)}, \quad \forall\, t \geq 0, x \in D(A^\alpha) \tag{1.27}$$

or, equivalently,

$$\|T(t)A^{-\alpha}\| \leq Mt^{-\beta}, \quad t \geq 0. \tag{1.28}$$

---

1 F. Huang. Characteristic condition for exponential stability of linear dynamical systems in Hilbert space. Ann. Differential. Equations 1 (1) (1985), 43–56.

2 A. Bátkai, K. J. Engel, J. Prüss and R. Schnaubelt. Polynomial stability of operator semigroups. Math. Nachr. 279 (2006), 1425–1440.

Theorem 1.3. [3] *Let A be the generator of a C_0-semigroup of bounded operators on a Hilbert space $\mathcal{H}$ such that $i\mathbb{R} \subset \rho(A)$. Then we have the polynomial decay*

$$\|e^{tA}U_0\| \leq \frac{C}{t^{1/\theta}}\|U_0\|, \quad t > 0 \tag{1.29}$$

if and only if

$$\limsup_{|\lambda|\to+\infty} \frac{1}{|\lambda|^{\theta}}\|(i\lambda - A)^{-1}\| < \infty. \tag{1.30}$$

1.2.2 Useful inequalities for decay rate analysis

Lemma 1.2 (Gronwall's inequality). *Let $T > 0$ and $C \geq 0$. Let*

$$G, \alpha, \beta : [0, T] \to \mathbb{R},$$

be continuous and nonnegative functions

$$0 \leq \mu(s) \leq C + 2\int_0^s \alpha(r)\sqrt{G(r)}\,dr + \int_0^s \beta(r)G(r)\,dr.$$

Then

$$\mu(t) \leq \left(\sqrt{C} + \int_0^t \alpha(s)\,ds\right)^2 e^{\int_0^t \beta(s)\,ds}.$$

Lemma 1.3 (Modified Gronwall inequality). *Let μ and f be nonnegative functions on $[0, +\infty)$ satisfying*

$$0 \leq \mu(t) \leq K + \int_0^t f(s)\mu(s)^{r+1}\,ds,$$

with $K > 0$ and $r > 0$. Then

$$\mu(t) \leq \left\{K^{-r} - r\int_0^t f(s)\,ds\right\}^{-1/r},$$

as long as the RHS exists.

---

3 A. Borichev, Y. Tomilov. Optimal polynomial decay of functions and operator semigroups. Math. Ann. 347 (2) (2010), 455–478.

To deal with stability, we need to use the next lemma.

Lemma 1.4. [4] *Let*

$$E : \mathbb{R}^+ \to \mathbb{R}^+,$$

be a nonincreasing differentiable function and

$$\Psi : \mathbb{R}^+ \to \mathbb{R}^+,$$

a convex and increasing function such that $\Psi(0) = 0$. *Assume that*

$$\int_S^T \Psi(E(t))\, dt \le E(s), \quad \forall 0 \le S \le T.$$

Then E satisfies the following estimate:

$$E(t) \le \psi^{-1}(h(t) + \psi(E(0))), \quad \forall t \ge 0 \tag{1.31}$$

where $\psi(t) = \int_t^1 \frac{1}{\Psi(s)}\, ds$ *for* $t > 0$, $h(t) = 0$ *for* $0 \le t \le \frac{E(0)}{\Psi(E(0))}$ *and*

$$h^{-1}(t) = t + \frac{\psi(t + \psi(E(0)))}{\Psi(\psi^{-1}(t + \psi(E(0))))}, \quad \forall t \ge \frac{E(0)}{\Psi(E(0))}.$$

Lemma 1.5. [5] *Let*

$$E : \mathbb{R}^+ \longrightarrow \mathbb{R}^+,$$

be a nonincreasing function and assume that there are two constants $\mu \ge 0$ *and* $\omega > 0$ *such that*

$$\int_t^{+\infty} E(s)^{\mu+1}\, ds \le \omega E(0)^{\mu} E(t), \quad \forall t \ge 0.$$

Then we have for every $t > 0$,

$$\begin{cases} E(t) \le E(0)\left(\frac{1+\mu}{1+\omega\mu t}\right)^{-\frac{1}{\mu}}, & \text{if } \mu > 0, \\ E(t) \le E(0)e^{1-\omega t}, & \text{if } \mu = 0. \end{cases}$$

---

4 Existence globale et stabilisation interne non lineaire d'un systeme de Petrovsky. Bell. Belg. Math. Soc. 5 (1998), 583–594.

5 A new method to obtain decay rate estimates for dissipative systems. ESAIM Control Optim. Calc. Var. 4 (1999), 419-444.

Lemma 1.6. [6] *Let*

$$E : \mathbb{R}_+ \to \mathbb{R}_+,$$

be a nonincreasing differentiable function, $\lambda \in \mathbb{R}_+$ *and*

$$\varphi : \mathbb{R}_+ \to \mathbb{R}_+,$$

a convex and increasing function such that

$$\varphi(0) = 0.$$

Assume that

$$\begin{cases} \int_s^{+\infty} \varphi(E(t))\,dt \le E(s), & \forall s \ge 0, \\ E'(t) \le \lambda E(t) & \forall t \ge 0. \end{cases}$$

Then E satisfies the following estimate:

$$E(t) \le e^{\tau_0 \lambda} d^{-1}(e^{\lambda(t-h(t))} \varphi(\psi^{-1}(h(t) + \psi(E(0)))), \quad \forall t \ge 0 \tag{1.32}$$

where

$$\psi(t) = \int_t^1 \frac{1}{\varphi(s)}\,ds, \quad \forall t \ge 0,$$

$$d(t) = \begin{cases} \psi(t), & \text{if } \lambda = 0, \\ \int_0^t \frac{\varphi(s)}{s}\,ds & \text{if } \lambda > 0, \end{cases}$$

$$h(t) = \begin{cases} K^{-1}(D(t)), & \forall t > T_0, \\ 0 & \forall t \in [0, T_0], \end{cases}$$

$$K(t) = D(t) + \frac{\psi^{-1}(t + \psi(E(0)))}{\varphi(\psi^{-1}(t + \psi(E(0))))} e^{\lambda t}, \quad \forall t \ge 0,$$

$$D(t) = \int_0^t e^{\lambda s}\,ds \quad \forall t \ge 0,$$

$$T_0 = D^{-1}\left(\frac{E(0)}{\varphi(E(0))}\right), \quad \tau_0 = \begin{cases} 0 & \forall t > T_0, \\ T_0 & \forall t \in [0, T_0]. \end{cases}$$

---

6 A difference inequality and its applications to nonlinear evolution equations. J. Math. Soc. Japan 30 (1978), 747–762.

Lemma 1.7 (Sobolev–Poincaré inequality). [7] *For every*

$$v \in \{u \in H^1(0, L) \mid u(0) = 0\},$$

we have

$$\int_0^L v^2\,dx \le P \int_0^L v_x^2\,dx, \quad \text{with } P = \frac{L^2}{2},$$

an important technical lemma.

1.2.2.1 Assumption on the viscoelastic term

The viscoelastic materials show a behavior, which is something between that of elastic solids and Newtonian fluids. Indeed, the stresses in these media depend on the entire history of their deformation, not only on their current state of deformation or their current state of motion. This is the reason why they are called materials with memory. The integral

$$\int_0^t \mu(t-s)\Delta_x u(s)ds,$$

is the viscoelastic term and the scalar function $\mu(x)$ is the relaxation kernel such that $\mu \in C^1(\mathbb{R}^+, \mathbb{R}^+)$ assumed to satisfy (to guarantee the hyperbolicity of systems):

$$1 - \overline{\mu} = \mu_0 > 0, \quad \forall t \in \mathbb{R}^+, \tag{1.33}$$

in the classical case and

$$m_0 - \overline{\mu} = \mu_0 > 0, \quad \forall t \in \mathbb{R}^+, \tag{1.34}$$

where m_0 is constant of the Kirchhoof-type function and

$$\overline{\mu} = \int_0^\infty \mu(t)dt.$$

In addition,

$$\mu(0) > 0, \quad \mu'(t) \le 0, \quad \forall t \in \mathbb{R}^+ \tag{1.35}$$

to let the systems dissipative. The condition

---

7 R. A. Adams. Sobolev spaces. Pure and Applied Mathematics, Vol. 65. Academic Press, 1978.

$$\mu'(t) + \kappa\mu(t) \le 0, \quad \kappa \in \mathbb{R}^+ \tag{1.36}$$

is taken to show the growth rate of μ. It means that

$$\mu(t) \le ce^{-\kappa t}.$$

For a coupled system, there exist nonincreasing continuous functions $\xi_1, \xi_2 \in C^1(\mathbb{R}^+, \mathbb{R}^+)$ satisfying

$$\xi'(t) \le 0, \quad \int_0^\infty \xi(t) = \infty, \quad \xi(t) = \min\{\xi_1(t), \xi_2(t)\}, \tag{1.37}$$

where

$$\mu'(t) + \xi(t)\mu(t) \le 0. \tag{1.38}$$

In order to investigate the decay structure based on the weak-memory, the integral

$$\alpha(t) \int_0^t \mu(t-s)\Delta_x u(s)ds,$$

is the weak viscoelastic term for a differentiable function $\alpha(t) \in C^1(\mathbb{R}^+, \mathbb{R}^+)$ satisfying

$$\mu'(t) + \alpha(t)\mu(t) \le 0 \quad \forall t \in \mathbb{R}^+, \tag{1.39}$$

which means that

$$\mu(t) \le ce^{-\int \alpha(s)ds},$$

where

$$\alpha(t) > 0, \quad \alpha'(t) \le 0, \quad \forall t > 0. \tag{1.40}$$

It is also considered in the following assumptions:

$$|\xi|^2 - \alpha(t) \int_0^t \mu(t)dt \ge k > 0, \quad \mu(0) > 0, \tag{1.41}$$

$$\overline{\mu} < \infty, \quad \alpha(t) > 0. \tag{1.42}$$

Here, ξ is the variable associated with the Fourier transform. To extend (1.36), for the general rate, there exists a function $\beta \in C^1(\mathbb{R}^+, \mathbb{R}^+)$ such that

$$\beta(0) > 0, \quad \mu'(t) + \beta(t)\mu(t) \le 0, \quad \forall t \ge 0, \quad \lim_{t\to\infty} = \frac{\alpha'(t)}{\beta(t)\alpha(t)} = 0. \tag{1.43}$$

In more general, there exists a positive function $H \in C^1(\mathbb{R}^+)$ such that

$$\mu'(t) + H(\mu(t)) \le 0, \quad t \ge 0, \quad H(0) = 0 \tag{1.44}$$

and H is linear or strictly increasing and strictly convex C^2 function on $(0, r], 1 > r$.

Remark 1.1. (A) We can deduce that there exists $t_1 > 0$ large enough such that:

1) $\forall t \ge t_1$: We have $\lim_{s\to+\infty} \mu(s) = 0$, which implies that $\lim_{s\to+\infty} -\mu'(s)$ cannot be positive, so $\lim_{s\to+\infty} -\mu'(s) = 0$. Then $\mu(t_1) > 0$ and

$$\max\{\mu(s), -\mu'(s)\} < \min\{r, H(r), H_0(r)\}, \tag{1.45}$$

where $H_0(t) = H(D(t))$ provided that D is a positive C^1 function, with $D(0) = 0$, for which H_0 is strictly increasing and strictly convex C^2 function on $(0, r]$ and

$$\int_0^{+\infty} \mu(s) H_0(-\mu'(s))ds < +\infty.$$

2) $\forall t \in [0, t_1]$: As μ is nonincreasing, $\mu(0) > 0$ and $\mu(t_1) > 0$ then $\mu(t) > 0$ and

$$\mu(0) \ge \mu(t) \ge \mu(t_1) > 0.$$

Therefore, since H is a positive continuous function, then

$$a \le H(\mu(t)) \le b$$

for some positive constants a and b. Consequently,

$$\mu'(t) \le -H(\mu(t)) \le -\kappa\mu(t), \quad \kappa > 0$$

which gives

$$\mu'(t) \le -\kappa\mu(t), \quad k > 0. \tag{1.46}$$

(B) Let H_0^* be the convex conjugate of H_0 in the sense of Young (see [1], pp. 61–64), then

$$H_0^*(s) = s(H_0')^{-1}(s) - H_0[(H_0')^{-1}(s)], \quad s \in (0, H_0'(r))$$

and satisfies the following Young's inequality:

$$AB \le H_0^*(A) + H_0(B), \quad A \in (0, H_0'(r)), B \in (0, r]. \tag{1.47}$$

Remark 1.2. For the degenerate Kirchooff type, we have that μ is a positive function and $m_0 - \bar{\mu} > 0$. Then $\int_0^\infty \mu(t)dt \geq 0$ and $\int_0^\infty \mu(t)dt \leq m_0$. Therefore, $\lim_{t\to\infty} \mu(t) = 0$. Because $\mu(t)$ is a positive function on $\mathbb{R}_+$, $\mu(0) > 0$ and $\lim_{t\to\infty} \mu(t) = 0$, we conclude that there is a $t_1 > 0$, large enough, so that $\mu(t)$ is a decreasing function on $[t_1, \infty)$ and $\mu(t_1) > 0$. Hence, $\mu'(t) \leq 0$ on $[t_1, \infty)$. Since $\mu(t) \to 0$ as $t \to \infty$, we get that $\mu'(t) \to 0$ as $t \to \infty$. We set

$$H_0(t) = H(D(t))$$

for a C^1-function D such that $D(0) = 0$, H_0 is strictly increasing and strictly convex C^2-function on $(0, r]$. If there is a need, we enlarge $t_1 > 0$ so that $\mu(t_1) > 0$ and

$$\max\{\mu(s), -\mu'(s)\} < \min\{r, H(r), H_0(r)\} \tag{1.48}$$

for any $s \in [t_1, \infty)$. Note that

$$\begin{aligned}
\int_0^\infty \mu(s)H_0(-\mu'(s))ds &= \int_0^{t_1} \mu(s)H_0(-\mu'(s))ds + \int_{t_1}^\infty \mu(s)H_0(-\mu'(s))ds \\
&\leq \int_0^{t_1} \mu(s)H_0(-\mu'(s))ds + \int_{t_1}^\infty \mu(s)H_0(r)ds \\
&= \int_0^{t_1} \mu(s)H_0(-\mu'(s))ds + H_0(r)\int_{t_1}^\infty \mu(s)ds \\
&< \infty.
\end{aligned}$$

Since μ is a positive function on $(0, t_1]$ and H is a positive continuous function on $\mathbb{R}_+$, we have that there exist positive constants α_1 and β_1 such that

$$\alpha_1 \leq H(\mu(t)) \leq \beta_1 \quad \text{for any } t \in [0, t_1].$$

Let

$$m_1 = \min_{T\in[0,t_1]} \mu(t), \quad M_1 = \max_{t\in[0,t_1]} \mu(t).$$

Then $m_1 > 0$ and $M_1 > 0$ and

$$-H(\mu(t)) \leq -\alpha_1 = -\frac{\alpha_1}{M_1} M_1 \leq -\frac{\alpha_1}{M_1}\mu(t)$$

for any $t \in [0, t_1]$. We set

$$\kappa = \frac{\alpha_1}{M_1}.$$

Then $\kappa > 0$ and

$$-H(\mu(t)) \leq -\kappa\mu(t) \quad \text{for any } t \in [0, t_1].$$

Hence,

$$\mu'(t) \leq -H(\mu(t)) \leq -\kappa\mu(t) \quad \text{for any } t \in [0, t_1]. \tag{1.49}$$

Let H_0^* be the convex conjugate of H_0 in the sense of Young (see [1], pp. 61–64), then

$$H_0^*(s) = s(H_0')^{-1}(s) - H_0[(H_0')^{-1}(s)], \quad s \in (0, H_0'(r))$$

and satisfies the following Young's inequality:

$$AB \leq H_0^*(A) + H_0(B), \quad A \in (0, H_0'(r)), B \in (0, r]. \tag{1.50}$$

Notation: For any real (complex)-valued function $h(t)$, we define

$$(\mu * h)(t) = \int_0^t \mu(t-\tau)h(\tau)d\tau,$$

$$(\mu \circ h)(t) = \int_0^t \mu(t-\tau)|h(t) - h(\tau)|^2 d\tau.$$

Lemma 1.8. *For any $v \in C^1(0, T, H^1(\mathbb{R}^n))$, we have for a self-adjoint operator A,*

$$\begin{aligned}
&-\int_{\mathbb{R}^n} \alpha(t) \int_0^t \mu(t-s)Av(s)\frac{\partial v}{\partial t} dsdx \\
&= \frac{1}{2}\frac{d}{dt}\alpha(t)(\mu \circ A^{1/2}v)(t) \\
&\quad - \frac{1}{2}\frac{d}{dt}\left[\alpha(t)\int_0^t \mu(s) \int_{\mathbb{R}^n} |A^{1/2}v(t)|^2 dxds\right] \\
&\quad - \frac{1}{2}\alpha(t)(\mu' \circ A^{1/2}v)(t) + \frac{1}{2}\alpha(t)\mu(t)\int_{\mathbb{R}^n} |A^{1/2}v(t)|^2 dxds \\
&\quad - \frac{1}{2}\alpha'(t)(\mu \circ A^{1/2}v)(t) + \frac{1}{2}\alpha'(t)\int_0^t \mu(s)ds \int_{\mathbb{R}^n} |A^{1/2}v(t)|^2 dxds.
\end{aligned}$$

Proof. It is not hard to see

$$
\begin{aligned}
&\int_{\mathbb{R}^n} \alpha(t) \int_0^t \mu(t-s)Av(s)\frac{\partial v}{\partial t} ds dx \\
&\quad = \alpha(t) \int_0^t \mu(t-s) \int_{\mathbb{R}^n} A^{1/2}\frac{\partial v}{\partial t} A^{1/2} v(s) dx ds \\
&\quad = \alpha(t) \int_0^t \mu(t-s) \int_{\mathbb{R}^n} A^{1/2}\frac{\partial v}{\partial t} [A^{1/2} v(s) - A^{1/2} v(t)] dx ds \\
&\qquad + \alpha(t) \int_0^t \mu(t-s) \int_{\mathbb{R}^n} A^{1/2}\frac{\partial v}{\partial t} A^{1/2} v(t) dx ds.
\end{aligned}
$$

Consequently,

$$
\begin{aligned}
&\int_{\mathbb{R}^n} \alpha(t) \int_0^t \mu(t-s)Av(s)\frac{\partial v}{\partial t} ds dx \\
&\quad = -\frac{1}{2}\alpha(t) \int_0^t \mu(t-s)\frac{d}{dt} \int_{\mathbb{R}^n} |A^{1/2} v(s) - A^{1/2} v(t)|^2 dx ds \\
&\qquad + \alpha(t) \int_0^t \mu(s)\left(\frac{d}{dt}\frac{1}{2} \int_{\mathbb{R}^n} |A^{1/2} v(t)|^2 dx\right) ds,
\end{aligned}
$$

which implies

$$
\begin{aligned}
&\int_{\mathbb{R}^n} \alpha(t) \int_0^t \mu(t-s)Av(s)\frac{\partial v}{\partial t} ds dx \\
&\quad = -\frac{1}{2}\frac{d}{dt}\left[\alpha(t) \int_0^t \mu(t-s) \int_{\mathbb{R}^n} |A^{1/2} v(s) - A^{1/2} v(t)|^2 dx ds\right] \\
&\qquad + \frac{1}{2}\frac{d}{dt}\left[\alpha(t) \int_0^t \mu(s) \int_{\mathbb{R}^n} |A^{1/2} v(t)|^2 dx ds\right] \\
&\qquad + \frac{1}{2}\alpha(t) \int_0^t \mu'(t-s) \int_{\mathbb{R}^n} |A^{1/2} v(s) - A^{1/2} v(t)|^2 dx ds \\
&\qquad - \frac{1}{2}\alpha(t)\mu(t) \int_{\mathbb{R}^n} |A^{1/2} v(t)|^2 dx ds.
\end{aligned}
$$

$$+\frac{1}{2}\alpha'(t)\int_0^t \mu(t-s)\int_{\mathbb{R}^n}|A^{1/2}v(s)-A^{1/2}v(t)|^2 dxds$$
$$-\frac{1}{2}\alpha'(t)\int_0^s \mu(s)ds\int_{\mathbb{R}^n}|A^{1/2}v(t)|^2 dxds.$$

This completes the proof. □

The following technical lemmas will play an important role in the sequel.

Lemma 1.9. *For any two functions $\mu, v \in C^1(\mathbb{R})$ and $\theta \in [0,1]$, we have*

$$\begin{aligned}\frac{\partial v}{\partial t}\int_0^t \mu(t-s)v(s)ds &= -\frac{1}{2}\frac{d}{dt}\int_0^t \mu(t-s)|v(t)-v(s)|^2 ds + \frac{1}{2}\frac{d}{dt}\left(\int_0^t \mu(s)ds\right)|v(t)|^2\\ &\quad+\frac{1}{2}\int_0^t \mu'(t-s)|v(t)-v(s)|^2 ds - \frac{1}{2}\mu(t)|v(t)|^2\end{aligned} \tag{1.51}$$

and

$$\left|\int_0^t \mu(t-s)(v(t)-v(s))ds\right|^2 \le \left(\int_0^t |\mu(s)|^{2(1-\theta)}ds\right)\left(\int_0^t |\mu(t-s)|^{2\theta}|v(t)-v(s)|^2 ds\right). \tag{1.52}$$

1.2.3 Useful inequalities for compactly supported initial data

Lemma 1.10 (Gagliardo–Nirenberg inequality).[8] *Let*

$$1 \le q \le \frac{2n}{[n-2]^2},$$

$2 \le q$ *and* $n \ge 1$. *Then*

$$\|u\|_q \le K_0\|\nabla u\|_2^{\theta}\|u\|_r^{1-\theta}, \quad u \in H^1(\mathbb{R}^n), K_0 > 0$$

holds and

$$\theta = \left(\frac{1}{r}-\frac{1}{q}\right)\left(\frac{1}{r}-\frac{1}{2}+\frac{1}{n}\right)^{-1},$$

provided that $\theta \in (0,1]$.

8 R. Ikehata and M. Ohta. Critical exponents for semilinear dissipative wave equations in $\mathbb{R}^n$. J. Math. Anal. Appl. 269 (2002), 87–97.

Lemma 1.11. [9] *Let the function ψ (will be called a modified weighted function) defined by*

$$\psi(x,t) = \frac{|x|^2}{4(1+t)}, \quad x \in \mathbb{R}^n, t \geq 0. \tag{1.53}$$

For each $\sigma > 0$, there exists a constant $C = C_\sigma > 0$ such that

$$\int_{\mathbb{R}^n} e^{-2p\sigma\psi(x,t)} dx \leq C(t+1)^{n/2},$$

where

$$\begin{aligned} &1 + \frac{2}{n} < p \leq \frac{n}{n-2}, \quad n \geq 3, \quad or \\ &1 + \frac{2}{n} < p \leq +\infty, \quad n = 1, 2. \end{aligned} \tag{1.54}$$

For $\sigma > 0$ and $t \geq 0$, we define a family of weighted function spaces $H^1_{\sigma\psi}(\mathbb{R}^n)$,

$$g \in H^1_{\sigma\psi}(\mathbb{R}^n) \equiv g \in H^1(\mathbb{R}^n),$$

and

$$\|e^{\sigma\psi} g\|_2^2 + \|e^{\sigma\psi} \nabla g\|_2^2 < +\infty, \quad t \geq 0.$$

Lemma 1.12 (Generalized Gagliardo–Nirenberg inequality in $\mathbb{R}^n$). *Let*

$$\theta(q) = n\left(\frac{1}{2} - \frac{1}{q}\right),$$

and

$$1 \leq \theta(q) \leq 1,$$

$\sigma \in (0,1]$ *and* $n \geq 1$. *If* $\sigma = 1$ *and let*

$$g \in H^1_\psi(\mathbb{R}^n), \quad t \geq 0,$$

then we have

$$\|e^{\sigma\psi} g\|_q \leq K_0 (t+1)^{(1-\theta(q))/2} \|\nabla g\|_2^{1-\sigma} \|e^{\psi} \nabla g\|_2^{\sigma}, \quad g \in H^1_\psi(\mathbb{R}^n), K_0 > 0,$$

---

9 R. Ikehata and K. Tanzawa. Global existence of solutions for semilinear damped wave equations in $\mathbb{R}^n$ with noncompactly supported initial data. Nonl. Anal. 61 (2005), 1189–1208.

Proof. Let

$$g \in H^1_\psi(\mathbb{R}^n).$$

If $\sigma \in (0,1]$ from the Hölder's inequality, it follows that

$$\begin{aligned}\|e^{\sigma\psi}\nabla g\|_2^2 &= \int_{\mathbb{R}^n} e^{2\sigma\psi}|\nabla g|^{2\sigma}|\nabla g|^{2(1-\sigma)} \\ &\le \|e^{2\sigma\psi}|\nabla g|^{2\sigma}\|_{1/\sigma}\||\nabla g|^{2(1-\sigma)}\|_{1/(1-\sigma)} \\ &= \|e^{\psi}|\nabla g|\|_2^{2\sigma}\|\nabla g\|_2^{2(1-\sigma)},\end{aligned} \tag{1.55}$$

and

$$\begin{aligned}\|e^{\sigma\psi}g\|_2^2 &= \int_{\mathbb{R}^n} e^{2\sigma\psi}|g|^{2\sigma}|g|^{2(1-\sigma)} \\ &\le \|e^{2\sigma\psi}|g|^{2\sigma}\|_{1/\sigma}\||g|^{2(1-\sigma)}\|_{1/(1-\sigma)} \\ &= \|e^{\psi}g\|_2^{2\sigma}\|g\|_2^{2(1-\sigma)}.\end{aligned}$$

So, we notice that

$$g \in H^1_{\sigma\psi}(\mathbb{R}^n), \quad \forall \sigma \in (0,1].$$

Thus, from Lemma 1.13, we obtain

$$f = e^{\sigma\psi}g \in H^1(\mathbb{R}^n),$$

and from Lemma 1.10, we have

$$\|f\|_q \le C\|f\|_2^{1-\theta(q)}\|\nabla f\|_2^{\theta(q)}.$$

Here,

$$\theta(q) = n\left(\frac{1}{2} - \frac{1}{q}\right) \in [0,1].$$

Furthermore, from Lemma 1.13, we have again

$$\|\nabla f\|_2 \le \|e^{\sigma\psi}\nabla g\|_2,$$

and

$$\|f\|_2 \le \left(\frac{2}{n\sigma}\right)^{1/2}(t+1)^{1/2}\|e^{\sigma\psi}\nabla g\|_2.$$

Thus,

$$\|f\|_q \le C_\sigma (t+1)^{(1-\theta(q))/2} \|e^{\sigma\psi} \nabla g\|_2.$$

Finally, by (1.55), we get the desired estimate. □

Lemma 1.13. *Let $\sigma > 0$ and $g \in H^1_{\sigma\psi}(\mathbb{R}^n)$. Then*

$$\frac{\sigma n}{2} \frac{1}{t+1} \|e^{\sigma\psi} g\|_2^2 + \|\nabla(e^{\sigma\psi} g)\|_2^2 \le \|e^{\sigma\psi} \nabla g\|_2^2, \quad t \ge 0.$$

Proof. Let

$$f = e^{\sigma\psi} g, \quad \sigma > 0.$$

Then

$$\nabla g = e^{-\sigma\psi}(\nabla f - \sigma f \nabla \psi),$$

so

$$e^{\sigma\psi} \nabla g = \nabla f - \sigma f \nabla \psi.$$

This implies

$$\begin{aligned} \|e^{\sigma\psi} \nabla g\|_2^2 &= \int_{\mathbb{R}^n} |\nabla f - \sigma f \nabla \psi|^2 dx \\ &= \int_{\mathbb{R}^n} |\nabla f|^2 dx + \sigma^2 \int_{\mathbb{R}^n} f^2 |\nabla \psi|^2 dx - 2\sigma \int_{\mathbb{R}^n} f(\nabla f \nabla \psi) dx. \end{aligned} \tag{1.56}$$

Since

$$f \nabla f = \frac{1}{2} \nabla f^2,$$

we have

$$\int_{\mathbb{R}^n} f(\nabla f \nabla \psi) dx = -\frac{1}{2} \int_{\mathbb{R}^n} f^2 \Delta \psi dx. \tag{1.57}$$

By (1.56) and (1.57), we have

$$\|e^{\sigma\psi} \nabla g\|_2^2 \ge \int_{\mathbb{R}^n} |\nabla f|^2 dx + \sigma \int_{\mathbb{R}^n} f^2 \Delta \psi dx,$$

using (2.21) to complete the proof. □

1.2.4 Results related to the weighted weak spaces

We consider the solutions in an appropriate spaces weighted by the density function $\rho(x)$ in order to compensate for the lack of Poincare's inequality, which play a decisive role in the proof. The density function satisfies

$$\rho : \mathbb{R}^n \to \mathbb{R}_+^*, \quad \rho(x) \in C^{0,\tilde{\gamma}}(\mathbb{R}^n) \tag{1.58}$$

with $\tilde{\gamma} \in (0,1)$ and $\rho \in L^s(\mathbb{R}^n) \cap L^\infty(\mathbb{R}^n)$, where $s = \frac{2n}{2n-qn+2q}$, $2 \le q \le \frac{2n}{n-2}$.

The space

$$\mathscr{H}(\mathbb{R}^n),$$

is defined as the closure of $C_0^\infty(\mathbb{R}^n)$ functions with respect to the norm

$$\|u\|^2_{\mathscr{H}(\mathbb{R}^n)} = \int_{\mathbb{R}^n} |\nabla_x u|^2 dx.$$

It is defined in the next definition.

Definition 1.3. [10] We define the function spaces of our problem and its norm as follows:

$$\mathscr{H}(\mathbb{R}^n) = \{f \in L^{2n/(n-2)}(\mathbb{R}^n) : \nabla_x f \in (L^2(\mathbb{R}^n))^n\}$$

and that $\mathscr{H}$ is embedded continuously in $L^{2n/(n-2)}$.

The space $L^2_\rho(\mathbb{R}^n)$ is to be the closure of $C_0^\infty(\mathbb{R}^n)$ functions with respect to the inner product

$$(f,h)_{L^2_\rho(\mathbb{R}^n)} = \int_{\mathbb{R}^n} \rho f h dx.$$

For $1 < q < \infty$, if f is a measurable function on $\mathbb{R}^n$, we define

$$\|f\|_{L^q_\rho(\mathbb{R}^n)} = \left(\int_{\mathbb{R}^n} \rho |f|^q dx \right)^{1/q}. \tag{1.59}$$

Remark 1.3. The space $L^2_\rho(\mathbb{R}^n)$ is a separable Hilbert space. So, we are able to construct the necessary *evolution triple* for the space setting of our problem, which is

10 P. G. Papadopulos and N. M. Stavrakakies. Global existence and blow-up results for an equation of Kirchhoff type on $\mathbb{R}^n$. Methods in Nonlinear Analysis 17 (2001), 91–109.

$$\mathscr{H}(\mathbb{R}^n) \subset L^2_\rho(\mathbb{R}^n) \subset \mathscr{H}^{-1}(\mathbb{R}^n), \tag{1.60}$$

where all the embedding are compact and dense.

The next lemma can be easily shown.

Lemma 1.14. [11] *Let ρ satisfy* (1.58), *then for any $u \in \mathscr{H}(\mathbb{R}^n)$,*

$$\|u\|_{L^q_\rho(\mathbb{R}^n)} \le \|\rho\|_{L^s(\mathbb{R}^n)} \|\nabla_x u\|_{L^2(\mathbb{R}^n)} \tag{1.61}$$

with $s = \frac{2n}{2n-qn+2q}$, $2 \le q \le \frac{2n}{n-2}$.

Corollary 1.1. *If $q = 2$, then Lemma* 1.14 *yields*

$$\|u\|_{L^2_\rho(\mathbb{R}^n)} \le \|\rho\|_{L^{n/2}(\mathbb{R}^n)} \|\nabla u\|_{L^2(\mathbb{R}^n)},$$

where we can assume $\|\rho\|_{L^{n/2}(\mathbb{R}^n)} = c > 0$ to get

$$\|u\|_{L^2_\rho(\mathbb{R}^n)} \le c\|\nabla u\|_{L^2(\mathbb{R}^n)}. \tag{1.62}$$

To study the properties of the operator $\phi\Delta$, we consider the equation

$$\phi(x)\Delta u(x) = \eta(x), \quad x \in \mathbb{R}^n, \tag{1.63}$$

without boundary conditions. Since for every u, v in $C_0^\infty(\mathbb{R}^n)$,

$$(\phi\Delta u, v)_{L^2_\rho} = \int_{\mathbb{R}^n} \nabla u \nabla v dx. \tag{1.64}$$

We may consider equation (1.63) as an operator equation:

$$\Delta_0 u = \eta, \quad \Delta_0 : \mathscr{H}(\Delta_0) \subseteq L^2_\rho(\mathbb{R}^n) \to L^2_\rho(\mathbb{R}^n), \quad \eta \in L^2_\rho(\mathbb{R}^n).$$

Relations (1.64) imply that the operators $\phi\Delta$ with domain of definition $\mathscr{H}(\Delta_0) = C_0^\infty(\mathbb{R}^n)$ being symmetric. Let us note that the operator $\phi\Delta$ is not symmetric in the standard Lebesgue space $L^2(\mathbb{R}^n)$, because of the appearance of $\phi(x)$ (see [5], pp. 185–187). From (1.62) and (1.64), we have

$$\|u\|_{L^2_\rho} \le c(\Delta_0 u, u)_{L^2_\rho}, \quad \text{for all } u \in \mathscr{H}(\Delta_0). \tag{1.65}$$

From (1.64) and (1.65), we conclude that Δ_0 is a symmetric, strongly monotone operator on $L^2_\rho(\mathbb{R}^n)$. The energy scalar product is given by

---

11 N. I. Karachalios and N. M. Stavrakakis. Existence of global attractor for semilinear dissipative wave equations on $\mathbb{R}^n$. J. Diff. Equ. 157 (1999), 183–205.

$$(u,v)_E = \int_{\mathbb{R}^n} \nabla u \nabla v dx,$$

and the energy space is the completion of $\mathscr{H}(\Delta_0)$ with respect to $(u,v)_E$. It is obvious that the energy space X_E is the homogeneous Sobolev space $\mathscr{H}(\mathbb{R}^n)$. The energy extension Δ_E, namely

$$\phi\Delta : \mathscr{H}(\mathbb{R}^n) \to D^{-1}(\mathbb{R}^n),$$

is defined to be the duality mapping of $\mathscr{H}(\mathbb{R}^n)$. For every $\eta \in \mathscr{H}^{-1}(\mathbb{R}^n)$, the equation (1.63) has a unique solution. Define $\mathscr{H}(\Delta_1)$ to be the set of all solutions of the equations (1.63) for arbitrary $\eta \in L^2_\rho(\mathbb{R}^n)$. The operator extension Δ_1 of Δ_0 is the restriction of the energy extension Δ_E to the set $\mathscr{H}(\Delta_1)$. The operator Δ_1 is self-adjoint and, therefore, graph closed. Its domain, is a Hilbert space with respect to the graph scalar product

$$(u,v)_{\mathscr{H}(\Delta_1)} = (u,v)_{L^2_\rho} + (\Delta_1 u, \Delta_1 v)_{L^2_\rho}, \quad \text{for all } u,v \in \mathscr{H}(\Delta_1).$$

The norm induced by the scalar product $(u,v)_{\mathscr{H}(\Delta_1)}$ is

$$\|u\|_{\mathscr{H}(\Delta_1)} = \left\{ \int_{\mathbb{R}^n} \rho |u|^2 dx + \int_{\mathbb{R}^n} \phi |\Delta u|^2 dx \right\}^{\frac{1}{2}},$$

which is equivalent to the norm

$$\|\Delta_1 u\|_{L^2_\rho} = \left\{ \int_{\mathbb{R}^n} \phi |\Delta u|^2 dx \right\}^{\frac{1}{2}}.$$

So, we have established the *evolution quartet*

$$\mathscr{K}(\Delta_1) \subset \mathscr{H}(\mathbb{R}^n) \subset L^2_\rho(\mathbb{R}^n) \subset \mathscr{H}^{-1}(\mathbb{R}^n), \tag{1.66}$$

where all the embedding are dense and compact. A consequence of the compactness of the embedding in (1.66) is that the eigenvalue problem

$$-\Delta u = \mu u, \quad x \in \mathbb{R}^n, \tag{1.67}$$

has a complete system of eigenfunctions $\{w_n, \mu_n\}$ with the following properties:

$$\begin{cases} -\Delta w_j = \mu w_j, \quad j = 1, 2 \cdots, \quad w_j \in \mathscr{H}(\mathbb{R}^n), \\ 0 < \mu_1 \le \mu_2 \le \cdots, \quad \mu_j \to \infty, \quad \text{as } j \to \infty. \end{cases} \tag{1.68}$$

It can be shown that every solution of (1.67) is such that

$$u(x) \longrightarrow 0, \quad \text{as } |x| \longrightarrow \infty, \tag{1.69}$$

uniformly with respect to x.

1.2.5 Results on the Fourier transform

The Fourier transform is used in the form

$$Ff(\xi) = \hat{f}(\xi) = \int e^{-2\pi x\xi} f(x)dx, \tag{1.70}$$

where

$$\langle \xi \rangle = (1 + |\xi|^{1/2}).$$

For $s \in \mathbb{R}$, we define

$$L_s^2(\mathbb{R}^n) = \{g \in S'(\mathbb{R}^n) : \langle \xi \rangle^s g(\xi) \in L^2(\mathbb{R}^n)\}. \tag{1.71}$$

The Sobolev spaces and local Sobolev spaces, respectively, are defined by

$$H^s(\mathbb{R}^n) = \{g \in S'(\mathbb{R}^n) : \hat{g} \in L_s^2(\mathbb{R}^n)\}, \tag{1.72}$$

and

$$H_{\text{loc}}^s(\mathbb{R}^n) = \{g \in S'(\mathbb{R}^n) : \psi g \in H^s(\mathbb{R}^n),\ \forall \psi \in \mathscr{D}(\mathbb{R}^n)\}. \tag{1.73}$$

The distribution

$$(s + i0)^\lambda \in S'(\mathbb{R}^n),$$

is defined as

$$\lim_{\epsilon \to 0} (x^2 + \epsilon^2)^{\lambda/2} e^{i\lambda \arg(x + i\epsilon)}, \quad \lambda \in \mathbb{C}.$$

Its Fourier transform is given by

$$F((x + i0)\lambda)(\xi) = \frac{(2\pi)^{-\lambda}}{\Gamma(-\lambda)} e^{i\lambda\pi/2} \xi_+^{-\lambda-1}, \tag{1.74}$$

where ξ_+^μ is a pseudo function.

Let $S, T \in S'(\mathbb{R}^n)$. The S'-convolution of S and T is said to exist, if

$$(\phi * S^-)T \in \mathscr{D}'_{L^1}(\mathbb{R}^n), \quad \forall \phi \in S(\mathbb{R}^n),$$

where $S^{-}(x) = S(-x)$. In this case, the convolution is defined by

$$\langle S * T, \phi\rangle = \langle((\phi * S^{-1})T, 1\rangle,$$

and

$$S * T \in S'(\mathbb{R}^n).$$

Let

$$u, v \in S'(\mathbb{R}^n).$$

If the S'-convolution of Fu and Fv exists, we may define the Fourier product

$$u.v = F^{-1}(Fu * Fv).$$

Let Γ be a closed, acute, convex cone in $\mathbb{R}^n$ and $s \in \mathbb{R}$. Define

$$H^s_\Gamma(\mathbb{R}^n) = \{g \in H^s(\mathbb{R}^n) : \operatorname{supp} \widehat{g} \subset \Gamma\} = S'(\mathbb{R}^n) \cap H^s(\mathbb{R}^n). \tag{1.75}$$

The product of two members $f \in H^s_\Gamma(\mathbb{R}^n)$ and $g \in H^s_\Gamma(\mathbb{R}^n)$ is understood in the sense of the Fourier product.

Proposition 1.1. 1. *Let*

$$s_1 \leq 0, \quad s_2 \leq \frac{n}{2}, \quad f \in H^{s_1}_\Gamma(\mathbb{R}^n), \quad g \in H^{s_2}_\Gamma(\mathbb{R}^n).$$

Then

$$f.g \in H^\sigma_\Gamma(\mathbb{R}^n), \quad \sigma < -\frac{n}{2} + s_1 + s_2.$$

2. *Let*

$$s_1 \geq 0, \quad s_2 \in \mathbb{R}, \quad f \in H^{s_1}_\Gamma(\mathbb{R}^n), \quad g \in H^{s_2}_\Gamma(\mathbb{R}^n).$$

Then

$$f.g \in H^\sigma_\Gamma(\mathbb{R}^n), \quad \sigma \leq s_1, \quad \sigma < -\frac{n}{2} + s_2.$$

Proof. Case 1: Assume that Γ is the positive coordinate cone.

1. Assume first that Γ is the positive coordinate cone

$$\Gamma = \{\xi \in \mathbb{R}^n : \xi_i \leq 0, i = 1, 2, \ldots n\},$$

Using Minkowski's inequality, noting that

$$1_{[0,\xi]}(\mu) = 1_{[\mu,\infty)}(\xi),$$

we have

$$\begin{aligned}
\|\widehat{f} * \widehat{g}\|_{L^2_\sigma} &= \left(\int_0^\infty \left|\int_0^\xi \widehat{f}(\xi-\mu)\widehat{g}(\mu)d\mu\right|^2 \langle\xi\rangle^{2\sigma} d\xi\right)^{1/2} \\
&= \left(\int_0^\infty \left|\int_0^\infty 1_{[0,\xi]}(\mu)\widehat{f}(\xi-\mu)\widehat{g}(\mu)d\mu\right|^2 \langle\xi\rangle^{\sigma} d\xi\right)^{1/2} \\
&\le \left(\int_0^\infty \int_0^\infty 1_{[0,\xi]}(\xi)|\widehat{f}(\xi-\mu)|^2 |\widehat{g}(\mu)|^2 \langle\xi\rangle^{2\sigma} d\xi\right)^{1/2} d\mu \\
&= \left(\int_0^\infty \int_\mu^\infty |\widehat{f}(\xi-\mu)|^2 |\widehat{g}(\mu)|^2 \langle\xi\rangle^{2\sigma} d\xi\right)^{1/2} d\mu \\
&= \left(\int_0^\infty \int_0^\infty |\widehat{f}(\xi)|^2 |\widehat{g}(\mu)|^2 \langle\xi+\mu\rangle^{2\sigma} d\xi\right)^{1/2} d\mu.
\end{aligned}$$

For

$$\xi \ge 0, \quad \mu \ge 0, \quad s_1 \le 0, \quad \sigma - s_1 \le 0,$$

we have

$$\begin{aligned}
\langle\xi+\mu\rangle^{2\sigma} &= \langle\xi+\mu\rangle^{2s_1}\langle\xi+\mu\rangle^{2\sigma-2s_1} \\
&\le \langle\xi\rangle^{2s_1}\langle\mu\rangle^{2\sigma-2s_1} \\
&= \langle\xi\rangle^{2s_1}\langle\mu\rangle^{2s_2+2\sigma-2s_1-2s_2}.
\end{aligned} \tag{1.76}$$

Thus,

$$\begin{aligned}
\|\widehat{f} * \widehat{g}\|_{L^2_\sigma} &\le \left(\int_0^\infty |\widehat{f}(\xi)|^2 \langle\xi\rangle^{2s_1} d\xi\right)^{1/2} \int_0^\infty |\widehat{g}(\mu)|\langle\mu\rangle^{s_2}\langle\mu\rangle^{\sigma-s_1-s_2} d\mu \\
&\le \|f\|_{H^{s_1}}\|g\|_{H^{s_2}}\left(\langle\mu\rangle^{2\sigma-2s_1-2s_2} d\mu\right)^{1/2}.
\end{aligned} \tag{1.77}$$

For

$$\sigma < -\frac{n}{2} + s_1 + s_2,$$

the integral in (1.77) is finite.

2. We use the fact that

$$s_1 \geq 0, \quad \sigma - s_1 \leq 0.$$

We have by (1.76) and thanks to Hölder's inequality to get

$$\begin{aligned}\|\widehat{f} * \widehat{g}\|_{L^2_\sigma} &\leq \left(\int_0^\infty |\widehat{f}(\xi)|^2 \langle\xi\rangle^{2s_1} d\xi\right)^{1/2} \int_0^\infty |\widehat{g}(\mu)| \langle\mu\rangle^{s_2} \langle\mu\rangle^{\sigma - s_2} d\mu \\ &\leq \|f\|_{H^{s_1}} \|g\|_{H^{s_2}} \left(\langle\mu\rangle^{2\sigma - 2s_2} d\mu\right)^{1/2}. \end{aligned} \tag{1.78}$$

For

$$\sigma < -\frac{n}{2} + s_2,$$

the integral in (1.78) is finite.

Case 2: Assume that Γ is an arbitrary convex cone, and note that if the supports of $\widehat{f}$ and $\widehat{g}$ are contained in Γ, it is the support of $\widehat{f} * \widehat{g}$. Since Γ is acute, one may assume (after rotation) that

$$\Gamma \subset \{(\xi', \xi_n) \in \mathbb{R}^n : \xi_n \geq \alpha|\xi'|\}, \quad \alpha > 0,$$

where

$$\xi' = (\xi_1, \xi_2, \ldots, \xi_n).$$

Using a dilation,

$$\tilde{\xi_n} = \beta\xi_n, \quad \tilde{\xi'} = \xi'.$$

We may make the opening angle arbitrarily small. After a further rotation, one may assume that

$$\Gamma \subset \{\xi \in \mathbb{R}^n : xi_n \geq 0, i = 1, 2, \ldots, n\},$$

which is case 1. Rotations and dilations do not change $H^s(\mathbb{R}^n)$. □

2 Semilinear dissipative wave equations in $\mathbb{R}^n$

2.1 Property solutions with compactly supported initial data

2.1.1 Wave equation with frictional damping

The linear problem will first concern

$$\frac{\partial^2 u}{\partial t^2} - \Delta u + \frac{\partial u}{\partial t} = 0, \tag{2.1}$$

with initial data

$$u(x, t = 0) - u_0(x), \quad \frac{\partial u}{\partial t}(x, t = 0) - u_1(x). \tag{2.2}$$

Proposition 2.1. *Let $n \geq 0$. Then for*

$$(u_0, u_1) \in (H^1(\mathbb{R}^n) \cap L^m(\mathbb{R}^n)) \times (L^2(\mathbb{R}^n) \cap L^m(\mathbb{R}^n)), \quad m \in [1, 2],$$

the weak solution of (2.1)–(2.2) *is given as*

$$u \in C([0, \infty), H^1(\mathbb{R}^n)) \cap C^1([0, \infty), L^2(\mathbb{R}^n))$$

and has the decay estimates

$$\|u\|_2^2 \leq C(\|u_0\|_2 + \|u_0\|_m + \|u_1\|_2 + \|u_1\|_m)^2 (1 + t)^{-n/(1/m - 1/2)},$$

and

$$\left\|\frac{\partial u}{\partial t}\right\|_2^2 + \|\nabla u\|_2^2 \leq C(\|u_0\|_2 + \|u_0\|_m + \|u_1\|_2 + \|u_1\|_m)^2 (1 + t)^{-n/(1/m\ 1/2)}.$$

Proof. Since

$$u_0 \in H^1(\mathbb{R}^n) \cap L^m(\mathbb{R}^n),$$

and

$$u_1 \in L^2(\mathbb{R}^n) \cap L^m(\mathbb{R}^n)$$

and can be approximated by smooth functions

$$\phi_n \subset C_0^\infty(\mathbb{R}^n),$$

and

https://doi.org/10.1515/9783111633787-002

$$\psi_n \subset C_0^\infty(\mathbb{R}^n),$$

satisfying

$$\|\phi_n - u_0\|_{\dot{H}^1} + \|\phi_n - u_0\|_m \to 0, \quad n \to \infty,$$

and

$$\|\psi_n - u_1\|_2 + \|\psi_n - u_1\|_m \to 0, \quad n \to \infty.$$

□

2.1.2 Semilinear dissipative wave equation

We will use the compacted support initial data to analyze the problems in an unbounded domain. We are concerned with the following Cauchy problem:

$$\frac{\partial^2 u}{\partial t^2} - \Delta u + \frac{\partial u}{\partial t} = |u|^{p-1}u, \tag{2.3}$$

with initial data

$$u(x, t = 0) - u_0(x), \quad \frac{\partial u}{\partial t}(x, t = 0) - u_1(x). \tag{2.4}$$

We will distinguish many cases according to the dimension of the space $\mathbb{R}^n$.

Theorem 2.1. *Let $n = 1, 2$ with the exponent p satisfy*

$$1 + \frac{2m}{n} \le p \le +\infty, \quad n = 1, 2, \ m \in [1, 2]. \tag{2.5}$$

Then there exists $\epsilon_0 > 0$ such that

$$(u_0, u_1) \in (H^1(\mathbb{R}^n) \cap L^m(\mathbb{R}^n)) \times (L^2(\mathbb{R}^n) \cap L^m(\mathbb{R}^n)), \quad m \in [1, 2]$$

further satisfy

$$I_{0,u} = \|u_0\|_{L^m} + \|u_0\|_{H^1} + \|u_1\|_{L^m} + \|u_1\|_{L^2} + \le \epsilon_0.$$

The problems (2.3)–(2.4) *admit a global solution*

$$u \in C([0, \infty), H^1(\mathbb{R}^n)) \cap C^1([0, \infty), L^2(\mathbb{R}^n)),$$

satisfying

$$\|u\|^2 \le CI_{0,u}^2(1+t)^{n(\frac{1}{m}-\frac{1}{2})},$$

$$\left\|\frac{\partial u}{\partial t}\right\|^2 + \|\nabla u\|^2 \le CI_{0,u}^2(1+t)^{-1-n(\frac{1}{m}-\frac{1}{2})}, \quad C>0.$$

Proof. Define a semigroup

$$S(t) : H^1(\mathbb{R}^n)\times L^2(\mathbb{R}^n) \to H^1(\mathbb{R}^n)\times L^2(\mathbb{R}^n),$$

by

$$S(t) : \begin{bmatrix} u_0 \\ u_1 \end{bmatrix} \to \begin{bmatrix} v \\ \frac{\partial v}{\partial t} \end{bmatrix},$$

where

$$v \in C([0,+\infty), H^1(\mathbb{R}^n)) \cap C^1([0,+\infty), L^2(\mathbb{R}^n))$$

is the unique solution to the linear problem (2.1).

Lemma 2.1. *If $\beta > 1$ and $\eta \le \beta$, then there exists a constant $C_{\beta,\eta} > 0$ depending only on β and η such that*

$$\int_0^t (1+t-s)^{-\eta}(1+s)^{-\beta}ds \le C_{\beta,\eta}(1+t)^{-\eta}, \quad t \ge 0.$$

By a standard semigroup theory, the nonlinear problems (2.3)–(2.4) are rewritten as the integral equation

$$U(t) = S(t)U_0 + \int_0^t S(t-s)F(s)ds, \tag{2.6}$$

where

$$U(t) = \begin{bmatrix} u \\ \frac{\partial u}{\partial t} \end{bmatrix}, \quad U_0 = \begin{bmatrix} u_0 \\ u_1 \end{bmatrix}, \quad F(s) = \begin{bmatrix} 0 \\ f(u) \end{bmatrix},$$

with $f(u) = |u|^{p-1}u$. In order to show global existence, it suffices to obtain the a priori estimates for

$$\left\|\frac{\partial u}{\partial t}\right\| + \|\nabla u\|.$$

By the way, in the occasion of the proof of the next proposition, we have the following.

Proposition 2.2. [1] *Let $1 < p \le \frac{n}{[n-2]^+}$, $n \ge 1$. For each*

$$(u_0, u_1) \in H^1(\mathbb{R}^n) \times L^2(\mathbb{R}^n),$$

there exists a maximal existence time $T_{\max} > 0$ such that the problems (2.3)–(2.4) have a unique solution

$$u \in C(0, T_{\max}, H^1(\mathbb{R}^n)) \cap C^1(0, T_{\max}, L^2(\mathbb{R}^n)).$$

If $T_{\max} < \infty$, then

$$\lim_{t \to T_{\max}} \left[\left\| \frac{\partial u}{\partial t} \right\|_{L^2} + \|\nabla u\|_{L^2} \right] = \infty.$$

Lemma 2.2. *Under the assumptions as in Theorem 2.1, we have*

$$\|S(t)U_0\|_E \le C I_{0,u} (1+t)^{-1/2-(n/2)(1/m-1/2)},$$

on $[0, T_{\max})$, where we set

$$\left\| \begin{bmatrix} u \\ v \end{bmatrix} \right\|_E = \|v\| + \|\nabla u\|.$$

Furthermore, if

$$I(s) = \|f\| + \|f\|_m < +\infty, \tag{2.7}$$

for each $s \in [0, t]$ with $t \in [0, T_{\max})$. Then from Proposition 2.1, we have

$$\|S(t-s)F(s)\|_E \le C I(s)(1+t-s)^{-1/2-(n/2)(1/m-1/2)}. \tag{2.8}$$

Thus, by (2.5) one can estimate $U(t)$ as follows:

$$\begin{aligned} \|U(t)\|_E \le{} & I_{0,u}(1+t)^{-1/2-(n/2)(1/m-1/2)} \\ & + C \int_0^t (1+t-s)^{-1/2-(n/2)(1/m-1/2)} I(s)\,ds. \end{aligned} \tag{2.9}$$

Take $K > 0$ so large and choose $T \in (0, T_m)$ so small such as

$$\|U\|_E \le K I_{0,u} (1+t)^{-1/2-(n/2)(1/m-1/2)} \tag{2.10}$$

1 W. A. Strauss. Nonlinear wave equations in: CBMS Regional Conf. Ser. in Math., American Mathematical Society, Providence, RI, 1989.

$$\|u\|_E \le KI_{0,u}(1+t)^{-(n/2)(1/m-1/2)}. \tag{2.11}$$

By Lemma 1.10 and (2.5), we have

$$\|f\|_m = \|u\|_{mp}^p \le K_0\|u\|^{p(1-\theta_1)}\|\nabla u\|^{p\theta_1},$$

with $\theta_1 = n(mp-2)/2mp \in (0,1]$. Note that (2.5) implies

$$mp > 2, \quad m \in [1,2]. \tag{2.12}$$

Similarly, we have

$$\|f\| \le K_0\|u\|^{p(1-\theta_2)}\|\nabla u\|^{p\theta_2},$$

with $\theta_2 = n(p-1)/2p \in (0,1]$. Therefore,

$$\begin{aligned}
\|f\|_m &\le K_0\{KI_{0,u}(1+s)^{-(n/2)(1/m-1/2)}\}^{p(1-\theta_1)} \\
&\quad\times \{KI_{0,u}(1+s)^{-1/2-(n/2)(1/m-1/2)}\}^{p\theta_1} \\
&= K_0K^pI_{0,u}^p(1+s)^{-p(\theta_1/2-(n/2)(1/m-1/2))}, \\
\|f\| &\le K_0\{KI_{0,u}(1+s)^{-(n/2)(1/m-1/2)}\}^{p(1-\theta_2)} \\
&\quad\times \{KI_{0,u}(1+s)^{-1/2-(n/2)(1/m-1/2)}\}^{p\theta_2} \\
&= K_0K^pI_{0,u}^p(1+s)^{-p(\theta_2/2-(n/2)(1/m-1/2))}.
\end{aligned}$$

Summarizing these calculations, we have the following lemma, which shows the validity of the condition (2.7).

Lemma 2.3. *Let* (2.10) *and* (2.11) *hold. Then we have*

$$\begin{aligned}
\|f(u)\| &\le K_0K^pI_{0,u}^p(1+t)^{-p(n/2)(1/m-1/2)-n(mp-2)/4m}, \\
\|f(u)\| &\le K_0K^pI_{0,u}^p(1+t)^{-p(n/2)(1/m-1/2)-n(p-1)/4m}.
\end{aligned}$$

Applying Lemma 2.2, (2.7) to (2.6), we get

$$\begin{aligned}
\|U(t)\|_E &\le I_{0,u}(1+t)^{-1/2-(n/2)(1/m-1/2)} \\
&\quad + C\int_0^t (1+t-s)^{-1/2-(n/2)(1/m-1/2)}K_0K^pI_{0,u}^p \\
&\quad \times \{(1+s)^{-\gamma_1} + (1+s)^{-\gamma_2}\}ds \\
&\le CI_{0,u}(1+t)^{-1/2-(n/2)(1/m-1/2)} \\
&\quad + CK_0K^pI_{0,u}^p\int_0^t (1+t-s)^{-1/2-(n/2)(1/m-1/2)}(1+s)^{-\gamma_1}ds,
\end{aligned}$$

where

$$\gamma_1 = \frac{n}{2}p\left(\frac{1}{m} - \frac{1}{2}\right) + \frac{n(mp-2)}{4m},$$
$$\gamma_2 = \frac{n}{2}p\left(\frac{1}{m} - \frac{1}{2}\right) + \frac{n(p-1)}{4}. \tag{2.13}$$

By (2.5), we have $\gamma_2 \geq \gamma_1 > 1$. Then from Lemma 2.1, it follows

$$\begin{aligned}\|U(t)\|_E &\leq CI_{0,u}(1+t)^{-1/2-(n/2)(1/m-1/2)} \\ &\quad + CK_0K^pI_{0,u}^p(1+t)^{-1/2-(n/2)(1/m-1/2)}, \quad C > 0.\end{aligned}$$

Set $Q_0(I_{0,u}, K) = C(1 + K_0K^pI_{0,u}^{p-1})$.

Lemma 2.4. *Let* (2.10) *and* (2.11) *hold. Then we have*

$$\|U(t)\|_E \leq I_{0,u}Q_0(I_{0,u}, K)(1+t)^{-1/2-(n/2)(1/m-1/2)}.$$

We derive the L^2 estimates for the local solution $u(t,x)$ to the problems (2.3)–(2.4). Indeed, we have from (2.6) that

$$\begin{aligned}\|u\| &\leq CI_{0,u}(1+t)^{-(n/2)(1/m-1/2)} \\ &\quad + C\int_0^t (1+t-s)^{-(n/2)(1/m-1/2)}I(s)ds.\end{aligned} \tag{2.14}$$

Therefore, it follows from Lemma 2.3 and the similar argument to Lemma 2.4 that

$$\begin{aligned}\|u\| &\leq CI_{0,u}(1+t)^{-(n/2)(1/m-1/2)} \\ &\quad + C\int_0^t (1+t-s)^{-(n/2)(1/m-1/2)}K_0K^pI_{0,u}^p \\ &\quad \times \left[(1+s)^{-n(mp-2)/4m-p(n/2)(1/m-1/2)}\right. \\ &\quad \left. + (1+s)^{-n(p-1)/4-p(n/2)(1/m-1/2)}\right]ds \\ &\leq CI_{0,u}(1+t)^{-(n/2)(1/m-1/2)} \\ &\quad + CK_0K^pI_{0,u}^p\int_0^t (1+t-s)^{-(n/2)(1/m-1/2)} \\ &\quad \times (1+s)^{-n(mp-2)/4m-p(n/2)(1/m-1/2)}ds\end{aligned}$$

with some generous constant $C > 0$, when

$$n(mp-2)/4m + p(n/2)(1/m-1/2) > 1.$$

This together with Lemma 2.1 implies

$$\|u\| \le CI_{0,u}(1+t)^{-(n/2)(1/m-1/2)} + CK_0K^pI_{0,u}^p(1+t)^{-(n/2)(1/m-1/2)}.$$

Thus, we have the following.

Lemma 2.5. *Let* (2.10) *and* (2.11) *hold. Then we have*

$$\|u(t)\| \le I_{0,u}Q_0(I_{0,u},K)(1+t)^{-(n/2)(1/m-1/2)}.$$

Take $K > C$ so large and take $I_{0,u}$ so small such that

$$CK_0K^pI_{0,u}^{p-1} < K - C. \tag{2.15}$$

For such $K > 0$ and $I_{0,u}$, we have

$$Q_0(I_{0,u},K) < K.$$

Therefore, by combining this with Lemma 2.4 and Lemma 2.5, we have

$$\|U(t)\|_E < KI_{0,u}(1+t)^{-1/2-(n/2)(1/m-1/2)}, \tag{2.16}$$
$$\|u\| < KI_{0,u}(1+t)^{-(n/2)(1/m-1/2)}. \tag{2.17}$$

Then (2.10), (2.11) and (2.16), (2.17) shows that under assumption (2.15), the local solution $u(t,\cdot)$ exists globally in time and these estimates hold in fact for all $t \ge 0$. Taking

$$\epsilon_0 = \left(\frac{K-c}{CK_0K^p}\right)^{1/(p-1)}.$$

The proof of Theorem 2.1 is now completed. □

Theorem 2.2. *Let* $3 \le n \le 6$ *and suppose that*

$$\frac{\sqrt{n^2+16n}-n}{4} \le m < \min\left\{2, \frac{n}{n-2}\right\},$$
$$1 + \frac{2m}{n} < p \le \frac{n}{n-2}.$$

Then there exists $\epsilon_0 > 0$ *such that*

$$(u_0,u_1) \in (H^1(\mathbb{R}^n) \cap L^m(\mathbb{R}^n)) \times (L^2(\mathbb{R}^n) \cap L^m(\mathbb{R}^n)), \quad m \in [1,2]$$

further satisfy

$$I_{0,u} = \|u_0\|_{L^m} + \|u_0\|_{H^1} + \|u_1\|_{L^m} + \|u_1\|_{L^2} \le \epsilon_0.$$

The problems (2.3)–(2.4) *admit a global solution*

$$u \in C([0,\infty), H^1(\mathbb{R}^n)) \cap C^1([0,\infty), L^2(\mathbb{R}^n)),$$

satisfying

$$\|u\|^2 \le C I_{0,u}^2 (1+t)^{n(\frac{1}{m}-\frac{1}{2})},$$

$$\left\|\frac{\partial u}{\partial t}\right\|^2 + \|\nabla u\|^2 \le C I_{0,u}^2 (1+t)^{-1-n(\frac{1}{m}-\frac{1}{2})}, \quad C > 0.$$

Proof. By (2.12) and (2.13), we have

$$mp > 2, \quad n(mp-2)/2mp \in (0,1],$$
$$\gamma_2 \ge \gamma_1 > 1.$$

So, the same argument as in Theorem 2.1 can proceed and we have the desired conclusion. □

Finally, let us prove Theorem 2.3. For this purpose, we set

$$E(u,v) = J(u) + \frac{1}{2}\|v\|^2,$$
$$J(u) = \frac{1}{2}\|\nabla u\|^2 - \frac{1}{p+1}\|u\|_{p+1}^{p+1}.$$

In the occasion of the proof of Theorem 2.3, the following blowup results.

Proposition 2.3. [2] *Let* $(u_0, u_1) \in H^1(\mathbb{R}^n) \times L^2(\mathbb{R}^n)$ *satisfy* $E(0) < 0$, *then the associated solution to the problems* (2.3)–(2.4) *do not exist globally in time.*

Theorem 2.3. *Let* $n \ge 1$ *and*

$$1 < p < 1 + \frac{2m}{m}, \quad m \in [1,2]. \tag{2.18}$$

Then for any $\epsilon > 0$, *there exists*

$$(u_0, u_1) \in (H^1(\mathbb{R}^n) \cap L^m(\mathbb{R}^n)) \times (L^2(\mathbb{R}^n) \cap L^m(\mathbb{R}^n)), \quad m \in [1,2]$$

further satisfy

$$I_{0,u} = \|u_0\|_{L^m} + \|u_0\|_{H^1} + \|u_1\|_{L^m} + \|u_1\|_{L^2} \le \epsilon,$$

2 H. A. Levine. Some additional remarks on the nonexistence of global solutions to nonlinear wave equations. SIAM J. Math. Anal. 5 (1974), 138–146.

such that the associated solution $u(t,x)$ to the problems (2.3)–(2.4) do not exist globally in time.

Proof. Let $\phi \in H^1(\mathbb{R}^n) \times L^m(\mathbb{R}^n)/\{0\}$ and $\alpha > 0$ such that

$$n/m < \alpha < 2/(p-1)$$

be arbitrarily fixed. This is guaranteed by an assumption (2.18). For $\lambda > 0$, set

$$\phi_\lambda(x) = \lambda^\alpha \phi(\lambda x).$$

Then

$$E(\phi_\lambda, 0) = J(\phi_\lambda) < 0, \tag{2.19}$$

for sufficiently small $\lambda > 0$. Indeed, first we have

$$J(\phi_\lambda) = \frac{1}{2}\lambda^{2\alpha+2-n}\|\nabla\phi\|^2 - \frac{1}{p+1}\lambda^{\alpha(p+1)-n}\|\phi\|_{p+1}^{p+1}.$$

Since $\alpha < 2/(p-1)$ implies

$$2\alpha + 2 - n > \alpha(p+1) - n,$$

the desired (2.19) can be derived if we take $\lambda > 0$ small enough. Therefore, the solution $u_\lambda(t,x)$ to the problems (2.3)–(2.4) with $(\phi_\lambda, 0)$ as the initial data blows up in a finite time because of Proposition 2.3. On the other hand, we see that

$$\begin{aligned}&\|\phi_\lambda\|_m + \|\phi_\lambda\| + \|\nabla\phi_\lambda\| \\ &\quad = \lambda^{\alpha-n/m}\|\phi\|_m + \lambda^{\alpha-n/2}\|\phi\| + \lambda^{\alpha+1-n/2}\|\nabla\phi\|.\end{aligned}$$

Since $m \le 2$ and $n/m < \alpha$ imply

$$0 < \alpha - n/m \le \alpha - n/2 < \alpha - n/2 + 1.$$

For $\epsilon > 0$, take $\lambda > 0$ further sufficiently small, we have

$$\|\phi_\lambda\|_m + \|\phi_\lambda\| + \|\nabla\phi_\lambda\| \le \epsilon,$$

which implies the desired statement and then Theorem 2.3 is proved. □

2.2 Noncompactly supported initial data

For noncompactly supported initial data, which has a small energy, we shall derive a global in time existence of (2.4)–(2.4) with $|u|^p$ instead of $|u|^{p-1}u$ in weighted function spaces. For this end, we assume the power type of nonlinearity $|u|^p$ satisfying (1.54) and let the space

$$X_1(0,T)(\mathbb{R}^n) = C([0,T), H^1(\mathbb{R}^n)) \cap C^1([0,T), L^2(\mathbb{R}^n)), \quad T \in [0,\infty). \tag{2.20}$$

We will prove that the energy associated to the unique global solution exists and concentrates in a ball much smaller than $|x| < t$ as t is large enough even if the solution does not have a compact support. The proof will be relied on in the next decay estimate on the linear case (2.1)–(2.2) in Proposition 2.1. For this end, let ψ in (2.6) be a weight function satisfy

$$\begin{aligned} \psi_t &< 0, \\ \psi_t + |\psi_t|^2 &= 0, \\ \Delta\psi &= \frac{n}{2(1+t)}. \end{aligned} \tag{2.21}$$

Proposition 2.4. *Let $n \geq 1$ and*

$$1 < p \leq \frac{n}{n-2}, \quad n \geq 3,$$

or

$$1 < p < +\infty, \quad n = 1, 2.$$

Then, for

$$(u_0, u_1) \in H^1(\mathbb{R}^n) \times L^2(\mathbb{R}^n),$$

satisfying

$$\int_{\mathbb{R}^n} e^{|x|^2}(|u_1|^2 + |\nabla u_0|^2 + |u_0|^2) < +\infty.$$

Then there exists a maximal existence time $T_m > 0$ such that problems (2.3)–(2.4) have a unique solution $u \in X_1$ and satisfy

$$\sup_{[0,T]}\left[\|e^{\psi(t)}\nabla u\|_2 + \left\|e^{\psi(t)}\frac{\partial u}{\partial t}\right\|_2 + \|e^{\psi(t)}u\|_2\right] < +\infty$$

for any $T < T_m$. If, in particular, $T_m < +\infty$, then it is true that

$$\lim_{t\to T_m}\sup_{[0,T]}\left[\|e^{\psi(t)}\nabla u\|_2 + \left\|e^{\psi(t)}\frac{\partial u}{\partial t}\right\|_2 + \|e^{\psi(t)}u\|_2\right] = +\infty.$$

Lemma 2.6. *Let u be a local solution to problems (2.3)–(2.4) on $[0, T_m)$, as in Proposition 2.4. Then for all $t \in [0, T_m)$, we have*

$$\|e^{\psi(t)}Du\|_2 \leq CI_0 + C\Big(\sup_{s\in[0,t]} (1+s)^\sigma \|e^{\gamma\psi(t)}u\|_{p+1}\Big)^{(p+1)/2}$$

where $\gamma \in (\frac{2}{p+1}, 1]$, $C > 0$, $\sigma > 0$ and $D = (\partial_t, \nabla)$.

Proof. We multiply (2.3) by $e^{2\psi}\frac{\partial u}{\partial t}$ and by (2.21) to get

$$\begin{aligned}\frac{-2\psi_t}{p+1}e^{2\psi}|u|^p u &= \frac{d}{dt}\Big(\frac{e^{2\psi}}{2}\Big(\Big|\frac{\partial u}{\partial t}\Big|^2 + |\nabla u|^2\Big) - \frac{e^{2\psi}}{p+1}|u|^p u\Big)\\ &\quad - \operatorname{div}\Big(e^{2\psi}\frac{\partial u}{\partial t}\nabla u\Big) - \frac{e^{2\psi}}{\psi_t}\Big|\psi_t\nabla u - \frac{\partial u}{\partial t}\nabla\psi\Big|^2 - \psi_t\frac{\partial u}{\partial t}^2 e^{2\psi}.\end{aligned}$$

Integrating over $[0, t] \times \mathbb{R}^n$ and recalling (2.21), we get

$$\begin{aligned}\frac{2}{p+1}\int_0^t\int_{\mathbb{R}^n}\psi_s e^{2\psi}|u|^p u\,dxds &\leq -\int_0^t\int_{\mathbb{R}^n}\frac{d}{ds}\Big(\frac{e^{2\psi}}{2}(|u_s|^2 + |\nabla u|^2) - \frac{e^{2\psi}}{p+1}|u|^p u\Big)dxds\\ &\quad + \int_0^t\int_{\mathbb{R}^n}\operatorname{div}(e^{2\psi}u_s\nabla u).\end{aligned}$$

Since

$$\int_0^t\int_{\mathbb{R}^n}\operatorname{div}(e^{2\psi}u_s\nabla u) = 0,$$

we find the following estimate with some constant $C > 0$:

$$\begin{aligned}\|e^{\psi}Du\|^2 &\leq CI_0^2 + C\|e^{(2/(p+1))\psi}u\|_{p+1}^{p+1}\\ &\quad + C\int_0^t\int_{\mathbb{R}^n}|\psi_s|e^{(2-\gamma(p+1))\psi}e^{\gamma(p+1)\psi}|u|^{p+1}dxds.\end{aligned}$$

Thus,

$$\begin{aligned}\|e^{\psi}Du\|^2 &\leq CI_0^2 + C\|e^{(2/(p+1))\psi}u\|_{p+1}^{p+1}\\ &\quad + C\int_0^t\Big(\max_{x\in\mathbb{R}^n}\phi(x,s)\Big)\|e^{\gamma\psi}u\|_{p+1}^{p+1}ds,\end{aligned}$$

where

$$\phi(x,s) = |\psi_s| e^{(2-\gamma(p+1))\psi},$$

with

$$\gamma > \frac{2}{p+1}.$$

Actually, the above estimate holds for any weak solution u by the density argument.

We shall show

$$\max_{x\in\mathbb{R}^n} \phi(x,t) \le \frac{C_\gamma}{t+1}. \tag{2.22}$$

Indeed, notice that by (1.53), we have

$$\psi_t = \frac{|x|^2}{1(t+1)^2}. \tag{2.23}$$

Thus, it follows from

$$\sup_{r\ge 0}(re^{(2-\gamma(p+1)/r)}) < +\infty,$$

$$\begin{aligned}\phi(x,s) &\le \frac{|x|^2}{4(1+s)^2} e^{(2-\gamma(p+1)|x|^2/4(s+1))} \\ &\le \frac{1}{s+1}\left\{\frac{|x|^2}{4(s+1)^2} e^{(2-\gamma(p+1)|x|^2/4(s+1))}\right\} \\ &\le \frac{C}{s+1}.\end{aligned} \tag{2.24}$$

In fact, from (2.11) and (2.22), we have

$$\begin{aligned}\|e^{\psi}Du\|^2 &\le CI_0^2 + C\|e^{(2/(p+1))\psi}u\|_{p+1}^{p+1} + C\int_0^t \frac{1}{s+1}\|e^{\gamma\psi}u\|_{p+1}^{p+1} ds \\ &\le CI_0^2 + C\Big\{\sup_{s\in[0,t]}(s+1)^{\sigma}\|e^{\sigma\psi}u\|_{p+1}\Big\}^{p+1} \\ &\quad + C\int_0^t \frac{1}{(s+1)^{1+\sigma(p+1)}}\Big\{\sup_{s\in[0,t]}(s+1)^{\sigma}\|e^{\sigma\psi}u\|_{p+1}\Big\}^{p+1} ds \\ &\le CI_0^2 + C\Big\{\sup_{s\in[0,t]}(s+1)^{\sigma}\|e^{\sigma\psi}u\|_{p+1}\Big\}^{p+1},\end{aligned} \tag{2.25}$$

where we have used

$$\int_0^t \frac{1}{(s+1)^{1+\sigma(p+1)}} ds < +\infty.$$

This completes the proof. □

Lemma 2.7. *Let u be a solution as in Proposition 2.4. Then*

$$\|u\|_p^p \le C(s+1)^{n/4} \|e^{\sigma\psi} u\|_{2p},$$

and

$$\|u\|_{2p}^p \le C(s+1)^{n/4} \|e^{\sigma\psi} u\|_{2p},$$

for any $s \in [0, T_m)$.

Proof. From the Schwarz inequality and Lemma 1.11 that

$$\|u\|_p^p \le \Big(\int_{\mathbb{R}^n} e^{-2p\sigma\psi} dx \Big)^{1/2} \Big(\int_{\mathbb{R}^n} e^{2p\sigma\psi} |u|^{2p} dx \Big)^{1/2},$$

which implies the desired estimate. The second one is similar. □

Lemma 2.8. *Let u be a local solution to the Cauchy problems* (2.3)–(2.4) *in* $[0, T_m)$ *as in Proposition* 2.4. *Then the following decay estimates hold: for*

$$D^1 = D, \quad D^0 = 1, \quad j = 0, 1,$$

we have

$$(t + 1^{n/4+j/2}) \|D^j u\|_2 \le C I_0 + C \Big[\sup_{s \subset [0,t]} (s+1)^\beta \|e^{\sigma\psi} u\|_{2p} \Big]^p.$$

Here,

$$\beta = \frac{n}{4p} + \frac{1}{p} + \frac{\eta}{p}, \quad \eta > 0.$$

Proof. In the linear problems (2.1)–(2.2), taking $u_0 = 0$, $u_1 = h \in L^2(\mathbb{R}^n)$. By $S(t)h$, we denote the unique solution of the linear problem and let u be the solution of the nonlinear system (2.3)–(2.4). Then

$$u = S(t)(u_0 + u_1) + \partial_t (S(t) u_0) + \int_0^t S(t-s) |u(s)|^p ds. \tag{2.26}$$

We should estimate $\|\frac{\partial u}{\partial t}\|_2$ and $\|\nabla u\|_2$. Set (linear solution)

$$u_L = S(t)(u_0 + u_1) + \partial_t(S(t)u_0).$$

Then by Proposition 2.1, we have for $m = 1$,

$$\|D^j u_l\|_2 \le C(t+1)^{-n/4-j/2} I_0. \tag{2.27}$$

Next, let us estimate the nonlinear term in (2.26), and we set (nonlinear solution)

$$u_N = S(t)(u_0 + u_1) + \partial_t(S(t)u_0).$$

We split the integral into two parts

$$\begin{aligned} u_N &= \int_0^{t/2} S(t-s)|u(s)|^p ds + \int_{t/2}^{t} S(t-s)|u(s)|^p ds \\ &= u_{N,1} + u_{N,2}. \end{aligned}$$

For the first integral, we apply Proposition 2.1 with $m = 1$, and obtain (for $j = 0, 1$)

$$\|D^j S(t-s)|u|^p\| \le C(1+t-s)^{-n/2-j/2}(\|u^p_{2p}+\|u^p_p). \tag{2.28}$$

Thus, it follows from Lemma 2.7 that for $j = 0, 1$,

$$\|D^j S(t-s)|u|^p\| \le C(1+t-s)^{-n/2-j/2}(s+1)^{n/4}\|e^{\sigma\psi}u\|^p_{2p}.$$

This implies

$$\begin{aligned} &\int_0^{t/2} \|D^j S(t-s)|u|^p\| ds \\ &\quad \le C\int_0^{t/2} (1+t-s)^{-n/2-j/2}(s+1)^{-1-\eta}\big((s+1)^{n/(4p)+1/p+\eta/p}\|e^{\sigma\psi}u\|_{2p}\big)^p ds. \\ &\quad \le C\Big(\sup_{s\in[0,t]}(s+1)^{\beta}\|e^{\sigma\psi}u\|_{2p}\Big)^p \int_0^{t/2}(1+t-s)^{-n/2-j/2}(s+1)^{-1-\eta} ds. \end{aligned}$$

Here,

$$\beta = \frac{n}{4p} + \frac{1+\eta}{p}, \quad \eta > 0.$$

Since

$$\int_0^{t/2}(1+t-s)^{-n/2-j/2}(s+1)^{-1-\eta} ds \le C(t+1)^{-n/4-j/2},$$

we get

$$\|D^j u_{N,1}\| \le \int_0^{t/2} \|D^j S(t-s)|u|^p\| ds \qquad (2.29)$$
$$\le C(t+1)^{-n/4-j/2}\Big(\sup_{s\in[0,t]} (s+1)^\beta \|e^{\sigma\psi} u\|_{2p}\Big)^p.$$

On the other hand, for the estimate of $u_{N,2}$ from Proposition 2.1 with $m = 2$, we have

$$\|D^j S(t-s)|u|^p\| \le C(t-s+1)^{-j/2}\|u\|_{2p}^p$$
$$\le C(t-s+1)^{-j/2}(s+1)^{-n/4-1}\Big(\sup_{s\in[0,t]} (s+1)^{n/(4p)+1/p}\|u\|_{2p}\Big)^p.$$

Thus, since $e^{\sigma\psi} \ge 1$, we have

$$\int_0^{t/2} \|D^j S(t-s)|u|^p\| ds \le C\int_0^{t/2} (t-s+1)^{-j/2}(s+1)^{-n/4-1} ds\Big(\sup_{s\in[0,t]} (s+1)^\beta \|u\|_{2p}\Big)^p$$
$$\le C(t+1)^{-n/4-j/2} ds\Big(\sup_{s\in[0,t]} (s+1)^\beta \|u\|_{2p}\Big)^p$$
$$\le C(t+1)^{-n/4-j/2} ds\Big(\sup_{s\in[0,t]} (s+1)^\beta \|e^{\sigma\psi} u\|_{2p}\Big)^p.$$

This implies

$$\|D^j u_{N,2}\| \le \int_0^{t/2} \|D^j S(t-s)|u|^p\| ds \qquad (2.30)$$
$$\le C(t+1)^{-n/4-j/2}\Big(\sup_{s\in[0,t]} (s+1)^\beta \|e^{\sigma\psi} u\|_{2p}\Big)^p.$$

Finally, (2.27), (2.29) and (2.30) imply the desired estimate. □

Theorem 2.4. *Let $n \le 1$ and*

$$(u_0, u_1) \in H^1(\mathbb{R}^n) \times L^2(\mathbb{R}^n),$$

and assume further (1.54). *Then there exists a positive number $\epsilon > 0$ such that if*

$$I_0^2 = \int_{\mathbb{R}^n} s^{|x|^2}(|u_1|^2 + |u_0|^2 + |\nabla u_0|^2) dx < \epsilon,$$

the corresponding problems (2.3)–(2.4) *have a unique global solution*

$$u \in C([0,T), H^1(\mathbb{R}^n)) \cap C^1([0,T), L^2(\mathbb{R}^n)), \quad T \in [0,\infty),$$

satisfying

$$\left\|\frac{\partial u}{\partial t}\right\|_2 + \|\nabla u\|_2 \le CI_0(1+t)^{-(1/2+n/4)},$$

and

$$\|u\|_2 \le CI_0(1+t)^{-n/4}.$$

In addition, the unique solution satisfy

$$\int\limits_{|x|\ge(1+t)^{\rho+1/2}} \left(\left|\frac{\partial u}{\partial t}\right|^2 + |\nabla u|^2\right)dx \le CI_0^2 e^{(1+t)^{\rho/2}}, \quad \rho > 0, \tag{2.31}$$

with some constant $C > 0$.

Proof. Let

$$W(t) = \|e^{\sigma\psi} Du\|_2 + (t+1)^{n/4+1/2}\|Du\|_2 + (t+1)^{n/4}\|u\|_2.$$

It follows from Lemma 2.6 and Lemma 2.8 that

$$\begin{aligned} W \le CI_0 &+ C\Big(\sup_{s\in[0,t]} (s+1)^{\delta}\|e^{\gamma\psi}u\|_{p+1}\Big)^{(p+1)/2} \\ &+ C\sup_{s\in[0,t]} (s+1)^{\beta}\|e^{\sigma\psi}u\|_{2p})^{p}, \end{aligned}$$

for

$$\gamma > \frac{2}{p+1}, \quad \delta > 0.$$

On the other hand, Lemma 1.12 with $q = 2p$ and $\delta = \sigma$ gives

$$\begin{aligned} \|e^{\sigma\psi}u\|_{2p} &\le C(s+1)^{(1-\theta(2p))/2}\|\nabla u\|_2^{1-\sigma}\|e^{\psi}\nabla u\|_2^{\sigma} \\ &\le C(s+1)^{(1-\theta(2p))/2}\{(1+s)^{n/4+1/2}\|\nabla u\|_2\}^{1-\sigma} \\ &\quad \times (s+1)^{-(1-\sigma)(n/4+1/2)}W(s)^{\sigma} \\ &= C(s+1)^{(1-\theta(2p))/2-(1-\sigma)/(n/4+1/2)}W(s). \end{aligned} \tag{2.32}$$

Similarly, we have

$$\|e^{\gamma\psi}u\|_{p+1} \le C(s+1)^{(1-\theta(p+1))/2-(1-\gamma)/(n/4+1/2)}W(s). \tag{2.33}$$

It follows from (2.32) and (2.33) that

$$W(t) \leq CI_0 + C\Big\{\sup_{s\in[0,1]} (s+1)^{\sigma+(1-\theta(p+1))/2-(1-\gamma)/(n/4+1/2)} W(s)\Big\}^{(p+1)/2}$$
$$+ C\Big\{\sup_{s\in[0,1]} (s+1)^{\beta+(1-\theta(2p))/2-(1-\sigma)/(n/4+1/2)} W(s)\Big\}^{p}.$$

Set

$$\gamma = \frac{2}{p+1} + \epsilon.$$

Then by definition of the quantities $\theta(2p)$ and $\theta(p+1)$, we obtain

$$\beta + \frac{1-\theta(2p)}{2} - (1-\sigma)\Big(\frac{n}{4} + \frac{1}{2}\Big) = \Big(\frac{n}{2p} + \frac{1}{p} - \frac{n}{2}\Big) + \Big(\sigma\Big(\frac{1}{4} + \frac{1}{2}\Big) + \frac{\eta}{p}\Big) < 0.$$

If we choose $\eta > 0$ and $\sigma > 0$ are small enough, because we have assumed

$$1 + \frac{1}{n} < p.$$

Similarly, because of

$$1 + \frac{2}{n} < p,$$

we see that

$$\sigma + \frac{(1-\theta(p+1))}{2} - (1-\gamma)\Big(\frac{n}{4} + \frac{1}{2}\Big) = \frac{2n-2pn+4}{4(p+1)} + \Big(\sigma + \epsilon\Big(\frac{1}{4} + \frac{1}{2}\Big)\Big) < 0,$$

if ϵ and σ are small enough. Note that the parameter $\eta > 0$, $\sigma > 0$ and $\epsilon > 0$ depend only on p and n. Let the parameters above be fixed. Then we have

$$\sup_{s\in[0,t]} W(s) \leq CI_0 + C\Big(\sup_{s\in[0,t]} W(s)\Big)^{(p+1)/2}\Big(\sup_{s\in[0,t]} W(s)\Big)^{p}. \tag{2.34}$$

Define

$$M(t) = \sup_{s\in[0,t]} W(s)$$

and (2.34) can be written as

$$M(t)CI_0 + CW(t)^{(p+1)/2} + CW(t)^{p}.$$

Therefore, by taking I_0 sufficiently small, we have arrived at the good a priori bound:

$$M(t) \le CI_0, \quad t \in [0, T_m).$$

This causes

$$\|e^{\psi} Du\|_2 + (t+1)^{n/4+1/2}\|Du\|_2 + (t+1)^{n/4}\|u\|_2 \le CI_0. \tag{2.35}$$

Furthermore, since

$$u = u_0 + \int_0^t u_s ds,$$

we have

$$e^{\psi} u = e^{\psi} u_0 + \int_0^t e^{\psi} u_s ds,$$

so

$$\begin{aligned} \|e^{\psi} u\| &\le \|e^{\psi} u_0\| + \int_0^t \|e^{\psi}(t) u_s\| ds \\ &\le \|e^{\psi} u_0\| + \int_0^t \|e^{\psi}(0) u_s\| ds, \end{aligned}$$

where we have used the monotonicity of function

$$t \to e^{|x|^2/4(t+1)}.$$

From (2.35), it follows that

$$\|e^{\psi} u\|_2 \le \|e^{\psi} u_0\|_2 + CI_0 t, \quad t \in [0, T_m). \tag{2.36}$$

Thus, if $T_m < +\infty$, (2.35) and (2.36) imply that

$$\lim_{t \to T_m} \left[\left\| e^{\psi} \frac{\partial u}{\partial t} \right\|_2 + \|e^{\psi} u\|_2 + \|e^{\psi} \nabla u\|_2 \right] < +\infty,$$

which contradicts the statement of the latter part of Proposition 2.4. This shows $T_m = +\infty$, and the desired decay estimates follow from (2.35). This completes the proof of a unique global solution.

It remains to prove (2.31). From (2.35), we have

$$CI_0 \ge \int_{|x| \ge (t+1)^{\rho+1/2}} e^{|x|^2/2(t+1)} \left(\left| \frac{\partial u}{\partial t} \right|^2 + |\nabla u|^2 \right) dx$$

$$\geq \int\limits_{|x|\geq(t+1)^{\rho+1/2}} e^{(t+1)^{\rho}/2}\left(\left|\frac{\partial u}{\partial t}\right|^{2} + |\nabla u|^{2}\right)dx,$$

which implies the desired estimate (2.31). The proof is now completed. □

3 Viscoelastic wave equation in $\mathbb{R}^n$

The models here considered are well-known ones and refer to materials with memory as they are termed in the wide literature, which is concerned about their physical, mechanical behavior and the many interesting analytical problems. The physical characteristic property of such materials is that their behavior depends on time not only through the present time but also through their past history.

3.1 Viscoelastic evolution equation

We are concerned with the following Cauchy problem:

$$\frac{\partial^2 u}{\partial t^2} - \Delta u + \int_0^t \mu(t-s)\Delta u(s)ds - \Delta \frac{\partial^2 u}{\partial t^2} = 0, \tag{3.1}$$

with initial data

$$u(x,0) = u_0, \quad \frac{\partial u}{\partial t}(x,0) = u_1. \tag{3.2}$$

The function μ is supposed to be defined in (1.33), (1.35) such that

$$\int_0^\infty \mu(s)e^{(\alpha+2)s}ds < \infty, \quad \alpha > 0. \tag{3.3}$$

Proposition 3.1. *Assume that* (1.33) *and* (3.3) *hold and*

$$u_0, u_1 \in H^1(\mathbb{R}^n),$$

with compact support, and then problems (3.1)–(3.2) *possess at least one global weak solution such that*

$$u \in L^\infty(0,\infty, H_0^1(\mathbb{R}^n)),$$
$$\frac{\partial u}{\partial t} \in L^2(0,\infty, H_0^1(\mathbb{R}^n)),$$
$$\frac{\partial^2 u}{\partial t^2} \in L^2(0,\infty, H_0^1(\mathbb{R}^n)).$$

Lemma 3.1. *If u is the solution of* (3.1)–(3.2), *then*

$$\|u\|_2 \le (L+t)\|\nabla u\|_2, \quad L > 0. \tag{3.4}$$

https://doi.org/10.1515/9783111633787-003

Proof. By Lemma 1.10, we have

$$\|u\|_{p^*} \leq c\|\nabla u\|_2, \quad p^* = \frac{2n}{n-2}, \quad n \leq 3.$$

By using the finite-speed propagation property and Hölder's inequality, we get

$$\begin{aligned}\int_{\mathbb{R}^n} |u|^2 dx &= \int_{B(L+t)} |u|^2 dx \\ &\leq \Big(\int_{B(L+t)} 1 dx\Big)^{1-2/p^*} \Big(\int_{B(L+t)} (|u|^2)^{p^*/2} dx\Big)^{2/p^*} \\ &\leq c(L+t)^2 \|u\|_{p^*}^2,\end{aligned}$$

where $L > 0$. Let

$$B(L) = \{x \in \mathbb{R}^n : |x| < L\},$$

where L is taken so that

$$\operatorname{Supp}\{u_0, u_1\} \subset B(L).$$

Hence,

$$\begin{aligned}\|u\|_2 &\leq c(L+t)\|u\|_{p^*} \\ &\leq c(L+t)\|\nabla u\|_2.\end{aligned}$$

□

Now, we introduce the energy functional

$$E(t) = \frac{1}{2}\Big[\int_{\mathbb{R}^n} \Big|\frac{\partial u}{\partial t}\Big|^2 dx + \Big(1 - \int_0^t \mu(s) ds\Big) \int_{\mathbb{R}^n} |\nabla u|^2 dx + \int_{\mathbb{R}^n} \Big|\nabla \frac{\partial u}{\partial t}\Big|^2 dx + (\mu \circ \nabla u)\Big].$$

If u is a solution of (3.1), then the energy satisfies

$$\begin{aligned}E'(t) &= \frac{1}{2}(\mu' \circ \nabla u) - \frac{1}{2}\mu(t)\|\nabla u\|_2^2 - \Big\|\nabla \frac{\partial u}{\partial t}\Big\|_2^2 \\ &\leq \frac{1}{2}(\mu' \circ \nabla u) \\ &\leq 0. \end{aligned} \tag{3.5}$$

By multiplying the equation in (3.1) by $\frac{\partial u}{\partial t}$ and integrating over $\mathbb{R}^n$, using integration by parts, using the assumptions on the relaxation function μ we obtain the result.

3.1.1 Stability of solution

We use the notation

$$\bar{f} = \int_0^\infty |f(s)|ds, \quad \bar{f}_a = \int_0^\infty e^{as}|f(s)|ds.$$

To deal with different terms in our system, we set the functional

$$\phi_1(t) = \frac{1}{1+t}\int_{\mathbb{R}^n}\int_0^t \mu_1(a, t-s)\left|\nabla u(t) - \nabla u(s)\right|^2 ddx, \tag{3.6}$$

$$\phi_2(t) = \frac{1}{1+t}\int_{\mathbb{R}^n}\int_0^t \mu_2(a, t-s)\left|\nabla u(t) - \nabla u(s)\right|^2 ddx, \tag{3.7}$$

$$\phi_3(t) = \frac{1}{1+t}\left(\int_{\mathbb{R}^n} u\frac{\partial u}{\partial t}dx + \int_{\mathbb{R}^n} \nabla u \nabla\frac{\partial u}{\partial t}dx\right), \tag{3.8}$$

$$\begin{aligned}\phi_4(t) = \frac{1}{1+t}\Bigg(&\int_{\mathbb{R}^n} \nabla\frac{\partial u}{\partial t}\int_0^t \mu(t-s)(\nabla u(t) - \nabla u(s))dsdx\\ &+ \int_{\mathbb{R}^n}\frac{\partial u}{\partial t}\int_0^t \mu(t-s)(u(t) - u(s))dsdx\Bigg),\end{aligned} \tag{3.9}$$

where

$$\mu_1(a, t) = e^{-at}\int_t^\infty e^{as}\mu(s)ds, \tag{3.10}$$

$$\mu_1(a, t) = -e^{-at}\int_t^\infty e^{as}(1+s)^2\mu'(s)ds. \tag{3.11}$$

To prepare to introduce an appropriate Lyapunov functional, we have the following series of lemmas.

Lemma 3.2. *Let* (1.33) *and* (3.3) *hold. Then the functional* (3.6) *satisfies*

$$\phi_1'(t) \leq -\frac{1}{1+t}\left[\left(1 + (1+t)\left(a - \frac{1}{\delta_1}\bar{\mu}_1\right)\right)\phi_1(t) + (\mu \circ \nabla u) - \delta_1\left|\frac{\partial u}{\partial t}\right|_2^2\right], \tag{3.12}$$

for some $\delta_1 > 0$.

Proof. Using (3.3) and integration by parts, we can easily find that

$$\begin{aligned}\bar{\mu}_1 &= \int_0^\infty \left(e^{-at} \int_t^\infty e^{as}\mu(s)ds \right) dt \\ &\le \frac{1}{a}\int_t^\infty e^{as}\mu(s)ds \\ &\le \frac{1}{a}\bar{\mu}_a \\ &< \infty.\end{aligned}$$

By differentiating $\phi_1(t)$ and using Young's inequality, we have

$$\begin{aligned}\phi_1'(t) &= -\frac{\phi_1(t)}{1+t} + \frac{1}{1+t}\Bigg[-\alpha(1+t)\phi_1(t) - (\mu \circ \nabla u) \\ &\quad + 2\int_{\mathbb{R}^n} \nabla\frac{\partial u}{\partial t} \int_0^t \mu_1(\alpha, t-s)|\nabla u(t) - \nabla u(s)|^2 ddx \Bigg] \\ &\le -\frac{\phi_1(t)}{1+t} + \frac{1}{1+t}\Bigg[-\alpha(1+t)\phi_1(t) - (\mu \circ \nabla u) \\ &\quad + \delta_1 \left\| \nabla \frac{\partial u}{\partial t} \right\|_2^2 + \frac{1+t}{\delta_1}\bar{\mu}_1\phi_1(t) \Bigg] \\ &\le -\frac{1}{1+t}\left[\left(1 + (1+t)\left(\alpha - \frac{1}{\delta_1}\bar{\mu}_1 \right) \right)\phi_1(t) + (\mu \circ \nabla u) - \delta_1 \left\| \nabla\frac{\partial u}{\partial t} \right\|_2^2 \right].\end{aligned}$$

This completes the proof. □

Lemma 3.3. *Let* (1.33) *and* (3.3) *hold. Then the functional* (3.7) *satisfies*

$$\begin{aligned}\phi_2'(t) &\le -\frac{1}{1+t}\left[\left(1 + (1+t)\left(\alpha - \frac{1}{\delta_1}\bar{\mu}_2 \right) \right)\phi_2(t) - \delta_1 \left\| \nabla\frac{\partial u}{\partial t} \right\|_2^2 \right] \\ &\quad + (1+t)(\mu \circ \nabla u),\end{aligned} \tag{3.13}$$

for some $\delta_1 > 0$.

Proof. Using (3.3) and the fact that

$$(1+t)^2 \le e^{2t}, \quad t \ge 0.$$

Integration by parts, we can easily find that

$$-\int_t^\infty e^{as}(1+s)^2\mu'(s)ds = [-e^{as}(1+t)^2\mu(s)]_t^\infty + (\alpha+2)\int_t^\infty e^{as}(1+s)^2\mu(s)ds$$
$$\leq e^{at}(1+t)^2\mu(t) + (\alpha+2)\int_t^\infty e^{as}(1+s)^2\mu(s)ds.$$

Now

$$\begin{aligned}
0 &\leq \bar{\mu}_2 \\
&= \int_0^\infty \left(e^{-\alpha t}\int_t^\infty e^{as}(1+s)^2(-\mu'(s))ds \right)dt \\
&\leq \int_0^\infty (1+t)^2\mu(t)dt + \left[\frac{-e^{\alpha t}}{\alpha}\int_0^\infty e^{as}(1+s)^2\mu(s)ds\right]_0^\infty - \frac{1}{\alpha}\int_0^\infty (1+t)^2\mu(t)dt \\
&\leq \frac{1}{\alpha}\int_0^\infty e^{(\alpha+2)s}\mu(s)ds - \left(\frac{1}{\alpha}-1\right)\int_0^\infty (1+t)^2\mu(t)dt,
\end{aligned}$$

where

$$\begin{aligned}
\int_0^\infty (1+t)^2\mu(t)dt &\leq \int_0^\infty e^{2t}\mu(t)dt \\
&< \int_0^\infty e^{(\alpha+2)t}\mu(t)dt \\
&< \infty.
\end{aligned}$$

Then

$$\mu_2 < \infty.$$

By differentiating ϕ_2 and using Young's inequality, we have

$$\begin{aligned}
\phi_2'(t) \leq &-\frac{\phi_2(t)}{1+t} + \frac{1}{1+t}\bigg[-\alpha(1+t)\phi_2(t) + (1+t)^2(\mu\circ\nabla u) \\
&+ 2\int_{\mathbb{R}^n}\nabla\frac{\partial u}{\partial t}\int_0^t \mu_2(\alpha, t-s)(\nabla u(t)-\nabla u(s))dsdx\bigg] \\
\leq &-\frac{1}{1+t}\left[\phi_2(t) + \alpha(1+t)\phi_2(t) - \delta_1\left\|\nabla\frac{\partial u}{\partial t}\right\|_2^2 - \frac{(1+t)}{\delta_1}\bar{\mu}_2\left\|\nabla\frac{\partial u}{\partial t}\right\|^2 - 2\right] \\
&+ (1+t)(\mu'\circ\nabla u)
\end{aligned}$$

$$\le -\frac{1}{1+t}\left[1+(1+t)\left(a-\frac{1}{\delta_1}\bar{\mu}_2\right)\phi_2(t)-\delta_1\left\|\nabla\frac{\partial u}{\partial t}\right\|_2^2\right]$$
$$+(1+t)(\mu'\circ\nabla u).$$

This completes the proof. □

Lemma 3.4. *Let* (1.33) *and* (3.3) *hold. Then the functional* (3.8) *satisfies, along with the solution of* (3.1),

$$\phi_3'(t)\le\frac{1}{1+t}\left(1+\frac{c}{4\delta_2}\right)\left\|\frac{\partial u}{\partial t}\right\|_2^2+\frac{1}{1+t}\left(1+\frac{(1+t)^{-1}}{4\delta_2}\right)\left\|\nabla\frac{\partial u}{\partial t}\right\|_2^2$$
$$-\frac{1}{1+t}\left[l-\delta_2\left(1+c+\frac{1}{1+t}\right)\right]\|\nabla u\|_2^2+\frac{\frac{\bar{\mu}}{1+t}}{4\delta_2}(\mu\circ\nabla u), \tag{3.14}$$

for some $\delta_2, c>0$

Proof. By differentiating $\phi_3(t)$, we have

$$\phi_3'(t)=-\frac{1}{(1+t)^2}\left(\int_{\mathbb{R}^n}u\frac{\partial u}{\partial t}dx+\int_{\mathbb{R}^n}\nabla u\nabla\frac{\partial u}{\partial t}dx\right)$$
$$+\frac{1}{1+t}\left(\int_{\mathbb{R}^n}\frac{\partial u^2}{\partial t}dx+\int_{\mathbb{R}^n}u\frac{\partial^2 u}{\partial t^2}dx+\int_{\mathbb{R}^n}\nabla\frac{\partial u^2}{\partial t}dx+\int_{\mathbb{R}^n}\nabla u\nabla\frac{\partial^2 u}{\partial t^2}dx\right).$$

By (3.1), we have

$$\int_{\mathbb{R}^n}u\frac{\partial^2 u}{\partial t^2}dx+\int_{\mathbb{R}^n}\nabla u\nabla\frac{\partial^2 u}{\partial t^2}dx=-\int_{\mathbb{R}^n}|\nabla u|^2dx+\int_{\mathbb{R}^n}\nabla u\int_0^t\mu(t-s)\nabla u(s)dsdx.$$

Using the estimates

$$\int_{\mathbb{R}^n}u\frac{\partial u}{\partial t}dx\le c(1+t)\left(\delta_2\|\nabla u\|_2^2+\frac{1}{4\delta_2}\left|\frac{\partial u}{\partial t}\right|_2^2\right),$$

and

$$\int_{\mathbb{R}^n}\nabla u\nabla\frac{\partial u}{\partial t}dx\le\delta_2\|\nabla u\|_2^2+\frac{1}{4\delta_2}\left\|\nabla\frac{\partial u}{\partial t}\right\|_2^2,$$

and

$$\int_{\mathbb{R}^n}\nabla u\int_0^t\mu(t-s)\nabla u(s)dsdx\le\delta_2\|\nabla u\|_2^2+\frac{\bar{\mu}}{4\delta_2}(\mu\circ\nabla u)+\bar{\mu}\|\nabla u\|_2^2,$$

and combining to get

$$\phi_3'(t) \le \frac{1}{1+t}\left(1+\frac{c}{4\delta_2}\right)\left\|\frac{\partial u}{\partial t}\right\|_2^2 + \frac{1}{1+t}\left(1+\frac{(1+t)^{-1}}{4\delta_2}\right)\left\|\nabla\frac{\partial u}{\partial t}\right\|_2^2$$
$$- \frac{1}{1+t}\left[l-\delta_2\left(1+c+\frac{1}{1+t}\right)\right]\|\nabla u\|_2^2 + \frac{\frac{\bar{\mu}}{1+t}}{4\delta_2}(\mu\circ\nabla u).$$

This completes the proof. □

Lemma 3.5. *Let* (1.33) *and* (3.3) *hold. Then the functional* (3.9) *satisfies, along with the solution of* (3.1),

$$\phi_4'(t) \le \frac{1}{1+t}\left(\int_0^t \mu(s)ds - c\delta_2 - \delta_2\right)\left\|\frac{\partial u}{\partial t}\right\|_2^2$$
$$+ \frac{1}{1+t}\left(\int_0^t \mu(s)ds - 2\delta_2\right)\left\|\nabla\frac{\partial u}{\partial t}\right\|_2^2$$
$$+ \frac{2\delta_3}{1+t}\|\nabla u\|_2^2 - \frac{1}{1+t}\frac{\mu(0)}{4\delta_2}(\mu'\circ\nabla u)$$
$$- \frac{1}{1+t}\left[\frac{1}{4\delta_2}+\frac{c}{4\delta_2}+\frac{1}{2\delta_3}+1\right]\bar{\mu}(\mu\circ\nabla u) - \frac{c\mu(0)}{4\delta_2}(1+t)(\mu'\circ\nabla u), \tag{3.15}$$

for some $\delta_2, \delta_3 > 0$.

Proof. By differentiating $\phi_4(t)$, we have

$$\phi_4'(t) = \frac{1}{(1+t)^2}\left(\int_{\mathbb{R}^n}\nabla\frac{\partial u}{\partial t}\int_0^t \mu(t-s)(\nabla u(t)-\nabla u(s))dsdx\right.$$
$$\left.+\int_{\mathbb{R}^n}\frac{\partial u}{\partial t}\int_0^t \mu(t-s)(u(t)-u(s))dsdx\right)$$
$$-\frac{1}{1+t}\left[\int_{\mathbb{R}^n}\nabla\frac{\partial^2 u}{\partial t^2}\int_0^t \mu(t-s)(\nabla u(t)-\nabla u(s))dsdx\right.$$
$$\times\int_{\mathbb{R}^n}\nabla\frac{\partial u}{\partial t}\int_0^t \mu'(t-s)(\nabla u(t)-\nabla u(s))dsdx + \int_{\mathbb{R}^n}\frac{\partial^2 u}{\partial t^2}\int_0^t \mu(t-s)(u(t)-u(s))dsdx$$
$$\left.+\int_{\mathbb{R}^n}\frac{\partial u}{\partial t}\int_0^t \mu'(t-s)(u(t)-u(s))dsdx + \left(\int_0^t \mu(s)ds\right)\left\|\frac{\partial u}{\partial t}\right\|_2^2 + \int_0^t \mu(t-s)\left\|\nabla\frac{\partial u}{\partial t}\right\|_2^2\right].$$

By (3.1), we have

$$\int_{\mathbb{R}^n} \frac{\partial^2 u}{\partial t^2} \int_0^t \mu(t-s)(u(t)-u(s))dsdx + \int_{\mathbb{R}^n} \nabla \frac{\partial^2 u}{\partial t^2} \int_0^t \mu(t-s)(\nabla u(t)-\nabla u(s))dsdx$$

$$= -\int_{\mathbb{R}^n} \nabla u \int_0^t \mu(t-s)(\nabla u(t)-\nabla u(s))dsdx$$

$$+ \int_{\mathbb{R}^n} \int_0^t \mu(t-s)\nabla u(s)ds \int_0^t \mu(t-s)(\nabla u(t)-\nabla u(s))dsdx.$$

Using the estimates,

$$\int_{\mathbb{R}^n} \nabla \frac{\partial u}{\partial t} \int_0^t \mu(t-s)(\nabla u(t)-\nabla u(s))dsdx \le \delta_2 \left\| \nabla \frac{\partial u}{\partial t} \right\|_2^2 - \frac{\bar{\mu}}{4\delta_2}(\mu \circ \nabla u),$$

$$\int_{\mathbb{R}^n} \frac{\partial u}{\partial t} \int_0^t \mu(t-s)(\nabla u(t)-\nabla u(s))dsdx \le c(1+t)\left(\delta_2 \left\| \frac{\partial u}{\partial t} \right\|_2^2 + \frac{\bar{\mu}}{4\delta_2}(\mu \circ \nabla u) \right),$$

$$\int_{\mathbb{R}^n} \nabla u \int_0^t \mu(t-s)(\nabla u(t)-\nabla u(s))dsdx \le \delta_3 \|\nabla u\|_2^2 + \frac{\bar{\mu}}{4\delta_3}(\mu \circ \nabla u),$$

$$-\int_{\mathbb{R}^n} \nabla \frac{\partial u}{\partial t} \int_0^t \mu'(t-s)(\nabla u(t)-\nabla u(s))dsdx \le \delta_2 \left\| \nabla \frac{\partial u}{\partial t} \right\|_2^2 - \frac{\mu(0)}{4\delta_2}(\mu' \circ \nabla u),$$

$$\int_{\mathbb{R}^n} \frac{\partial u}{\partial t} \int_0^t \mu'(t-s)(u(t)-u(s))dsdx \le \delta_2 \left| \frac{\partial u}{\partial t} \right|_2^2 - \frac{c\mu(0)}{4\delta_2}(1+t)^2(\mu' \circ \nabla u).$$

Also, exploiting the estimate

$$\int_{\mathbb{R}^n} \left| \int_0^t \mu(t-s)(\nabla u(t)-\nabla u(s))ds \right|^2 dx \le \bar{\mu}(\mu \circ \nabla u),$$

then

$$\int_{\mathbb{R}^n} \int_0^t \mu(t-s)\nabla u(s)ds \int_0^t \mu(t-s)(\nabla u(t)-\nabla u(s))dsdx$$

$$\le \bar{\mu} \int_{\mathbb{R}^n} \nabla u(t) \int_0^t \mu(t-s)(\nabla u(t)-\nabla u(s))dsdx$$

$$+ \int_{\mathbb{R}^n} \left| \int_0^t \mu(t-s)(\nabla u(t)-\nabla u(s))ds \right|^2 dx$$

$$\leq \delta_3\|\nabla u\|_2^2 + \frac{\bar{\mu}}{4\delta_3}(\mu \circ \nabla u) + \bar{\mu}(\mu \circ \nabla u),$$

combining to obtain the desired results. □

Let us define the Lyapunov functional as

$$L(t) = E(t) + \sum_{i=1}^{i=4} \gamma_i \phi_i, \quad \gamma_i > 0,\ t \geq 0. \tag{3.16}$$

We prove the equivalence between the Lyapunvov functional and energy function in the next lemma.

Lemma 3.6. *Let* (1.33) *and* (3.3) *hold. Then there exists a positive constant* γ_0 *such that* $\gamma_0 \geq \gamma_3, \gamma_4$ *and* (3.16) *satisfy*

$$\xi_1 E(t) \leq L(t) \leq \xi_2[E(t) + \Phi_1 + \phi_2], \tag{3.17}$$

where $\xi_1, \xi_2 > 0$ *depending only on* γ_1, γ_2 *and* γ_3.

Proof. We estimate terms in the above functional using Cauchy–Schwarz, Young's inequalities and (3.5) as follows:

$$\begin{aligned}\int_{\mathbb{R}^n} u\frac{\partial u}{\partial t}dx &\leq \Big(\int_{\mathbb{R}^n} |u|^2 dx\Big)^{1/2}\Big(\int_{\mathbb{R}^n}\Big|\frac{\partial u}{\partial t}\Big|^2 dx\Big)^{1/2}\\ &\leq c(1+t)\|\nabla u\|_2\Big\|\frac{\partial u}{\partial t}\Big\|_2\\ &\leq c(1+t)\Big(\frac{1}{4}\|\nabla u\|_2^2 + \Big\|\frac{\partial u}{\partial t}\Big\|_2^2\Big),\end{aligned} \tag{3.18}$$

and

$$\int_{\mathbb{R}^n} \nabla u \nabla \frac{\partial u}{\partial t}dx \leq \frac{1}{4}\Big\|\nabla\frac{\partial u}{\partial t}\Big\|_2^2 + \|\nabla u\|_2^2,$$

and

$$\int_{\mathbb{R}^n} \nabla\frac{\partial u}{\partial t}\int_0^t \mu(t-s)(\nabla u(t) - \nabla u(s))ds dx \leq \frac{\bar{\mu}}{4}(\mu \circ \nabla u) + \Big\|\nabla\frac{\partial u}{\partial t}\Big\|_2^2,$$

and

$$\int_{\mathbb{R}^n} \frac{\partial u}{\partial t}\int_0^t \mu(t-s)(u(t) - u(s))ds dx \leq c(1+t)\Big(\frac{\bar{\mu}}{4}(\mu \circ \nabla u) + \Big\|\frac{\partial u}{\partial t}\Big\|_2^2\Big).$$

Inserting the above inequalities in (3.16), we get

$$L(t) \geq \left(\frac{1}{2} - c(\gamma_3 + \gamma_4)\right)\left|\frac{\partial u}{\partial t}\right|_2^2 + \left(\frac{l}{2} - \gamma_3\left(\frac{c}{4} + \frac{1}{1+t}\right)\right)\|\nabla u\|_2^2$$
$$+ \left[\frac{1}{2} - \frac{\gamma_3}{4}\frac{1}{1+t} - \gamma_4\frac{1}{1+t}\right]\left\|\nabla\frac{\partial u}{\partial t}\right\|_2^2 + \left[\frac{1}{2} - \frac{\gamma_4\bar{\mu}}{4}\left(c + \frac{1}{1+t}\right)\right](\mu \circ \nabla u).$$

Since

$$\frac{1}{1+t} < 1, \quad \forall t > 0,$$

if

$$\gamma_0 \geq \gamma_3, \gamma_4,$$

we have

$$L(t) \geq \left(\frac{1}{2} - 2c\gamma_0\right)\left|\frac{\partial u}{\partial t}\right|_2^2 + \left(\frac{l}{2} - \gamma_0\left(\frac{c}{4} + 1\right)\right)\|\nabla u\|_2^2$$
$$+ \left[\frac{1}{2} - \frac{5\gamma_0}{4}\right]\left\|\nabla\frac{\partial u}{\partial t}\right\|_2^2 + \left[\frac{1}{2} - \frac{\gamma_0\bar{\mu}}{4}(c+1)\right](\mu \circ \nabla u),$$

choosing γ_0 small enough to satisfy

$$L(t) \geq \xi_1 E(t),$$

where

$$\xi_1 = \min\left\{\frac{1}{2} - 2c\gamma_0, \frac{l}{2} - \gamma_0\left(\frac{c}{4} + 1\right), \frac{1}{2} - \frac{5\gamma_0}{4}, \frac{1}{2} - \frac{\gamma_0\bar{\mu}}{4}(c+1)\right\}.$$

Similarly, using the same estimates, we can easily see that

$$L(t) \leq \gamma_1\phi_1 + \gamma_2\phi_2 + \left(\frac{1}{2} - 2c\gamma_0\right)\left|\frac{\partial u}{\partial t}\right|_2^2 + \left(\frac{l}{2} - \gamma_0\left(\frac{c}{4} + 1\right)\right)\|\nabla u\|_2^2$$
$$+ \left[\frac{1}{2} - \frac{5\gamma_0}{4}\right]\left\|\nabla\frac{\partial u}{\partial t}\right\|_2^2 + \left[\frac{1}{2} - \frac{\gamma_0\bar{\mu}}{4}(c+1)\right](\mu \circ \nabla u)$$
$$\leq \xi_2[E(t) + \phi_1 + \phi_2]. \tag{3.19}$$

The proof in now completed. □

Theorem 3.1. *Let* (1.33) *and* (3.3) *hold. Then for any* $t_0 > 0$, *there exist two positive constants* K *and* k *such that*

$$E(t) \leq \frac{K}{(1+t)^k}.$$

Proof. Direct differentiation of (3.16) yields

$$\begin{aligned} L'(t) &= E'(t) + \sum_{i=1}^{i=4} \gamma_i \phi_i' \\ &\leq \frac{1}{2}(\mu' \circ \nabla u) + \sum_{i=1}^{i=4} \gamma_i \phi_i', \end{aligned} \tag{3.20}$$

since μ is continuous and $\mu(0) > 0$, then for any $t \geq t_0$, we have

$$\int_0^t \mu(s)ds \geq \int_0^{t_0} \mu(s) = \mu_0 > 0.$$

By Lemma 3.2–Lemma 3.5, we have

$$\begin{aligned} L'(t) \leq & -\frac{1}{1+t}\left(1 + (1+t)\left(\alpha - \frac{1}{\delta_1}\bar{\mu}_1\right)\right)\gamma_1\phi_1(t) \\ & - \frac{1}{1+t}[\left(1 + (1+t)\left(\alpha - \frac{1}{\delta_1}\bar{\mu}_2\right)\right)\gamma_2\phi_2(t) \\ & + \left[\frac{1}{2} - \gamma_4\frac{c\mu(0)}{4\delta_2}\right](\mu' \circ \nabla u) + \left[\gamma_2 - \gamma_4\frac{c\mu(0)}{4\delta_2}\right](1+t)(\mu' \circ \nabla u) \\ & - \frac{1}{1+t}\left[\gamma_1 - \bar{\mu}\left(\frac{\gamma_3}{4\delta_2} + \gamma_4\left(\frac{1}{4\delta_2}\frac{c}{4\delta_2} + \frac{1}{2\delta_3} + 1\right)\right)\right](\mu \circ \nabla u) \\ & - \frac{1}{1+t}[\gamma_3(l - \delta_2(c+2)) - \gamma_4(2\delta_3)]\|\nabla u\|_2^2 \\ & - \frac{1}{1+t}\left[\gamma_4(\mu_0 - 2\delta_2) - \gamma_3\left(1 + \frac{1}{4\delta_2}\right) - \delta_1(\gamma_1 + \gamma_2)\right]\left\|\nabla\frac{\partial u}{\partial t}\right\|_2^2 \\ & - \frac{1}{1+t}\left[\gamma_4(\mu_0 - \delta_2(c+1)) - \gamma_3\left(1 + \frac{1}{4\delta_2}\right)\right]\left\|\frac{\partial u}{\partial t}\right\|_2^2. \end{aligned}$$

At this point, we choose δ_2 and δ_3 so that

$$\delta_2 < \min\left\{\frac{\mu_0}{c+1}, \frac{\mu_0}{2}, \frac{l}{c+2}\right\}$$

and

$$\delta_3 < \frac{[l - \delta_2(c+2)]\min\{\mu_0 - 2\delta_2, \mu_0 - \delta_2(c+1)\}}{2(1 + \frac{1}{4\delta_2})}.$$

Whence δ_2 and δ_3 are fixed, any choice of γ_3, γ_4 satisfying

$$\frac{3\delta_3}{l - \delta_2(c+2)}\gamma_4 < \gamma_3 < \gamma_4\frac{\min\{\mu_0 - 2\delta_2, \mu_0 - \delta_2(c+1)\}}{\frac{1}{4\delta_2}},$$

then

$$\gamma_3(l - \delta_2(c + 2)) - \gamma_4(3\delta_3) > 0,$$
$$\gamma_4(\mu_0 - 2\delta_2) - \gamma_3\left(1 + \frac{1}{4\delta_2}\right) = k_1 > 0,$$
$$\gamma_4(\mu_0 - \delta_2(c + 1)) - \gamma_3\left(1 + \frac{1}{4\delta_2}\right) > 0.$$

So, we choose $\gamma_3 < \gamma_0$ and $\gamma_4 < \gamma_0$ so small, then

$$\frac{1}{2} - \gamma_4\left(\frac{\mu(0)}{4\delta_2}\right) > 0.$$

Then

$$\begin{aligned} L'(t) &\leq \frac{-c_1}{1+t}[E(t) + \phi_1 + \phi_2] \\ &\leq \frac{-c_1}{\xi_2(1+t)} L(t). \end{aligned} \tag{3.21}$$

Integration of (3.21) over (t_0, t) yields

$$L(t) \leq \frac{L(t_0)(1 + t_0)^{c_1/\xi_2}}{(1 + t)^{c_1/\xi_2}}, \quad c_1 > 0. \tag{3.22}$$

This completes the proof. □

3.2 Nonlinear damping wave equation: finite speed of propagation principle

For the nonlinear damping wave equation, we consider as in [3],

$$\frac{\partial^2 u}{\partial t^2} - \Delta u + a\left|\frac{\partial u}{\partial t}\right|^{m-1}\frac{\partial u}{\partial t} = |u|^{p-1}u - q(x)^2 u, \tag{3.23}$$

where the function $q(x)$ is a locally bounded measurable, with initial data

$$u(x, 0) = u_0(x), \quad \frac{\partial u}{\partial t}(x, 0) = u_1(x). \tag{3.24}$$

For $q(x) = 1$, we define the energy functional as

$$E(t) = \frac{1}{2}\left|\frac{\partial u}{\partial t}\right|_2^2 + \frac{1}{2}\|\nabla u\|_2^2 + \frac{1}{2}\|u\|_2^2 - \frac{1}{p+1}\|u\|_{p+1}^{p+1}. \tag{3.25}$$

Insofar as local existence is concerned, we have the next theorem.

Theorem 3.2. [1] *Let*

$$1 \le p < \frac{n}{n-2}, \quad \text{if } n > 2, \quad \text{or} \quad 1 \le p, \quad \text{if } n = 1, 2.$$

Then, for any compactly supported data,

$$u_0 \in W^{1,2}(\mathbb{R}^n), \quad u_1 \in L^2(\mathbb{R}^n). \tag{3.26}$$

The Cauchy problems (3.23)–(3.24) *have a unique solution such that*

$$u \in C([0,T), W^{1,2}(\mathbb{R}^n)),$$

and

$$\frac{\partial u}{\partial t} \in C([0,T), L^2(\mathbb{R}^n)) \cap L^{m+1}([0,T) \times \mathbb{R}^n),$$

for sufficiently small $T > 0$.

3.2.1 Global nonexistence

In this subsection, we shall consider the blow up of solution via unstable sets approach.

Theorem 3.3. *Let*

$$1 \le m < p < \frac{n}{n-2}, \quad \text{if } n > 2, \quad \text{or} \quad 1 \le p, \quad \text{if } n = 1, 2.$$

There exist a compacted support initial data

$$u_0 \in H^1(\mathbb{R}^n), \quad u_1 \in L^2(\mathbb{R}^n), \tag{3.27}$$

such that

$$E(0) = \lambda^2 > 0.$$

Then:

1. *If* $q(x) \neq 0$ *and*

$$q(x) \ge s(1 + \|x\|)^{-\mu},$$

1 H. D. A. Levine and G. Todorova. Blow-up of solutions of the Cauchy problem for a wave equation with nonlinear damping and source terms and positive initial energy. Proceedings of the American Mathematical Society 129 (3) (2000), 793–805.

where

$$0 \le \mu \le \frac{(p-1)m}{2(p-m)}.$$

Then every local solution of (3.23)–(3.24) *is not global in time.*

2. *If* $q(x) = 0$ *and* $m = 1$ *or*

$$p > m > 1, \quad m > p - 2.$$

Then the local solution of (3.23)–(3.24) *blows up in finite time.*

Proof. 1. The case $q(x) = 0$: Suppose to the contrary that for given $E(0) > 0$, u is a global solution of (3.23)–(3.24). Then, from energy identity, we have

$$E(0) - E(t) = \int_0^t \left\| \frac{\partial u}{\partial \tau} \right\|_{m+1}^{m+1} d\tau, \quad t \ge 0,$$

where $E(t)$ is defined in (3.25). It suffices to find t_1 such that

$$E(0) - E_{cr} < \int_0^{t_1} \left\| \frac{\partial u}{\partial \tau} \right\|_{m+1}^{m+1} d.$$

Then from the energy identity we have

$$E_{cr} > E(t_1).$$

We decompose the solution u into linear and nonlinear terms as follows:

$$u = w + v, \tag{3.28}$$

where

$$\begin{aligned} &\frac{\partial^2 w}{\partial t^2} - \Delta w = 0, \\ &w(x,0) = u_0(x), \quad \frac{\partial w}{\partial t}(x,0) = u_1(x), \end{aligned} \tag{3.29}$$

and

$$\begin{aligned} &\frac{\partial^2 v}{\partial t^2} - \Delta v = F(x,t), \\ &v(x,0) = 0, \quad \frac{\partial v}{\partial t}(x,0) = 0, \end{aligned} \tag{3.30}$$

where

$$F(x,t) = |u|^{p-1}u - q(x)^2 u - \left|\frac{\partial u}{\partial t}\right|^{p-1}\frac{\partial u}{\partial t}.$$

The idea of the proof is to show that for an appropriate data, the total energy E_{cr} falls below in finite time. We use negative Sobolev norms to accomplish this. First, we will need a good lower bound on $\|\frac{\partial u}{\partial t}\|_{m+1}^{m+1}$. To do this, we first observe from Hölder's inequality and the finite speed of propagation property of solutions of (3.24) that

$$\left\|\frac{\partial u}{\partial t}\right\|_2 \le C(t+L)^{n/q}\left\|\frac{\partial u}{\partial t}\right\|_{m+1},$$

where

$$\frac{1}{q} + \frac{1}{m+1} = \frac{1}{2}.$$

Additionally, we need the inequality

$$\left\|\frac{\partial u}{\partial t}\right\|_2 \ge C\left\|\frac{\partial u}{\partial t}\right\|_{H^{-s}}, \quad \frac{s}{n} \ge 0.$$

Combining these with the decomposition of the solution (3.28), we have

$$C\left[\left\|\frac{\partial w}{\partial t}\right\|_{H^{-s}}^{m+1} - \left\|\frac{\partial v}{\partial t}\right\|_{H^{-s}}^{m+1}\right] \le \left\|\frac{\partial u}{\partial t}\right\|_2^{m+1} \le \left[(t+L)^{n/q}\left\|\frac{\partial u}{\partial t}\right\|_{m+1}\right]^{m+1}, \tag{3.31}$$

where C is another constant. This gives us the desired lower bound for

$$\left\|\frac{\partial u}{\partial t}\right\|_{m+1}^{m+1},$$

on any interval $[0, T)$ of fixed, finite length, $T \le L$, say, provided we can find good estimates for the terms on the extreme right-hand side of (3.31). We now proceed to estimate these terms. We obtain, in turn:

1. An estimate for

$$\left\|\frac{\partial w}{\partial t}\right\|_{H^{-s}}^{m+1},$$

from below.

2. An estimate for

$$\left\|\frac{\partial v}{\partial t}\right\|_{H^{-s}}^{m+1},$$

from above.

3. An estimate for

$$E(t) \equiv \|u\|_2^2 + \|\nabla u\|_2^2,$$

from above.

These estimates hold for a small time interval. Using them, we are able to estimate the rate of decrease of the energy $E(t)$ from below in this time interval. Then we show that for any initial energy $E(0)$ we can choose v_0, u_0 with this initial energy such that for some t_0 in the above time interval, the energy

$$E(t_0) < E_{cr}.$$

We begin with the linear term w. In terms of the Fourier transform variable ξ, we can write the solution of the free equation (3.29) as follows:

$$\widehat{w}(\xi, t) = \cos(t|\xi|)\widehat{u}_0(\xi) + \frac{\sin(t|\xi|)}{|\xi|}\widehat{v}_0(\xi). \tag{3.32}$$

Computing $\widehat{w}_t$, squaring and using the inequalities

$$\begin{aligned}
(a+b)^2 &\geq \frac{a^2}{2} - b^2,\\
\sin^2 a &\leq a^2,\\
\cos^2 a &\geq 1 - 2a^2,
\end{aligned}$$

to find

$$|\widehat{w}|^2(\xi, t) \geq |\widehat{v}_0(\xi)|^2 - t^2|\xi|^2\big[|\xi|^2|\widehat{u}_0(\xi)|^2 + |\widehat{v}_0(\xi)|^2\big]. \tag{3.33}$$

We have after multiplying by

$$(1 + |\xi|^2)^{-s},$$

for $s > 2$, and integrating over $\xi \in \mathbb{R}^n$,

$$\left\|\frac{\partial w}{\partial t}\right\|_{H^{-s}}^2 + S(L,n)t^2E(0) \geq \frac{1}{2}\|v_0\|_{H^{-s}}^2, \tag{3.34}$$

for $S(L,n) > 0$. Raising both sides of this inequality to the $(m+1)/2$ power, we find for some suitable $C > 0$,

$$\left\|\frac{\partial w}{\partial t}\right\|_{H^{-s}}^{m+1} \geq C\big[\|v_0\|_{H^{-s}}^2 - t^{(m+1)/2}E^{(m+1)/2}(0)\big]. \tag{3.35}$$

We turn next to the second term. Taking the H^{-s} scalar product of both sides of (3.30) with $\frac{\partial v}{\partial t}$, it follows in the usual manner that there is a constant C such that

$$\left\|\frac{\partial v}{\partial t}\right\|_{H^{-s}} \le C\int_0^t \|F(x,\tau)\|_{H^{-s}}\,d\tau, \tag{3.36}$$

where F is defined in (3.30). We have that

$$\|g\|_{H^{-s}} \le C\|g\|_r, \quad r \le 2,$$

and

$$s \ge n\left(\frac{1}{r} - \frac{1}{2}\right).$$

We use this estimate twice, once with

$$g = u|u|^{p-1},$$

and $r = 2$ and with

$$g = \frac{\partial u}{\partial t}\left|\frac{\partial u}{\partial t}\right|^{p-1},$$

and

$$r = \frac{m+1}{m},$$

in

$$\|F\|_{H^{-s}} \le \|u|u|^{p-1}\|_{H^{-s}} + \left\|\frac{\partial u}{\partial t}\left|\frac{\partial u}{\partial t}\right|^{p-1}\right\|_{H^{-s}}, \tag{3.37}$$

Now we choose

$$s > \max\left\{2, \frac{n(m-1)}{2(m+1)}\right\}.$$

We obtain

$$\begin{aligned}\|F(x,t)\|_{H^{-s}} &\le C\left(\|u\|_{2p}^p + \left|\frac{\partial u}{\partial t}\right|_{m+1}^m\right)\\ &\le C\left((t+L)^{p_n}\|\nabla u\|_2^p + \left|\frac{\partial u}{\partial t}\right|_{m+1}^m\right),\end{aligned} \tag{3.38}$$

where

$$p_n \equiv \frac{n-(n-2)p}{2p},$$

which is strictly positive for $n > 2$ in view of our restrictions on p. In the above estimate, we used the finite propagation speed property of the solution and the Hölder and Poincare inequalities after noting that

$$2p \leq \frac{2n}{n-2}, \quad n > 2.$$

After integrating both sides of (3.38) and using the Hölder's inequality with respect to t in the integral

$$\int_0^t \left\|\frac{\partial u}{\partial \tau}\right\|_{m+1}^m d\tau,$$

to get

$$\int_0^t \|F(x,\tau)\|_{H^{-s}} d\tau \leq (t+L)^{p_n} \int_0^t E^{p/2}(\tau)d\tau + t^{1/(m+1)}\left(\int_0^t \left\|\frac{\partial u}{\partial \tau}\right\|_{m+1}^{m+1} d\tau\right)^{m/(m+1)}.$$

Since we are assuming that $E(t) > E_{cr}$, we have

$$\int_0^t \left\|\frac{\partial u}{\partial \tau}\right\|_{m+1}^{m+1} d\tau \leq E(0) - E_{cr},$$

and then

$$\int_0^t \|F(x,\tau)\|_{H^{-s}} d\tau \leq C(L) \int_0^t E^{p/2}(\tau)d\tau + t^{1/(m+1)}(E(0)-E_{cr})^{m/(m+1)}. \tag{3.39}$$

In order to complete our estimate for $\|\frac{\partial v}{\partial t}\|_{H^{-s}}$, we will need an upper bound for $E(t)$, which we obtain as follows:

$$\begin{aligned}
E'(t) &\leq \left(\frac{\partial u}{\partial t}, u|u|^{p-1}\right) \\
&\leq C\left\|\frac{\partial u}{\partial t}\right\|_2 \|u\|_{2p}^p \\
&\leq C(t+L)^{p_n}\left\|\frac{\partial u}{\partial t}\right\|_2 \|\nabla u\|_2^p \\
&\leq C(t+L)^{p_n} E^{(p+1)/2}(t),
\end{aligned} \tag{3.40}$$

where the constant C changes as we move from left to right. Let $t < L$. From (3.40) and a quadrature, we obtain

$$E(t) \leq \frac{E(0)}{(1 - CE(0)^{(p-2)/2}t)^{2/(p-1)}}.$$

We see that there exists

$$T = CE(0)^{(p-2)/2},$$

such that as long as

$$t < T_m = \min\{L, T_1\},$$

we have

$$E(t) < 2E(0).$$

Combining these observations with (3.36) and (3.39), we find that for $t \in [0, T_m)$,

$$\left\| \frac{\partial v}{\partial t} \right\|_{H^{-s}}^{m+1} \leq C[t^{m+1}E(0)^{p(m+1)/2} + t[E(0) - E_{cr}]^m]. \tag{3.41}$$

Let

$$C_1 = \max(1, C).$$

Using (3.35) and (3.41) we have, for some constant C and $t \in [0, T)$,

$$\begin{aligned} \left| \frac{\partial u}{\partial t} \right|_{m+1}^{m+1} \geq C[\|v_0\|_{H^{-s}}^{m+1} &- C_1 t^{m+1}E(0)^{p(m+1)/2} \\ &- t^{m+1}E(0)^{(m+1)/2} - C_1 t[E(0) - E_{cr}]^m]. \end{aligned} \tag{3.42}$$

From now on, we require that $E(0) > 1$. Then we have

$$C_1 t^{m+1}E(0)^{p(m+1)/2} \geq t^{m+1}E(0)^{(m+1)/2},$$

and the above inequality may be weakened to

$$\left| \frac{\partial u}{\partial t} \right|_{m+1}^{m+1} \geq C[\|v_0\|_{H^{-s}}^{m+1} - 2C_1 t^{m+1}E(0)^{p(m+1)/2} - C_1 t[E(0) - E_{cr}]^m], \tag{3.43}$$

which holds for $t \in [0, Tm)$. Integrating both sides of this inequality, we obtain

$$\begin{aligned} \int_0^t \left\| \frac{\partial u}{\partial \tau} \right\|_{m+1}^{m+1} d\tau \geq Ct\Big[\|v_0\|_{H^{-s}}^{m+1} &- \frac{2C_1}{m+2} t^{m+1}E(0)^{p(m+1)/2} \\ &- \frac{C_1}{2} t[E(0) - E_{cr}]^m \Big], \end{aligned} \tag{3.44}$$

for $t \in [0, T_m)$. If we can find $t \in [0, T_m)$ and

$$t > (E(0) - E_{cr})E(0)^{-p(m+1)/2m},$$

then

$$t^{m+1}E(0)^{p(m+1)/2} \geq t\big[E(0) - E_{cr}\big]^m.$$

Consequently, for such values of t we have, from (3.44), we have

$$\int_0^t \left\|\frac{\partial u}{\partial \tau}\right\|_{m+1}^{m+1} d\tau \geq Ct\left[\|v_0\|_{H^{-s}}^{m+1} - \frac{(m+6))C_1}{2(m+2)} t^{m+1}E(0)^{p(m+1)/2}\right]. \tag{3.45}$$

Set

$$t_1 = \|v_0\|_{H^{-s}}^{m+1}E(0)^{-p/2-\sigma},$$

where

$$0 < \sigma < \min\left\{\frac{p-m}{2m}, \frac{m+2-p}{2}\right\},$$

since $p > m$ and $m + 2 > p$. We will show that

$$T_m > t_1 > (E(0) - E_{cr})E(0)^{-p(m+1)/2m}.$$

Let

$$v_0 = \rho\phi, \quad \rho > 0,$$

and

$$u_0 = \sigma\psi, \quad \sigma > 0,$$

such that

$$E(0) = \lambda^2$$

is independent of these variables (this implies that

$$\sigma^{p+1}/\rho^2 = o(1) \quad \text{as } \rho \to +\infty.$$

Since

$$\|v_0\|_{H^{-s}}^{m+1} \leq \|v_0\|_2^{m+1} \leq \sqrt{E(0)},$$

then

$$t_1 \le T_m = \min\{L, T_1\},$$

will hold provided

$$E(0)^{-(p-1)/2-\sigma} < CE(0)^{-(p-1)/2},$$

and

$$E(0)^{-(p-1)/2-\sigma} < L.$$

These later inequalities clearly hold for ρ so large that

$$E(0) > \max\{(1/C)^1/\sigma,\ (1/L)^{1[(p-1)/2+\sigma]}\}.$$

To show that

$$t_1 > (E(0) - E_{cr})E(0)^{-p(m+1)/2m},$$

we define

$$\epsilon \equiv \frac{\|\phi\|_{H^{-s}}}{\|\phi\|_2}.$$

Then

$$\|v_0\|_{H^{-s}}E^{-\sigma}(0) > E^{-p/2m}(0)[E(0) - E_{cr}],$$

if

$$\epsilon\|v_0\|_2 E^{-\sigma}(0) > E^{-p/2m}(0)[E(0) - E_{cr}].$$

This last inequality will hold for our choice of initial data and for all sufficiently large ρ because as $\rho \to +\infty$, the left-hand side grows like $\rho^{1-2\sigma}$ while the right-hand side grows like $\rho^{2-p/m}$ while

$$1 - 2\sigma > 2 - p/m,$$

in view of our choice of σ. Setting $t = t_1$ in (3.46), we will be done if

$$\begin{aligned}\int_0^t \left\|\frac{\partial u}{\partial \tau}\right\|_{m+1}^{m+1} d\tau &\ge C\|v_0\|_{H^{-s}}^{m+2}E(0)^{-p/2-\sigma}\left[1 - \frac{(m+6))C_1}{2(m+2)}E(0)^{-\sigma(m+1)}\right] \\ &\ge E(0) - E_{cr}.\end{aligned} \tag{3.46}$$

For sufficiently large ρ, we have

$$\frac{(m+6))C_1}{2(m+2)}E(0)^{-\sigma(m+1)} < \frac{1}{2}.$$

Thus,

$$\begin{aligned}\int_0^t \left\|\frac{\partial u}{\partial \tau}\right\|_{m+1}^{m+1} d\tau &\geq \frac{\nu}{2\epsilon^{m+2}}\|v_0\|_2^{m+2}E(0)^{-p/2-\sigma} \\ &> E(0) - E_{cr}, \end{aligned} \tag{3.47}$$

and this inequality will in turn hold for sufficiently large ρ provided

$$m + 2 > p + 2\sigma.$$

The latter condition is fulfilled due to the choice of σ. Finally, when

$$m < \frac{np}{(n+p+1)},$$

we can guarantee the last blow-up condition, namely

$$(u_0, v_0)_{L^2(\mathbb{R}^n)} \geq 0.$$

This completes the proof when $q(x) = 0$.

2. The case $q(x) \neq 0$: The proof follows the same lines as before with slight modifications. Using (3.30), the estimates (3.37) and (3.38) must be written as follows:

$$\|F(x,t)\|_{H^{-s}} \leq \|u|u|^{p-1}\|_{H^{s}} + \left\|\frac{\partial u}{\partial t}\left|\frac{\partial u}{\partial t}\right|^{m-1}\right\|_{H^{-s}} + \|qu\|_{H^{-s}}, \tag{3.48}$$

and

$$\begin{aligned}\|F(x,t)\|_{H^{-s}} &\leq C\left(\|u\|_{2p}^p + \left|\frac{\partial u}{\partial t}\right|_{m+1}^m + \|qu\|_2\right) \\ &\leq C\left[(t+L)^{p_n}\|\nabla u\|_2^p + \left|\frac{\partial u}{\partial t}\right|_{m+1}^m + \|qu\|_2\right]. \end{aligned} \tag{3.49}$$

Then after a quadrature and an application of Hölder's inequality with respect to t in the integral

$$\int_0^t \|u_\tau\|_{m+1}^m d\tau,$$

we find that

$$\int_0^t \|F(x,\tau)\|_{H^{-s}} d\tau$$

$$\leq C(L)\left[\int_0^t [E^{p/2}(\tau) + E^{1/2}(\tau)]d\tau + t^{1/(m+1)}E(0)^{m/(m+1)}\right], \tag{3.50}$$

where

$$E(t) = \|\nabla u\|_2^2 + \left\|\frac{\partial u}{\partial t}\right\|_2^2 + \|qu\|_2^2.$$

In the above inequality, we have used the fact that if $q(x)$ satisfies the conditions (3.49), the solution of the Cauchy problem (3.23) is not global for any negative initial energy. Therefore, the constant E_{cr} from the mass-free case is now zero. We obtain the same upper bound for $E(t)$ as before on the same time interval $[0, T_m)$. We can then rewrite the estimate (3.42) as follows:

$$\left|\frac{\partial u}{\partial t}\right|_{m+1}^{m+1} \geq C[\|v_0\|_{H^{-s}}^{m+1} - C_1 t^{m+1}E(0)^{p(m+1)/2} - t^{m+1}E(0)^{(m+1)/2} - C_1 tE(0)^m]. \tag{3.51}$$

As before, since $E(0) > 1$ and $C_1 \geq 1$, the third term on the right-hand side of the above inequality dominates the second one. The remainder of the proof coincides with the remainder of the proof in the mass-free case, but is simplified since we have now $E_{cr} = 0$. □

4 Nonlinear viscoelastic wave equations in weighted spaces

4.1 Viscoelastic wave equation with nonlinear source

In weighted spaces, for wider class of relaxation functions, we establish the existence and decay of solution of semilinear viscoelastic wave equation in any spaces dimension.

We consider the following equation:

$$\frac{\partial^2 u}{\partial t^2} - \phi(x)\left(\Delta_x u - \int_0^t \mu(t-s)\Delta_x u(s)ds\right) + a\frac{\partial u}{\partial t} = u|u|^{p-1}, \tag{4.1}$$

where $p > 1$, $a > 0$ and the scalar function $\mu(s)$ is assumed to satisfy (1.33). Equation (4.1) is equipped by the following initial data:

$$u(0,x) = u_0(x) \in \mathcal{H}(\mathbb{R}^n), \quad \frac{\partial u}{\partial t}(0,x) = u_1(x) \in L^2_\rho(\mathbb{R}^n), \tag{4.2}$$

where the weighted spaces $\mathcal{H}$ is given in Definition 1.3 and the density function $\phi(x) > 0$, $\forall x \in \mathbb{R}^n$, $(\phi(x))^{-1} = \rho(x)$ satisfy (1.58). Multiplying the equation (4.1) by $\rho(x)\frac{\partial u}{\partial t}$ and integrating by parts over $\mathbb{R}^n$, we have that the energy of u at time t is given by

$$E(t) = \frac{1}{2}\left(\left\|\frac{\partial u}{\partial t}\right\|^2_{L^2_\rho} + \left(1 - \int_0^t \mu(s)ds\right)\|\nabla_x u\|_2^2 + (\mu \circ \nabla_x u) - \frac{2}{p+1}\|u\|^{p+1}_{L^{p+1}_\rho}\right) \tag{4.3}$$

and the following energy functional law holds:

$$E'(t) = \frac{1}{2}(\mu' \circ \nabla_x u) - \frac{1}{2}\mu(t)\|\nabla_x u\|_2^2 - a\left\|\frac{\partial u}{\partial t}\right\|^2_{L^2_\rho}, \quad \text{for all } t \geq 0, \tag{4.4}$$

which means that our energy is uniformly bounded and decreasing along the trajectories.

4.1.1 Existence of solution

First, to prove the local existence (in time), we consider the space

$$\mathfrak{I} = C([0,T], \mathcal{H}(\mathbb{R}^n)) \cap C^1([0,T], L^2_\rho(\mathbb{R}^n)),$$

endowed with the norm

$$\|u\|^2_{\mathfrak{I}} = \max_{t\in[0,T]}\left(\|u\|^2_{\mathcal{H}} + \|u\|^2_{L^2_\rho}\right).$$

https://doi.org/10.1515/9783111633787-004

Theorem 4.1 (Local existence). *Let*

$$u_0(x) \in \mathscr{H}(\mathbb{R}^n), \quad u_1(x) \in L^2_\rho(\mathbb{R}^n)$$

be given. Suppose that

$$1 < p < \frac{n+2}{n-2}, \quad n \geq 3. \tag{4.5}$$

Then, under hypothesis (1.33), (1.44) *and* (1.58), *the problem* (4.1) *has a unique local solution*

$$u \in C([0,T], \mathscr{H}(\mathbb{R}^n)) \cap C^1([0,T], L^2_\rho(\mathbb{R}^n)).$$

The proof of Theorem 4.1 will be established through several steps. By using the Galerkin method, we shall prove the next lemma.

Lemma 4.1. *Let $u \in \mathfrak{I}$. Suppose that*

$$u_0(x) \in \mathscr{H}(\mathbb{R}^n), \quad u_1(x) \in L^2_\rho(\mathbb{R}^n).$$

Then, under hypothesis (1.33), (1.44) *and* (1.58), *there exists a unique solution v of the problem*

$$\begin{aligned} \frac{\partial^2 v}{\partial t^2} - \phi(x)\left(\Delta_x v - \int_0^t \mu(t-s)\Delta_x v(s)ds\right) + a\frac{\partial v}{\partial t} &= u|u|^{p-1}, \\ v(x,0) = u_0(x), \quad \frac{\partial v}{\partial t}(x,0) &= u_1(x), \end{aligned} \tag{4.6}$$

such that

$$\begin{aligned} &v \in \mathfrak{I} \cap C^2([0,T], \mathscr{H}^{-1}(\mathbb{R}^n)), \\ &\frac{\partial v}{\partial t} \in L^2([0,T], \mathscr{H}(\mathbb{R}^n)). \end{aligned}$$

Proof. For any $m \geq 1$, let

$$W_m = \operatorname{span}\{w_1, \dots, w_m\},$$

where $\{w_j\}$, is the orthogonal complete system of eigenfunctions of $-\Delta\phi$ in $\mathscr{H}$ such that $\|w_j\|_{L^2_\rho} = 1$, $\forall j$. We construct approximations of the solutions of (4.6)

$$v_m(x,t) = \sum_{j=1}^m \varphi_j^m(t) w_j(x), \quad m = 1, 2, \dots \tag{4.7}$$

solves the problem

$$\begin{cases} \int_{\mathbb{R}^n} \rho(x)u|u|^{p-1}v dx = \int_{\mathbb{R}^n} \rho(x)v(\frac{\partial^2 v_m}{\partial t^2} + \frac{\partial v_m}{\partial t})dx \\ \qquad + \int_{\mathbb{R}^n}(\nabla_x v_m \nabla_x v - \int_0^t \mu(t-s)\nabla_x v_m(s)ds\nabla_x v)dx, \\ v_m(x,0) = u_0^m(x) = \sum_{j=1}^m (\int_{\mathbb{R}^n} \nabla_x u_0 . \nabla_x w_j)w_j \to u_0 \quad \text{in } \mathscr{H}(\mathbb{R}^n) \text{ as } m \to \infty, \\ \frac{\partial v_m}{\partial t}(x,0) = u_1^m(x) = \sum_{j=1}^m (\int_{\mathbb{R}^n} \rho u_0 \nabla_x w_j)w_j \to u_1 \quad \text{in } L^2_\rho(\mathbb{R}^n) \text{ as } m \to \infty, \end{cases} \tag{4.8}$$

for any $v \in \mathscr{H}(\mathbb{R}^n)$. Taking $v = w_j$, in (4.8) yields the following Cauchy problem for an ordinary differential equation with unknown $\varphi_j^m, j = 1, 2, \dots$

$$\begin{cases} \varphi_j''^m(t) + a\varphi_j'^m(t) + \lambda_j \varphi_j^m(t) - \lambda_j \int_0^t \mu(t-s)\varphi_j^m(s)ds = f_j(t), \\ \varphi_j^m(x,0) = \int_{\mathbb{R}^n} \nabla_x u_0 \nabla_x w_j dx, \\ \varphi_j'^m(x,0) = \int_{\mathbb{R}^n} \rho u_0 w_j dx, \end{cases} \tag{4.9}$$

where

$$f_j(t) = \int_{\mathbb{R}^n} \rho(x)u|u|^{p-1}w_j dx \in C[0,T].$$

By using the theorem for an ordinary differential equation, we deduce that the solution of O. D. E is local in time.

Substituting $\mu = \frac{\partial v_m}{\partial t}(t)$ in (4.8), integrating over $[0,t]$ to get

$$\frac{1}{2}\left(\left\|\frac{\partial v_m}{\partial t}\right\|^2_{L^2_\rho} + \left(1 - \int_0^t \mu(s)ds\right)\|\nabla_x v_m\|_2^2 + (\mu \circ \nabla_x u_m)\right)$$
$$= \frac{1}{2}\|v_1^m\|^2_{L^2_\rho} + \frac{1}{2}\|\nabla_x v_0^m\|_2^2$$
$$+ \frac{1}{2}\int_0^t \left((\mu' \circ \nabla_x u)(\iota) - \mu(\iota)\|\nabla_x u(t)\|_2^2 + 2\int_{\mathbb{R}^n} \rho u|u|^{p-1}\frac{\partial v_m}{\partial t}dx - 2a\int_{\mathbb{R}^n} \rho|v|^2 dx\right)ds.$$

Since

$$p < \frac{n+2}{n-2},$$

we have for

$$\frac{2np}{n-2} \le 2^* = \frac{2n}{n-2},$$

$$\int_0^t \int_{\mathbb{R}^n} \rho u|u|^{p-1}\frac{\partial v_m}{\partial t}dxds \le \int_0^t \int_{\mathbb{R}^n} \rho^{1-\frac{1}{2^*}}|u|^p \rho^{\frac{1}{2^*}}\left\|\frac{\partial v_m}{\partial t}\right\|dxds$$
$$\le \int_0^t \|u\|^p_{L_\rho^{\frac{2np}{n-2}}}\left\|\frac{\partial v_m}{\partial t}\right\|_{L^{2^*}_\rho} dxds$$

$$\leq c\int_0^t \|u\|^p_{\mathscr{H}} \left\|\frac{\partial v_m}{\partial t}\right\|_{L^{2^*}_\rho} dxds$$

$$\leq c_T + \int_0^T \left\|\frac{\partial v_m}{\partial t}\right\|_* dxds.$$

Then for c small enough $c_T > 0$ independent of m, we have

$$\frac{1}{2}\left(\left\|\frac{\partial v_m}{\partial t}\right\|^2_{L^2_\rho} + \left(1 - \int_0^t \mu(s)ds\right)\|\nabla_x v_m\|_2^2 + (\mu \circ \nabla_x u_m)\right) + (2a - c)\int_0^T \left\|\frac{\partial v_m}{\partial t}\right\|^2_{L^2_\rho} ds \leq c_T.$$

We obtain the next regularity, for all $m > 0$,

$$v_m \text{ is bounded in } L^\infty([0,T], \mathscr{H}(\mathbb{R}^n)),$$
$$\frac{\partial v_m}{\partial t} \text{ is bounded in } L^\infty([0,T], L^2_\rho(\mathbb{R}^n)).$$

Therefore, up to a subsequence, we observe that there exists a subsequence v_τ of v_m and a function v that we may pass to the limit in (4.8), we obtain a weak solution v of (4.6) with the above regularity:

$$v_\tau \rightharpoonup^* v \quad \text{in } L^\infty([0,T], \mathscr{H}(\mathbb{R}^n)),$$
$$\frac{\partial v_\tau}{\partial t} \rightharpoonup^* \frac{\partial v}{\partial t} \quad \text{in } L^\infty([0,T], L^2_\rho(\mathbb{R}^n)),$$

Hence, v defines a weak solution.

Concerning the uniqueness, we suppose two solutions v_1, v_2 with the same initial data, and after a classical calculus, we obtain $v_1 = v_2$. □

Proof of Theorem 4.1. Once the local solution v exists, we will use the contraction mapping theorem to show the local existence of our initial problem (4.1). This will be done under the assumption that T is required to be small enough. □

Now, we prove that the unique solution obtained in the above theorem (Local solution) is global in time. Let the potential of the problem (4.1) be

$$J(t) = \frac{1}{2}\left(\left(1 - \int_0^t \mu(s)ds\right)\|\nabla_x u\|_2^2 + (\mu \circ \nabla_x u) - \frac{2}{p+1}\|u\|^{p+1}_{L^{p+1}_\rho(\mathbb{R}^n)}\right)$$

and we set

$$I(t) = \left(1 - \int_0^t \mu(s)ds\right)\|\nabla_x u\|_2^2 + (\mu \circ \nabla_x u) - \|u\|^{p+1}_{L^{p+1}_\rho(\mathbb{R}^n)}.$$

Let

$$A = C^{-2/(p-1)},$$
$$B = \frac{p-1}{2(p+1)} C^{-2(p+1)/(p-1)},$$

with

$$C = c\|\rho\|_{L^s}^{1/(p+1)},$$

for

$$s = 2n/(n+2-p(n-2)), \quad 1 < p \leq (n+2)/(n-2).$$

From (4.3), (4.4), we have

$$E(t) < E(0).$$

Lemma 4.2. *Assume* $0 \leq E(0) < B$. *Then the local solution satisfies:*
(i) *If* $\|u_0\|_{L_\rho^{p+1}} < A$, *then* $\|u\|_{L_\rho^{p+1}} < A$ *for all* $t \in [0, T)$.
(ii) *If* $\|u_0\|_{L_\rho^{p+1}} > A$, *then* $\|u\|_{L_\rho^{p+1}} > A$ *for all* $t \in [0, T)$.

Theorem 4.2 (Global existence). *Let*

$$u_0(x) \in \mathscr{H}(\mathbb{R}^n), \quad u_1(x) \in L_\rho^2(\mathbb{R}^n),$$

and

$$\beta = \frac{c\|\rho\|_{L^s}}{\mu_0}\left(\frac{2(p+1)}{(p-1)\mu_0}E(0)\right)^{(p-1)/2} < 1, \tag{4.10}$$

assume that

$$0 \leq E(0) < B, \quad \|u_0\|_{L_\rho^{p+1}} < A.$$

Then, under hypothesis (1.33), (1.44) *and conditions* (1.58), (4.5), *the problem* (4.1) *has a global solution in time.*

Proof. We have by (4.3),

$$\left\|\frac{\partial u}{\partial t}\right\|_{L_\rho^2}^2 + \left(1 - \int_0^t \mu(s)ds\right)\|\nabla_x u\|_2^2 + (\mu \circ \nabla_x u) = 2E(t) + \frac{2}{p+1}\|u\|_{L_\rho^{p+1}}^{p+1}.$$

Then, using Lemma 4.2, to get

$$\left\|\frac{\partial u}{\partial t}\right\|_{L^2_\rho}^2 + \left(1 - \int_0^t \mu(s)ds\right)\|\nabla_x u\|_2^2 + (\mu \circ \nabla_x u)$$
$$< \frac{p-1}{p+1}C^{-2(p+1)/(p-1)} + \frac{2}{p+1}C^{-2(p+1)/(p-1)} = C^{-2(p+1)/(p-1)}. \tag{4.11}$$

By the definition of $E(t), J(t)$ and by the fact that

$$\int_0^t \mu(s)ds \le \int_0^\infty \mu(s)ds,$$

we easily see that

$$\begin{aligned}\|\nabla_x u\|_2^2 &\le \frac{2(p+1)}{\mu_0(p-1)}J(t)\\ &\le \frac{2(p+1)}{\mu_0(p-1)}E(t)\\ &\le \frac{2(p+1)}{\mu_0(p-1)}E(0), \quad \forall t \in [0,T).\end{aligned}$$

By the continuous embedding

$$\mathscr{H}(\mathbb{R}^n) \subset L^p_\rho(\mathbb{R}^n),$$

for

$$1 \le p \le \frac{2n}{n-2},$$

we have

$$\begin{aligned}\|u\|_{L^{p+1}_\rho}^{p+1} &\le c\|\rho\|_{L^s}^{p+1}\|\nabla_x u\|_2^{p+1}\\ &\le c\|\rho\|_{L^s}^{p+1}\|\nabla_x u\|_2^{p-1}\|\nabla_x u\|_2^2\\ &\le \frac{c\|\rho\|_{L^s}^{p+1}}{\mu_0}\|\nabla_x u\|_2^{p-1}\mu_0\|\nabla_x u\|_2^2\\ &< \frac{c\|\rho\|_{L^s}}{\mu_0}\left(\frac{2(p+1)}{(p-1)\mu_0}E(0)\right)^{(p-1)/2}\mu_0\|\nabla_x u\|_2^2,\end{aligned} \tag{4.12}$$

which means by the definition of μ_0

$$\begin{aligned}\|u\|_{L^{p+1}_\rho}^{p+1} &\le \beta\mu_0\|\nabla u\|_2^2\\ &< \left(\mu_0 - \int_0^t \mu(s)ds\right)\|\nabla u\|_2^2.\end{aligned} \tag{4.13}$$

Therefore,

$$I(t) = \left(1 - \int_0^t \mu(s)ds\right)\|\nabla_x u\|_2^2 + (\mu \circ \nabla_x u) - \|u\|_{L_\rho^{p+1}}^{p+1} > 0.$$

Then

$$\begin{aligned} E(t) &= \frac{1}{2}\left\|\frac{\partial u}{\partial t}\right\|_{L_\rho^2}^2 + \left(\frac{1}{2} - \frac{1}{p+1}\right)\left(\left(1 - \int_0^t \mu(s)ds\right)\|\nabla_x u\|_2^2 + \frac{1}{2}(\mu \circ \nabla_x u)\right) \\ &\quad + \frac{1}{p+1} I(t) \\ &> \frac{1}{2}\left\|\frac{\partial u}{\partial t}\right\|_{L_\rho^2}^2 + \left(\frac{1}{2} - \frac{1}{p+1}\right)\left(\left(1 - \int_0^t \mu(s)ds\right)\|\nabla_x u\|_2^2 + \frac{1}{2}(\mu \circ \nabla_x u)\right), \end{aligned}$$

which implies since $p > 1$ that $E(t) > 0$, $\forall t \geq 0$. Then we deduce the result. □

4.1.2 Behavior of solution at ∞

We apply the multiplier techniques to obtain useful estimates and prepare some functional associated with the nature of our problem to introduce an appropriate Lyapunov function. For this purpose, we introduce two functional

$$\psi_1(t) = \int_{\mathbb{R}^n} \rho(x) u \frac{\partial u}{\partial t} dx, \tag{4.14}$$

$$\psi_2(t) = -\int_{\mathbb{R}^n} \rho(x) \frac{\partial u}{\partial t} \int_0^t \mu(t-s)(u(t) - u(s)) ds dx. \tag{4.15}$$

Lemma 4.3. *Under the hypothesis* (1.33), (1.44), *the functional* ψ_1 *satisfies, along the solution of* (4.1),

$$\begin{aligned} \psi_1'(t) &\leq \left(1 + \frac{|a|}{4\sigma}\right)\left\|\frac{\partial u}{\partial t}\right\|_{L_\rho^2}^2 + \frac{(1-\mu_0)}{4\sigma}(\mu \circ \nabla_x u) \\ &\quad + \sigma(|a| + \|\rho\|_{L^2})\|\nabla_x u\|_2^2. \end{aligned} \tag{4.16}$$

Proof. From (4.14), integrate over $\mathbb{R}^n$, we have

$$\psi_1'(t) = \int_{\mathbb{R}^n} \rho(x)\left\|\frac{\partial u}{\partial t}\right\|^2 dx + \int_{\mathbb{R}^n} \rho(x) u \frac{\partial^2 u}{\partial t^2} dx$$

$$= \int_{\mathbb{R}^n} \left(\rho(x) \left\| \frac{\partial u}{\partial t} \right\|^2 + u\Delta_x u - u \int_0^t \mu(t-s)\Delta_x u(s,x)ds \right) dx$$
$$+ \int_{\mathbb{R}^n} \rho(x)u^2|u|^{p-1}dx - a \int_{\mathbb{R}^n} \rho(x)u\frac{\partial u}{\partial t}dx$$
$$\leq \left\| \frac{\partial u}{\partial t} \right\|^2_{L^2_\rho(\mathbb{R}^n)} - \mu_0 \|\nabla_x u\|_2^2 + \|u\|^{p+1}_{L^{p+1}_\rho} dx - a \int_{\mathbb{R}^n} \rho(x)u\frac{\partial u}{\partial t}dx$$
$$+ \int_{\mathbb{R}^n} \nabla_x u \int_0^t \mu(t-s)(\nabla_x u(s) - \nabla_x u(t))dsdx.$$

We have by using the generalized version of Poincare's inequality in Lemma 1.14,

$$\int_{\mathbb{R}^n} \rho(x)u\frac{\partial u}{\partial t}dx \leq \sigma\|u\|^2_{L^2_\rho} + \frac{1}{4\sigma}\left\| \frac{\partial u}{\partial t} \right\|^2_{L^2_\rho},$$
$$\leq \sigma\|\rho\|^2_{L^2}\|\nabla_x u\|_2^2 + \frac{1}{4\sigma}\left\| \frac{\partial u}{\partial t} \right\|^2_{L^2_\rho}, \quad \forall \sigma > 0. \tag{4.17}$$

Using Young's inequality and Lemma 1.9 for $\theta = 1/2$, we obtain

$$\psi_1'(t) \leq \left(1 + \frac{|a|}{4\sigma}\right)\left\| \frac{\partial u}{\partial t} \right\|^2_{L^2_\rho} + (|a|\sigma\|\rho\|_{L^2} - \mu_0)\|\nabla_x u\|_2^2 + \|u\|^{p+1}_{L^{p+1}_\rho}$$
$$+ \sigma\|\nabla_x u\|_2^2 + \frac{1}{4\sigma}\int_{\mathbb{R}^n}\left(\int_0^t \mu(t-s)|\nabla_x u(s) - \nabla_x u(t)|ds \right)^2 dx \tag{4.18}$$
$$\leq \left(1 + \frac{|a|}{4\sigma}\right)\left\| \frac{\partial u}{\partial t} \right\|^2_{L^2_\rho} + (\sigma(1 + |a|\|\rho\|_{L^2}) - \mu_0)\|\nabla_x u\|_2^2$$
$$+ \frac{(1-\mu_0)}{4\sigma}(\mu \circ \nabla_x u) + \|u\|^{p+1}_{L^{p+1}_\rho}.$$

By the continuous embedding,

$$\mathscr{H}(\mathbb{R}^n) \subset L^p_\rho(\mathbb{R}^n),$$

for

$$1 \leq p \leq \frac{2n}{n-2},$$

we have

$$\|u\|^{p+1}_{L^{p+1}_\rho} \leq c\|\rho\|^{p+1}_{L^s}\|\nabla_x u\|_2^{p+1}$$

$$\leq c\|\rho\|_{L^s}^{p+1}\|\nabla_x u\|_2^{p-1}\|\nabla_x u\|_2^2$$
$$< c\|\rho\|_{L^s}\left(\frac{2(p+1)}{(p-1)\mu_0}E(0)\right)^{(p-1)/2}\|\nabla_x u\|_2^2, \tag{4.19}$$

which means

$$\psi_1'(t) \leq \left(1+\frac{|a|}{4\sigma}\right)\left\|\frac{\partial u}{\partial t}\right\|_{L^2_\rho}^2 + \frac{(1-\mu_0)}{4\sigma}(\mu\circ\nabla_x u)$$
$$+\left(c\|\rho\|_{L^s}\left(\frac{2(p+1)}{(p-1)\mu_0}E(0)\right)^{(p-1)/2} + \sigma(1+|a|\|\rho\|_{L^2}) - \mu_0\right)\|\nabla_x u\|_2^2.$$

By the fact that

$$\beta = \frac{c\|\rho\|_{L^s}}{\mu_0}\left(\frac{2(p+1)}{(p-1)\mu_0}E(0)\right)^{(p-1)/2} < 1,$$

we have

$$\psi_1'(t) \leq \left(1+\frac{|a|}{4\sigma}\right)\left\|\frac{\partial u}{\partial t}\right\|_{L^2_\rho}^2 + \frac{(1-\mu_0)}{4\sigma}(\mu\circ\nabla_x u)$$
$$+\sigma(1+|a|\|\rho\|_{L^2})\|\nabla_x u\|_2^2.$$

□

Lemma 4.4. *Under the hypothesis* (1.33), (1.44), *the functional* ψ_2 *satisfies, along the solution of* (4.1), *for any* $\sigma \in (0,1)$,

$$\psi_2'(t) \leq \sigma(1+\|\rho\|_{L^s}^{p+1})\|\nabla_x u\|_2^2$$
$$+ c_\sigma(1+a\|\rho\|_{L^2}^2 + \|\rho\|_{L^s}^{p+1})(\mu\circ\nabla_x u)^{(p+1)/2} - c_\sigma\|\rho\|_{L^2}^2(\mu'\circ\nabla_x u)$$
$$+\left(\sigma(1+a) - \int_0^t \mu(s)ds\right)\left\|\frac{\partial u}{\partial t}\right\|_{L^2_\rho(\mathbb{R}^n)}^2.$$

Proof. Exploiting (4.1), (4.15) to get

$$\psi_2'(t) = -\int_{\mathbb{R}^n}\rho(x)\frac{\partial^2 u}{\partial t^2}\int_0^t \mu(t-s)(u(t)-u(s))dsdx$$
$$-\int_{\mathbb{R}^n}\rho(x)\frac{\partial u}{\partial t}\int_0^t \mu'(t-s)(u(t)-u(s))dsdx$$
$$-\int_0^t \mu(s)ds\left\|\frac{\partial u}{\partial t}\right\|_{L^2_\rho}^2$$

$$
\begin{aligned}
&= \int_{\mathbb{R}^n} \nabla_x u \int_0^t \mu(t-s)(\nabla_x u(t) - \nabla_x u(s))dsdx \\
&\quad + a \int_{\mathbb{R}^n} \rho(x)\frac{\partial u}{\partial t} \int_0^t \mu(t-s)(u(t)-u(s))dsdx \\
&\quad - \int_{\mathbb{R}^n} \rho(x)u|u|^{p-1} \int_0^t \mu(t-s)(u(t)-u(s))dsdx \\
&\quad - \int_{\mathbb{R}^n} \left(\int_0^t \mu(t-s)\nabla_x u(s,x)ds\right)\left(\int_0^t \mu(t-s)(\nabla_x u(t) - \nabla_x u(s))ds\right)dx \\
&\quad - \int_{\mathbb{R}^n} \rho(x)\frac{\partial u}{\partial t} \int_0^t \mu'(t-s)(u(t)-u(s))dsdx \\
&\quad - \int_0^t \mu(s)ds \left\|\frac{\partial u}{\partial t}\right\|_{L^2_\rho}^2.
\end{aligned}
$$

By (1.33), we have

$$
\begin{aligned}
\psi_2'(t) &= \left(1 - \int_0^t \mu(s)ds\right) \int_{\mathbb{R}^n} \nabla_x u \int_0^t \mu(t-s)(\nabla_x u(t) - \nabla_x u(s))dsdx \\
&\quad + \int_{\mathbb{R}^n} \left(\int_0^t \mu(t-s)(\nabla_x u(t) - \nabla_x u(s))ds\right)^2 dx \\
&\quad + a \int_{\mathbb{R}^n} \rho(x)\frac{\partial u}{\partial t} \int_0^t \mu(t-s)(u(t)-u(s))dsdx \\
&\quad - \int_{\mathbb{R}^n} \rho(x)u|u|^{p-1} \int_0^t \mu(t-s)(u(t)-u(s))dsdx \\
&\quad - \int_{\mathbb{R}^n} \rho(x)\frac{\partial u}{\partial t} \int_0^t \mu'(t-s)(u(t)-u(s))dsdx \\
&\quad - \int_0^t \mu(s)ds \left\|\frac{\partial u}{\partial t}\right\|_{L^2_\rho}^2 + c(\mu \circ \nabla_x u)(t).
\end{aligned}
$$

By Hölder's and Young's inequalities and Lemma 1.14, we estimate

$$
\begin{aligned}
&-\int_{\mathbb{R}^n} \rho(x)\frac{\partial u}{\partial t}\int_0^t \mu'(t-s)(u(t)-u(s))dsdx \\
&\quad \le \left(\int_{\mathbb{R}^n} \rho(x)\left\|\frac{\partial u}{\partial t}\right\|^2 dx\right)^{1/2} \\
&\quad \times \left(\int_{\mathbb{R}^n} \rho(x)\left|\int_0^t \mu'(t-s)(u(t)-u(s))ds\right|^2\right)^{1/2} \\
&\quad \le \sigma\left\|\frac{\partial u}{\partial t}\right\|^2_{L^2_\rho} + c_\sigma\left\|\int_0^t -\mu'(t-s)(u(t)-u(s))ds\right\|^2_{L^2_\rho} \\
&\quad \le \sigma\left\|\frac{\partial u}{\partial t}\right\|^2_{L^2_\rho} - c_\sigma\|\rho\|^2_{L^2}(\mu'\circ\nabla_x u)(t),
\end{aligned}
$$

and

$$
\begin{aligned}
&\int_{\mathbb{R}^n} \rho(x)\frac{\partial u}{\partial t}\int_0^t \mu(t-s)(u(t)-u(s))dsdx \\
&\qquad \le \sigma\left\|\frac{\partial u}{\partial t}\right\|^2_{L^2_\rho} + c_\sigma\|\rho\|^2_{L^2}(\mu\circ\nabla_x u)(t),
\end{aligned}
$$

and

$$
\begin{aligned}
&-\int_{\mathbb{R}^n} \rho(x)|u|^{p-1}u\int_0^t \mu(t-s)(u(t)-u(s))dsdx \\
&\quad \le \left(\int_{\mathbb{R}^n} \rho(x)|u|^{p+1}dx\right)^{p/(p+1)} \\
&\quad \times \left(\int_{\mathbb{R}^n} \rho(x)\left|\int_0^t \mu(t-s)(u(t)-u(s))ds\right|^{p+1}\right)^{1/(p+1)} \\
&\quad \le \sigma\|u\|^{p+1}_{L^{p+1}_\rho} + c_\sigma\left\|\int_0^t \mu(t-s)(u(t)-u(s))ds\right\|^{p+1}_{L^{p+1}_\rho} \\
&\quad \le \sigma\|\rho\|^{p+1}_{L^s}\|\nabla_x u\|^2_2 + c_\sigma\|\rho\|^{p+1}_{L^s}(\mu\circ\nabla_x u)^{(p+1)/2}.
\end{aligned}
$$

Using Young and Poincare's inequalities and Lemma 1.9 for $\theta = 1/2$, we obtain

$$
\psi_2'(t) \le \sigma(1+\|\rho\|^{p+1}_{L^s})\|\nabla_x u\|^2_2
$$

$$+ c_\sigma(1 + a\|\rho\|_{L^2}^2 + \|\rho\|_{L^s}^{p+1})(\mu \circ \nabla_x u)^{(p+1)/2} - c_\sigma\|\rho\|_{L^2}^2(\mu' \circ \nabla_x u)$$
$$+ \left(\sigma(1 + a) - \int_0^t \mu(s)ds\right)\left\|\frac{\partial u}{\partial t}\right\|_{L_\rho^2(\mathbb{R}^n)}^2. \qquad \square$$

Now, let us define

$$L(t) = \xi_1 E(t) + \psi_1(t) + \xi_2\psi_2(t) \tag{4.20}$$

for $\xi_1, \xi_2 > 1$. We need the next lemma, which means that there is an equivalent between the Lyapunov and energy functions, i. e., for $\xi_1, \xi_2 > 1$, we have

$$\beta_1 L(t) \le E(t) \le \beta_2 L(t) \tag{4.21}$$

holds for two positive constants β_1 and β_2.

Lemma 4.5. *For $\xi_1, \xi_2 > 1$, we have*

$$L(t) \sim E(t). \tag{4.22}$$

Proof. By (4.20), we have

$$\begin{aligned}|L(t) - \xi_1 E(t)| &\le |\psi_1(t)| + \xi_2|\psi_2(t)| \\ &\le \int_{\mathbb{R}^n}\left|\rho(x)u\frac{\partial u}{\partial t}\right|dx \\ &\quad + \xi_2\int_{\mathbb{R}^n}\left|\rho(x)\frac{\partial u}{\partial t}\int_0^t \mu(t-s)(u(t) - u(s))ds\right|dx.\end{aligned}$$

Thanks to Hölder and Young's inequalities, we have by using Lemma 1.14,

$$\begin{aligned}\int_{\mathbb{R}^n}\left|\rho(x)u\frac{\partial u}{\partial t}\right|dx &\le \left(\int_{\mathbb{R}^n}\rho(x)|u|^2dx\right)^{1/2}\left(\int_{\mathbb{R}^n}\rho(x)\left\|\frac{\partial u}{\partial t}\right\|^2dx\right)^{1/2} \\ &\le \frac{1}{2}\left(\int_{\mathbb{R}^n}\rho(x)|u|^2dx\right) + \frac{1}{2}\left(\int_{\mathbb{R}^n}\rho(x)\left\|\frac{\partial u}{\partial t}\right\|^2dx\right) \\ &\le c\left\|\frac{\partial u}{\partial t}\right\|_{L_\rho^2}^2 + c\|\rho\|_{L^2}^2\|\nabla_x u\|_2^2\end{aligned} \tag{4.23}$$

and

$$\begin{aligned}
&\int_{\mathbb{R}^n}\left|\left(\rho(x)^{\frac{1}{2}}\frac{\partial u}{\partial t}\right)\left(\rho(x)^{\frac{1}{2}}\int_0^t \mu(t-s)(u(t)-u(s))ds\right)\right|dx \\
&\quad\le\left(\int_{\mathbb{R}^n}\rho(x)\left\|\frac{\partial u}{\partial t}\right\|^2 dx\right)^{1/2}\times\left(\int_{\mathbb{R}^n}\rho(x)\left|\int_0^t \mu(t-s)(u(t)-u(s))ds\right|^2 dx\right)^{1/2} \\
&\quad\le\frac{1}{2}\left\|\frac{\partial u}{\partial t}\right\|^2_{L^2_\rho}+\frac{1}{2}\left\|\int_0^t \mu(t-s)(u(t)-u(s))ds\right\|^2_{L^2_\rho} \\
&\quad\le\frac{1}{2}\left\|\frac{\partial u}{\partial t}\right\|^2_{L^2_\rho}+\frac{1}{2}\|\rho\|^2_{L^2}(\mu\circ\nabla_x u).
\end{aligned}$$

Then

$$|L(t)-\xi_1 E(t)|\le cE(t).$$

Therefore, we can choose ξ_1 so that

$$L(t)\sim E(t). \tag{4.24}$$

□

Lemma 4.6. *For all $t\ge t_1>0$, we have*

$$\int_{t_1}^t(\mu\circ\nabla_x u)(s)ds\le H_0^{-1}\left(-\int_{t_1}^t H_0(-\mu'(s))\mu'(s)\int_{\mathbb{R}^n}\mu(s)|\nabla_x u(t)-\nabla_x u(t-s)|^2 dxds\right),$$

where H_0 introduced in (1.45).

Proof. By (4.4) and Remark 1.1, we have for all $t\ge t_1$,

$$\begin{aligned}
\int_{\mathbb{R}^n}\int_0^{t_1}\mu(t-s)|\nabla_x u(t)-\nabla_x u(s)|^2 dsdx &\le -\frac{1}{k}\int_{\mathbb{R}^n}\int_0^{t_1}\mu(t-s)|\nabla_x u(t)-\nabla_x u(s)|^2 dsdx \\
&\le -cE'(t).
\end{aligned}$$

Now, we define

$$I(t)=\int_{t_1}^t H_0(-\mu'(s))(\mu\circ\nabla_x u)(t)ds. \tag{4.25}$$

Since

$$\int_0^{+\infty} H_0(-\mu'(s))\mu(s)ds<+\infty,$$

from (4.4) we have

$$
\begin{aligned}
I(t) &= \int_{t_1}^{t} H_0(-\mu'(s)) \int_{\mathbb{R}^n} \mu(s)|\nabla_x u(t) - \nabla_x u(t-s)|^2 dxds \\
&\le 2\int_{t_1}^{t} H_0(-\mu'(s))\mu(s) \int_{\mathbb{R}^n} |\nabla_x u(t)|^2 + |\nabla_x u(t-s)|^2 dxds \\
&\le cE(0)\int_{t_1}^{t} H_0(-\mu'(s))\mu(s)ds < 1.
\end{aligned} \tag{4.26}
$$

Now, we define again a new functional $\lambda(t)$ related with $I(t)$ as

$$
\lambda(t) = -\int_{t_1}^{t} H_0(-\mu'(s))\mu'(s) \int_{\mathbb{R}^n} \mu(s)|\nabla_x u(t) - \nabla_x u(t-s)|^2 dxds. \tag{4.27}
$$

From (1.33), (1.44) and Remark 1.1, we get

$$
H_0(-\mu'(s))\mu(s) \le H_0(H(\mu(s)))\mu(s) = D(\mu(s))\mu(s) \le k_0,
$$

for some positive constant k_0. Then, for all $t \ge t_1$,

$$
\begin{aligned}
\lambda(t) &\le -k_0 \int_{t_1}^{t} \mu'(s) \int_{\mathbb{R}^n} |\nabla_x u(t) - \nabla_x u(t-s)|^2 dxds \\
&\le -k_0 \int_{t_1}^{t} \mu'(s) \int_{\mathbb{R}^n} |\nabla_x u(t)|^2 + |\nabla_x u(t-s)|^2 dxds \\
&\le -cE(0)\int_{t_1}^{t} \mu'(s)ds \\
&\le cE(0)\mu(t_1) \\
&< \min\{r, H(r), H_0(r)\}.
\end{aligned} \tag{4.28}
$$

Using the properties of H_0 (strictly convex in $(0,r]$, $H_0(0) = 0$), then for $x \in (0,r]$, $\theta \in [0,1]$,

$$
H_0(\theta x) \le \theta H_0(x).
$$

Using Remark 1.1, (4.26), (4.28) and Jensen's inequality leads to

$$\begin{aligned}\lambda(t) &= \frac{1}{I(t)}\int_{t_1}^{t} I(t)H_0[H_0^{-1}(-\mu'(s))]H_0(-\mu'(s))\mu'(s)\int_{\mathbb{R}^n}\mu(s)|\nabla_x u(t)-\nabla_x u(t-s)|^2dxds\\ &\geq \frac{1}{I(t)}\int_{t_1}^{t} H_0[I(t)H_0^{-1}(-\mu'(s))]H_0(-\mu'(s))\mu'(s)\int_{\mathbb{R}^n}\mu(s)|\nabla_x u(t)-\nabla_x u(t-s)|^2dxds\\ &\geq H_0\left(\frac{1}{I(t)}\int_{t_1}^{t} I(t)H_0^{-1}(-\mu'(s))H_0(-\mu'(s))\mu'(s)\int_{\mathbb{R}^n}\mu(s)|\nabla_x u(t)-\nabla_x u(t-s)|^2dxds\right)\\ &\geq H_0\left(\int_{t_1}^{t}\int_{\mathbb{R}^n}\mu(s)|\nabla_x u(t)-\nabla_x u(t-s)|^2dxds\right),\end{aligned}$$

which implies

$$\int_{t_1}^{t}\int_{\mathbb{R}^n}\mu(s)|\nabla_x u(t)-\nabla_x u(t-s)|^2dxds \leq H_0^{-1}(\lambda(t)).$$ □

Theorem 4.3. *Let*

$$(u_0, u_1) \in \mathscr{H}(\mathbb{R}^n)\times L^2_\rho(\mathbb{R}^n),$$

and suppose that (1.33), (1.44) *hold. Then there exist positive constants*

$$\alpha_0, \alpha_1, \alpha_2, \alpha_3,$$

such that the energy of solution of (4.1) *satisfies*

$$E(t) \leq \alpha_3 H_1^{-1}(\alpha_1 t + \alpha_2), \quad \textit{for all } t \geq 0,$$

where

$$H_1(t) = \int_t^1 (sH_0'(\alpha_0 s))^{-1}ds.$$

Proof. From (4.4), Lemma 4.3 and Lemma 4.4, we have

$$\begin{aligned}L'(t) &= \xi_1 E'(t) + \psi_1'(t) + \xi_2\psi_2'(t)\\ &\leq \left(\frac{1}{2}\xi_1 - c_\sigma\|\rho\|^2_{L^2}\xi_2\right)(\mu'\circ\nabla_x u) + M_0(\mu\circ\nabla_x u)^{(p+1)/2}\\ &\quad - M_1\left\|\frac{\partial u}{\partial t}\right\|^2_{L^2_\rho} - M_2\|\nabla_x u\|_2^2,\end{aligned}$$

where

$$M_0 = \left(\xi_2 c_\sigma (1 + a\|\rho\|_{L^2}^2 + \|\rho\|_{L^s}^{p+1}) + \frac{(1-\mu_0)}{4\sigma} \right) > 0,$$

$$M_1 = \left(a\xi_1 + \xi_2 \left(\int_0^{t_1} \mu(s)ds - \sigma(1+a) \right) - 1 - \frac{|a|}{4\sigma} \right),$$

$$M_2 = \left(\frac{1}{2}\xi_1 \mu(t_1) - \sigma(|a| + \|\rho\|_{L^2}) - \xi_2 \sigma (1 + \|\rho\|_{L^s}^{p+1}) \right),$$

and t_1 was introduced in Remark 1.1.

We choose σ so small that

$$\xi_1 > 2c_\sigma \|\rho\|_{L^2}^2 \xi_2.$$

Whence σ is fixed, we can choose ξ_1, ξ_2 large enough so that $M_1, M_2 > 0$, which yields

$$\begin{aligned} L'(t) &\leq M_0 (\mu \circ \nabla_x u)^{(p+1)/2} - cE'(t) \\ &\leq M_0 (\mu \circ \nabla_x u) E(t)^{(p-1)/2} - cE'(t) \\ &\leq M_0 (\mu \circ \nabla_x u) E(0)^{(p-1)/2} - cE'(t), \quad \forall t \geq t_1. \end{aligned} \tag{4.29}$$

Now we set

$$F(t) = L(t) + cE(t),$$

which is equivalent to $E(t)$. Then by (4.29), we get

$$\begin{aligned} F'(t) &= L'(t) + cE'(t) \\ &\leq -cE(t) + c \int_{\mathbb{R}^n} \int_{t_1}^{t} \mu(t-s) |\nabla_x u(t) - \nabla_x u(s)|^2 ds dx, \quad \text{for all } t \geq t_1. \end{aligned} \tag{4.30}$$

Using Lemma 4.6, we obtain

$$F'(t) \leq -cE(t) + cH_0^{-1}(\lambda(t)), \quad \text{for all } t \geq t_1.$$

Now, using the fact that

$$E' \leq 0, \quad 0 < H_0', \quad 0 < H_0'',$$

on $(0, r]$ to define the functional

$$F_1(t) = H_0'\left(\alpha_0 \frac{E(t)}{E(0)} \right) F(t) + cE(t), \quad \alpha_0 < r, 0 < c,$$

where

$$F_1(t) \sim E(t),$$

and

$$\begin{aligned} F_1'(t) &= \alpha_0 \frac{E'(t)}{E(0)} H_0''\left(\alpha_0 \frac{E(t)}{E(0)}\right) F(t) + H_0'\left(\alpha_0 \frac{E(t)}{E(0)}\right) F'(t) + cE'(t) \\ &\leq -cE(t) H_0'\left(\alpha_0 \frac{E(t)}{E(0)}\right) + cH_0'\left(\alpha_0 \frac{E(t)}{E(0)}\right) H_0^{-1}(\lambda(t)) + cE'(t). \end{aligned}$$

Let H_0^* given in Remark 1.1 and using Young's inequality (1.47) with

$$\begin{aligned} A &= H_0'\left(\alpha_0 \frac{E(t)}{E(0)}\right), \\ B &= H_0^{-1}(\lambda(t)), \end{aligned}$$

to get

$$\begin{aligned} F_1'(t) &\leq -cE(t) H_0'\left(\alpha_0 \frac{E(t)}{E(0)}\right) + cH_0^*\left(H_0'\left(\alpha_0 \frac{E(t)}{E(0)}\right)\right) + c\lambda(t) + cE'(t) \\ &\leq -cE(t) H_0'\left(\alpha_0 \frac{E(t)}{E(0)}\right) + c\alpha_0 \frac{E(t)}{E(0)} H_0'\left(\alpha_0 \frac{E(t)}{E(0)}\right) - c'E'(t) + cE'(t). \end{aligned}$$

Choosing α_0, c, c', such that for all $t \geq t_1$ we have

$$\begin{aligned} F_1'(t) &\leq -k \frac{E(t)}{E(0)} H_0'\left(\alpha_0 \frac{E(t)}{E(0)}\right) \\ &= -kH_2\left(\frac{E(t)}{E(0)}\right), \end{aligned}$$

where

$$H_2(t) = tH_0'(\alpha_0 t).$$

Using the strict convexity of H_0 on $(0, r]$ to find that H_2', H_2 are strict positives on $(0, 1]$, then

$$R(t) = \tau \frac{k_1 F_1(t)}{E(0)} \sim E(t), \quad \tau \in (0, 1) \tag{4.31}$$

and

$$R'(t) \leq -\tau k_0 H_2(R(t)), \quad k_0 \in (0, +\infty), t \geq t_1.$$

Then a simple integration and a suitable choice of τ yield

$$R(t) \le H_1^{-1}(\alpha_1 t + \alpha_2), \quad \alpha_1, \alpha_2 \in (0, +\infty), t \ge t_1.$$

Here,

$$H_1(t) = \int_t^1 H_2^{-1}(s)ds.$$

From (4.31), for a positive constant α_3, we have

$$E(t) \le \alpha_3 H_1^{-1}(\alpha_1 t + \alpha_2), \quad \alpha_1, \alpha_2 \in (0, +\infty), t \ge t_1.$$

The fact that H_1 is strictly decreasing function on $(0,1]$ and due to properties of H_2, we have

$$\lim_{t \to 0} H_1(t) = +\infty.$$

Then

$$E(t) \le \alpha_3 H_1^{-1}(\alpha_1 t + \alpha_2), \quad \text{for all } t \ge 0.$$

This completes the proof of Theorem 4.3. □

4.2 Viscoelastic wave equations with strong damping in $\mathbb{R}^n$

We consider the following problem:

$$\begin{cases} \frac{\partial}{\partial t}(|\frac{\partial u_1}{\partial t}|^{l-2}\frac{\partial u_1}{\partial t}) + au_2 - \phi(x)\Delta(u_1 - \int_0^t \mu_1(t-s)u_1(s,x)ds + \frac{\partial u_1}{\partial t}) = 0, \\ \frac{\partial}{\partial t}(|\frac{\partial u_2}{\partial t}|^{l-2}\frac{\partial u_2}{\partial t}) + au_1 - \phi(x)\Delta(u_2 - \int_0^t \mu_2(t-s)u_2(s,x)ds + \frac{\partial u_2}{\partial t}) = 0, \\ (u_1(0,x), u_2(0,x)) = (u_{10}(x), u_{20}(x)) \in (\mathcal{H}(\mathbb{R}^n))^2, \\ (\frac{\partial u_1}{\partial t}(0,x), \frac{\partial u_2}{\partial t}(0,x)) = (u_{11}(x), u_{21}(x)) \in (L_\rho^l(\mathbb{R}^n))^2, \end{cases} \tag{4.32}$$

where $a \ne 0$ where the space $\mathcal{H}(\mathbb{R}^n)$ defined in Definition 1.3 and $l, n \ge 2$, $\phi(x) > 0$, $\forall x \in \mathbb{R}^n$, $(\phi(x))^{-1} = \rho(x)$ defined in (1.58) and the functions μ_1, μ_2 defined as in (1.33).

The energy of (u_1, u_2) at time t is defined by

$$\begin{aligned} E(t) = \frac{(l-1)}{l}\sum_{i=1}^{2}\left\|\frac{\partial u_i}{\partial t}\right\|_{L_\rho^l(\mathbb{R}^n)}^l + \frac{1}{2}\sum_{i=1}^{2}\left(1 - \int_0^t \mu_i(s)ds\right)\|\nabla u_i\|_2^2 \\ + \frac{1}{2}\sum_{i=1}^{2}(\mu_i \circ \nabla u_i) + a\int_{\mathbb{R}^n} \rho u_1 u_2 dx. \end{aligned} \tag{4.33}$$

For a small enough, we use Lemma 1.14 to deduce that

$$E(t) \geq \frac{1}{2}(1 - c|\alpha|\|\rho\|_{L^{n/2}}^{-1})\left[\frac{2(l-1)}{l}\sum_{i=1}^{2}\left\|\frac{\partial u_i}{\partial t}\right\|_{L_\rho^l}^{l} + \sum_{i=1}^{2}\left(1 - \int_0^t \mu_i(s)ds\right)\|\nabla u_i\|_2^2 + \sum_{i=1}^{2}(\mu_i \circ \nabla u_i)\right], \tag{4.34}$$

and the following energy functional law holds, which means that our energy is uniformly bounded and decreasing along the trajectories:

$$E'(t) \leq \frac{1}{2}\sum_{i=1}^{2}(\mu_i' \circ \nabla u_i)(t) - \sum_{i=1}^{2}\left\|\nabla\frac{\partial u_i}{\partial t}\right\|_2^2, \quad \forall t \geq 0. \tag{4.35}$$

Finally, we give the definition of *weak solutions* for the problem (4.32).

Definition 4.1. A weak solution of (4.32) is (u_1, u_2) such that

- $(u_1, u_2) \in (L^2[0,T;\mathscr{H}(\mathbb{R}^n)])^2$, $(\frac{\partial u_1}{\partial t}, \frac{\partial u_2}{\partial t}) \in (L^2[0,T;L_\rho^l(\mathbb{R}^n)])^2$ and $(\frac{\partial^2 u_1}{\partial t^2}, \frac{\partial^2 u_2}{\partial t^2}) \in (L^2[0,T;D^{-1}(\mathbb{R}^n)])^2$,
- For all $(v,w) \in (C_0^\infty([0,T]\times\mathbb{R}^n))^2$, (u_1,u_2) satisfies the generalized formula:

$$\begin{cases} \int_0^T (\frac{\partial}{\partial t}(|\frac{\partial u_1}{\partial t}|^{l-2}\frac{\partial u_1}{\partial t}), v)_{L_\rho^l} ds + \alpha\int_0^T (u_2, v)_{L_\rho^2} ds + \int_0^T\int_{\mathbb{R}^n}\nabla u_1 \nabla v dx ds \\ \quad + \int_0^T\int_{\mathbb{R}^n}\nabla\frac{\partial u_1}{\partial t}\nabla v dx ds - \int_0^T\int_{\mathbb{R}^n}\int_0^s \mu_1(s-\tau)\nabla u_1(\tau)d\tau\nabla v(s) dx ds = 0, \\ \int_0^T (\frac{\partial}{\partial t}(|\frac{\partial u_2}{\partial t}|^{l-2}\frac{\partial u_2}{\partial t}), w)_{L_\rho^l} ds + \alpha\int_0^T (u_1, w)_{L_\rho^2} ds + \int_0^T\int_{\mathbb{R}^n}\nabla u_2 \nabla w dx ds \\ \quad + \int_0^T\int_{\mathbb{R}^n}\nabla\frac{\partial u}{\partial t}_2\nabla w dx ds - \int_0^T\int_{\mathbb{R}^n}\int_0^s \mu_2(s-\tau)\nabla u_2(\tau)d\tau\nabla w(s) dx ds = 0, \end{cases}$$

- (u_1,u_2) satisfies the initial conditions $(u_{10}(x), u_{20}(x)) \in (\mathscr{H}(\mathbb{R}^n))^2$, $(u_{11}(x), u_{21}(x)) \in (L_\rho^l(\mathbb{R}^n))^2$.

We are now ready to state and prove our existence results.

4.2.1 Well-posedness result for the nonlinear case

This section is devoted to prove the existence and uniqueness of solutions to system (4.32) taking account the nonlinear case in the terms responsible on the relation between tow equations, i. e., replacing $\alpha u_1, \alpha u_2$ by $f_1(u_1,u_2), f_2(u_1,u_2)$ introduced as

$$\begin{cases} \frac{\partial}{\partial t}(|\frac{\partial u_1}{\partial t}|^{l-2}\frac{\partial u_1}{\partial t}) + f_1(u_1,u_2) - \phi(x)\Delta(u_1 + \int_0^t \mu_1(s)u_1(t-s,x)ds + \frac{\partial u_1}{\partial t}) = 0, \\ \frac{\partial}{\partial t}(|\frac{\partial u_2}{\partial t}|^{l-2}\frac{\partial u_2}{\partial t}) + f_2(u_1,u_2) - \phi(x)\Delta(u_2 + \int_0^t \mu_2(s)u_2(t-s,x)ds + \frac{\partial u_2}{\partial t}) = 0, \\ (u_1(0,x), u_2(0,x)) = (u_{10}(x), u_{20}(x)) \in (D(\mathbb{R}^n))^2, \\ (\frac{\partial u_1}{\partial t}(0,x), \frac{\partial u_2}{\partial t}(0,x)) = (u_{11}(x), u_{21}(x)) \in (L_\rho^l(\mathbb{R}^n))^2, \end{cases} \tag{4.36}$$

where our nonlinearity is given by the functions f_1, f_2 satisfying the next assumptions:

1. The functions $f_i : \mathbb{R}^2 \to \mathbb{R}$ (for $i = 1, 2$) is of class C^1 and there exists a function F such that

$$f_1(x,y) = \frac{\partial F}{\partial x}, \quad f_2(x,y) = \frac{\partial F}{\partial y}, \tag{4.37}$$

$$F \geq 0, \quad xf_1(x,y) + yf_2(x,y) - F(x,y) \geq 0. \tag{4.38}$$

and

$$\left|\frac{\partial f_i}{\partial x}(x,y)\right| + \left|\frac{\partial f_i}{\partial y}(x,y)\right| \leq d(1 + |x|^{\beta_{i1}-1} + |y|^{\beta_{i2}-1}) \quad \forall (x,y) \in \mathbb{R}^2, \tag{4.39}$$

for some constant $d > 0$ and $1 \leq \beta_{ij} \leq \frac{n}{n-2}$ for $i,j = 1, 2$.

2. There exists a positive constant k such that

$$|f_i(x,y)| \leq k(|x| + |y| + |x|^{\beta_{i1}} + |y|^{\beta_{i2}}), \tag{4.40}$$

and

$$\begin{aligned} &|f_i(x,y) - f_i(r,s)| \\ &\quad \leq k(1 + |x|^{\beta_{i1}-1} + |y|^{\beta_{i2}-1} + |r|^{\beta_{i1}-1} + |s|^{\beta_{i2}-1})(|x - r| + |y - s|), \end{aligned} \tag{4.41}$$

for all $(x,y), (r,s) \in \mathbb{R}^2$ and $i = 1, 2$. Noting that we follow the same steps in the linear cases with the same perturbed function and some calculations related with the presence of f_1, f_2.

First, we prove the existence of unique solution of the restricted problem on

$$B_R = \{x \in \mathbb{R}^n : |x| < R\}.$$

The main ingredient used here is the Galerkin approximations.

Lemma 4.7. *Assume that* (1.37)–(1.38), *and* (1.58) *with* (4.37)–(4.41) *are satisfied. Suppose that the constants* $T > 0$, $R > 0$ *and the initial conditions*

$$(u_{10}, u_{20}) \in (D(B_R))^2 = (u_{11}, u_{21}) \in (L_p^l(B_R))^2,$$

are given. Then there exists a unique solution for the problem (4.32), *such that*

$$u_i \in C[0, T; D(B_R)] \quad \text{and} \quad \frac{\partial u_i}{\partial t} \in C[0, T; L_p^l(B_R)].$$

Proof. The existence is proved by using the Galerkin method, which consists of constructing approximations of the solution. Then we obtain a priori estimates necessary to guarantee the convergence of these approximations. So, we take $\{w_i\}_{i=1}^{\infty}$ be the eigenfunctions of operator $-\Delta$. Then $\{w_i\}_{i=1}^{\infty}$ is orthogonal basis of $D(B_R)$, which is orthonormal in $L^2_\rho(B_R)$.

Let

$$V_m = \operatorname{span}\{w_1, w_2, \ldots, w_m\},$$

and the projection of the initial data on the finite-dimensional subspace V_m is given by

$$u_{10}^m = \sum_{j=0}^m a_j w_j, \quad u_{20}^m = \sum_{j=0}^m b_j w_j, \quad u_{11}^m = \sum_{j=0}^m c_j w_j, \quad u_{21}^m = \sum_{j=0}^m d_j w_j.$$

We search the approximate solutions

$$u_1^m(x,t) = \sum_{j=0}^m h_j^m(t) w_j(x), \quad u_2^m(x,t) = \sum_{j=0}^m k_j^m(t) w_j(x)$$

of the approximate problem in V_m,

$$\begin{cases} \int_{B_R} (\rho(x) \frac{\partial}{\partial t} (|\frac{\partial u_1^m}{\partial t}|^{l-2} \frac{\partial u_1}{\partial t}^m) w - \int_0^t \mu_1(t-s) \nabla u_1^m(s,x) \nabla w ds) dx \\ \quad + \int_{B_R} (\rho(x) f_1(u_1^m, u_2^m) w + \nabla u_1^m \nabla w + \nabla \frac{\partial u_1}{\partial t}^m \nabla w) dx = 0, \\ \int_{B_R} (\rho(x) \frac{\partial}{\partial t} (|\frac{\partial u_2^m}{\partial t}|^{l-2} \frac{\partial u_2^m}{\partial t}) w - \int_0^t \mu_2(t-s) \nabla u_2^m(s,x) \nabla w ds) dx \\ \quad + \int_{B_R} (\rho(x) f_2(u_1^m, u_2^m) w + \nabla u_2^m \nabla w + \nabla \frac{\partial u_2}{\partial t}^m \nabla w) dx = 0, \\ u_1^m(0) = u_{10}^m, \quad \frac{\partial u_1}{\partial t}^m(0) = u_{11}^m, \quad u_2^m(0) = u_{20}^m, \quad \frac{\partial u_2}{\partial t}^m(0) = u_{21}^m. \end{cases} \tag{4.42}$$

Based on standard existence theory for differential equations, one can conclude the existence of solution (u_1^m, u_2^m) of (4.42) on a maximal time interval $[0, t_m)$, for each $m \in \mathbb{N}$.

– (A priori estimate 1): In (4.42), let

$$w = \frac{\partial u_1^m}{dt},$$

in the first equation and

$$w = \frac{\partial u_2^m}{dt},$$

in the second equation, add the resulting equations and integrate by parts to obtain

$$\frac{d}{dt} E^m(t) = \frac{1}{2} \sum_{i=1}^2 (\mu_i' \circ \nabla u_i^m)(t) - \frac{1}{2} \sum_{i=1}^2 \mu_i(t) \|\nabla u_i^m(t)\|_2^2 - \sum_{i=1}^2 \left\|\nabla \frac{\partial u_i}{\partial t}^m\right\|_2^2. \tag{4.43}$$

This means, using (1.37), that for some positive constant C independent of t and m, we have

$$E^m(t) \le E^m(0) \le C. \tag{4.44}$$

– (A priori estimate 2): In (4.42), let

$$w = -\Delta \frac{\partial u_1}{\partial t}^m,$$

in the first equation and $w = -\Delta \frac{\partial u_2}{\partial t}^m$ in the second equation, add the resulting equations, integrate by parts and use (1.33) to obtain

$$\begin{aligned}
&\frac{d}{dt} \sum_{i=1}^{2} \left(\frac{l-1}{l} \left\| \Delta \frac{\partial u_i}{\partial t}^m \right\|_{L_\rho^l}^l + \frac{1}{2}\left(1 - \int_0^t \mu_i(s)ds\right) \|\Delta u_i^m\|_2^2 + \frac{1}{2}(g_i \circ \Delta u_i^m) \right) \\
&= \sum_{i=1}^{2} \left(\frac{1}{2}(\mu' \circ \Delta u_i^m) - \frac{1}{2} g_i(t) \|\Delta u_i^m\|_2^2 + \left\| \Delta \frac{\partial u_i}{\partial t}^m \right\|_2^2 \right) \\
&\quad - \sum_{i=1}^{2} \int_{B_R} \rho(x) f_i(u_1^m, u_2^m) \Delta \frac{\partial u_i}{\partial t}^m dx \\
&\le - \sum_{i=1}^{2} \int_{B_R} \rho(x) f_i(u_1^m, u_2^m) \Delta \frac{\partial u_i}{\partial t}^m dx.
\end{aligned} \tag{4.45}$$

Then, integrating over $(0, t)$ yields

$$\begin{aligned}
&\sum_{i=1}^{2} \left(\frac{l-1}{l} \left\| \Delta \frac{\partial u_i}{\partial t}^m \right\|_{L_\rho^l}^l + \frac{1}{2}\left(1 - \int_0^t \mu_i(s)ds\right) \|\Delta u_i^m\|_2^2 + \frac{1}{2}(g_i \circ \Delta u_i^m) \right) \\
&\le \sum_{i=1}^{2} \left(\|\Delta u_{i1}^m\|_{L_\rho^l}^l + \|\Delta u_{i0}^m\|_2^2 - \int_{B_R} \rho(x) f_i(u_1^m, u_2^m) \Delta u_i^m dx \right) \\
&\quad + \sum_{i=1}^{2} \int_{B_R} \rho(x)(f_i(u_{10}^m, u_{20}^m) \Delta u_{i0}^m) dx \\
&\quad + \int_0^t \int_{B_R} \rho(x) \left(\frac{\partial f_1}{\partial u_2} \frac{\partial u_2}{\partial t}^m \Delta u_1^m + \frac{\partial f_2}{\partial u_1} \frac{\partial u_1}{\partial t}^m \Delta u_2^m \right) dxds.
\end{aligned} \tag{4.46}$$

To estimate the terms on the right-hand side of (4.46), we use (4.37)–(4.39), Young's inequality and (1.62) and take (4.44) into account to get

$$\begin{aligned}
\int_{B_R} \rho(x) f_i(u_1^m, u_2^m) \Delta u_i^m &\le k \int_{B_R} \rho(x)(|u_1^m| + |u_2^m| + |u_1^m|^{\beta_{i1}} + |u_2^m|^{\beta_{i2}}) \Delta u_i^m \\
&\le \delta \|\Delta u_i^m\|_{L_\rho^2}^2 + \frac{c}{\delta} \int_{B_R} \rho(x)(|u_1^m|^2 + |u_2^m|^2 + |u_1^m|^{2\beta_{i1}} + |u_2^m|^{2\beta_{i2}})
\end{aligned}$$

$$
\begin{aligned}
&\le \delta\|\Delta u_i^m\|_{L^2_\rho}^2 + \frac{c}{\delta}(\|u_1^m\|_{L^2_\rho}^2 + \|u_2^m\|_{L^2_\rho}^2 + \|u_1^m\|_{L^2_\rho}^{2\beta_{i1}} + \|u_2^m\|_{L^2_\rho}^{2\beta_{i2}}) \\
&\le \delta\|\Delta u_i^m\|_{L^2_\rho}^2 + \frac{c}{\delta}(\|\nabla u_1^m\|_{L^2_\rho}^2 + \|\nabla u_2^m\|_{L^2_\rho}^2 + \|\nabla u_1^m\|_{L^2_\rho}^{2\beta_{i1}} + \|\nabla u_2^m\|_{L^2_\rho}^{2\beta_{i2}}) \\
&\le \delta\|\Delta u_i^m\|_{L^2_\rho}^2 + \frac{c}{\delta}E^m(0)E^m(t) \\
&\le \delta\|\Delta u_i^m\|_{L^2_\rho}^2 + \frac{c}{\delta}.
\end{aligned}
\tag{4.47}
$$

Since $1 \le \beta_{ij}$, $i,j = 1,2$. Now, we estimate

$$
I = \int_{B_R} \rho(x)\frac{\partial f_i}{\partial u_1}\frac{\partial u_i}{\partial t}^m \Delta u_i^m.
$$

First, we observe that

$$
\frac{\beta_{1j}-1}{2\beta_{1j}} + \frac{1}{2\beta_{1j}} + \frac{1}{2} = 1,
$$

and use (1.37)–(1.38) and the generalized Hölder's inequality to infer

$$
\begin{aligned}
|I| &\le d\int_{B_R} \rho(x)(1 + |u_1^m|^{\beta_{11}-1} + |u_2^m|^{\beta_{12}-1})\frac{\partial u_i}{\partial t}^m \Delta u_i^m \\
&\le d\left(\left\|\frac{\partial u_i^m}{\partial t}\right\|_{L^2_\rho} + \left\|\frac{\partial u_i^m}{\partial t}\right\|_{L^{2\beta_{11}}_\rho}\|u_1^m\|_{L^{2\beta_{11}}_\rho}^{\beta_{11}-1} + \left\|\frac{\partial u_i^m}{\partial t}\right\|_{L^{2\beta_{12}}_\rho}\|u_2^m\|_{L^{2\beta_{12}}_\rho}^{\beta_{12}-1}\right)\|\Delta u_i^m\|_{L^2_\rho}.
\end{aligned}
$$

Then, by (1.62), (4.44) and Young's inequality, we arrive at

$$
\begin{aligned}
|I| &\le c(1 + \|\nabla u_1^m\|_2^{\beta_{11}-1} + \|\nabla u_2^m\|_2^{\beta_{12}-1})\left\|\nabla\frac{\partial u_i^m}{\partial t}\right\|_{L^2_\rho}\|\Delta u_i^m\|_{L^2_\rho} \\
&\le c\left(\left\|\nabla\frac{\partial u_i^m}{\partial t}\right\|_{L^2_\rho}\cdot\|\Delta u_i^m\|_{L^2_\rho}\right) \le c\left\|\nabla\frac{\partial u_i^m}{\partial t}\right\|_{L^2_\rho}^2 + c\|\Delta u_i^m\|_{L^2_\rho}^2.
\end{aligned}
\tag{4.48}
$$

Since the other terms in (4.46) can be similarly treated and the norms of the initial data are uniformly bounded, we combine (4.46), (4.48), and use (1.37), (1.38) and take δ small enough to end up with

$$
\sum_{i=1}^{2}\left(\left\|\nabla\frac{\partial u_i^m}{\partial t}\right\|_{L^l_\rho}^l + \|\Delta u_i^m\|_2^2\right) \le c + c\sum_{i=1}^{2}\int_0^t\left(\left\|\nabla\frac{\partial u_i^m}{\partial t}\right\|_{L^l_\rho}^l + \|\Delta u_i^m\|_2^2\right)ds.
$$

Using Gronwall's inequality, this implies that

$$
\sum_{i=1}^{2}\left(\left\|\nabla\frac{\partial u_i^m}{\partial t}\right\|_{L^l_\rho}^l + \|\Delta u_i^m\|_2^2\right) \le C, \quad \forall t \in [0,T] \text{ and } m \in \mathbb{N}.
\tag{4.49}
$$

– (A priori estimate 3): In (4.42), let $w = (u_1^m)''$ in the first equation and $w = (u_2^m)''$ in the second equation. Then, by exploiting the previous estimates and using similar arguments, we find

$$\sum_{i=1}^{2}\left\|\frac{\partial u_i^m}{\partial t}\right\|_2^2 \le C, \quad \forall t \in [0,T] \text{ and } m \in \mathbb{N}. \tag{4.50}$$

From (4.44), (4.49) and (4.50), we conclude that

$$\begin{aligned}
&u_i^m \text{ are uniformly bounded in } L^\infty(0,T;D(B_R)),\\
&u_i^{m'} \text{ are uniformly bounded in } L^\infty(0,T;L_\rho^l(B_R)),\\
&u_i^{m''} \text{ are uniformly bounded in } L^2(0,T;D^{-1}(B_R)),
\end{aligned}$$

which implies that there exists subsequences of $\{u_i^m\}$, which we still denote in the same way, such that

$$\begin{aligned}
&u_i^m \overset{*}{\longrightarrow} \text{weak } u_i \quad \text{in } L^\infty(0,T;D(B_R)),\\
&u_i^{m'} \overset{*}{\longrightarrow} \text{weak } \frac{\partial u_i}{\partial t} \quad \text{in } L^\infty(0,T;L_\rho^l(B_R)),\\
&u_i^{m''} \overset{*}{\longrightarrow} \text{weak } \frac{\partial^2 u_i}{\partial t^2} \quad \text{in } L^2(0,T;D^{-1}(B_R)).
\end{aligned} \tag{4.51}$$

In the sequel, we will deal with the nonlinear term. By Aubin's lemma (see [4]), we find up to a subsequence that

$$u_i^m \to u_i \quad \text{strongly in } L^2(0,T;L_\rho^l(B_R)). \tag{4.52}$$

Then

$$u_i^m \to u_i \quad \text{almost everywhere in } (0,T)\times B_R \tag{4.53}$$

and, therefore, from (4.40), (4.41),

$$f_i(u_1^m,u_2^m) \to f_i(u_1,u_2) \quad \text{almost everywhere in } (0,T)\times B_R, \text{ for } i=1,2. \tag{4.54}$$

Also, as u_i^m are bounded in $L^\infty(0,T;L_\rho^2(B_R))$, then the use of (4.37)–(4.41) gives that $f_i(u_1^m,u_2^m)$ is bounded in $L^\infty(0,T;L_\rho^2(B_R))$. From this (4.54), we can deduce that

$$f_i(u_1^m,u_2^m) \rightharpoonup f_i(u_1,u_2) \quad \text{in } L^2(0,T;L_\rho^2(B_R)), \text{ for } i=1,2.$$

Combining the results obtained above, we can pass to the limit and conclude that (u_1,u_2) is the solution of system (4.32) restricted in B_R. □

In the next result, we will extend our solutions to $\mathbb{R}^n$.

Theorem 4.4. *Assume that* (1.37)–(1.38), *and* (1.58) *with* (4.37)–(4.41) *are satisfied. Suppose that the initial conditions*

$$(u_{10}, u_{11}) \in (C_0^\infty(B_R))^2, \quad (u_{20}, u_{21}) \in (C_0^\infty(B_R))^2,$$

are given. Then for the problem (4.32), *there exists a unique solution such that*

$$(u_1, u_2) \in (C[0,T;\mathscr{H}(\mathbb{R}^n)])^2 \quad \textit{and} \quad \left(\frac{\partial u_1}{\partial t}, \frac{\partial u_2}{\partial t}\right) \in (C[0,T;L_\rho^l(\mathbb{R}^n)])^2.$$

Proof. (a) *Existence.* Let $R_0 > 0$ such that

$$\operatorname{supp}(u_{10}, u_{20}) \subset B_{R_0},$$

and

$$\operatorname{supp}(u_{11}, u_{21}) \subset B_{R_0}.$$

Then, for $R \geq R_0$, $R \in \mathbb{N}$, we consider the approximating problem

$$\begin{cases} \frac{\partial}{\partial t}(|\frac{\partial u_1^R}{\partial t}|^{l-2}\frac{\partial u_1^R}{\partial t}) + f_1(u_1^R, u_2^R) \\ \quad - \phi(x)\Delta(u_1^R + \int_0^t \mu_1(s)u_1^R(s-t,x)ds + \frac{\partial u_1^R}{\partial t}) = 0, \quad x \in B_R \times \mathbb{R}^+, \\ \frac{\partial}{\partial t}(|\frac{\partial u_2^R}{\partial t}|^{l-2}\frac{\partial u_2^R}{\partial t}) + f_2(u_1^R, u_2^R) \\ \quad - \phi(x)\Delta(u_2^R + \int_0^t \mu_2(s)u_2^R(s-t,x)ds + \frac{\partial u_2^R}{\partial t}) = 0, \quad x \in B_R \times \mathbb{R}^+, \\ (u_1^R(0,x), u_2^R(0,x)) = (u_1^0(x), u_2^0(x)) \in (C_0^\infty(B_R))^2, \\ (\frac{\partial u_1^R}{\partial t}(0,x), \frac{\partial u_2^R}{\partial t}(0,x)) = (u_1^1(x), u_2^2(x)) \in (C_0^\infty(B_R))^2. \end{cases} \tag{4.55}$$

By Lemma (4.7), problem (4.55) has a unique solution u_i^R such that

$$(u_1^R, u_2^R) \in (C[0,T;D(B_R)])^2 \quad \text{and} \quad \left(\frac{\partial u_1^R}{\partial t}, \frac{\partial u_2^R}{\partial t}\right) \in (C[0,T;L_\rho^l(B_R)])^2.$$

We extend the solution of the problem (4.55) as

$$(\tilde{u}_1^R, \tilde{u}_2^R) =: \begin{cases} (u_1^R, u_2^R), & \text{if } |x| \leq R, \\ 0, & \text{otherwise.} \end{cases}$$

The solution (u_1^R, u_2^R) satisfies the estimates

$$\begin{aligned} &\|\tilde{u}_i^R\|_{L^\infty(0,T;D(\mathbb{R}^n))} \leq K, \quad \|f(\tilde{u}_i^R)\|_{L^\infty(0,T;\mathscr{H}(\mathbb{R}^n))} \leq K, \\ &\left\|\frac{\partial \tilde{u}_i^R}{\partial t}\right\|_{L^\infty(0,T;L_\rho^l(\mathbb{R}^n))} \leq K, \quad \left\|\frac{\partial^2 \tilde{u}_i^R}{\partial t^2}\right\|_{L^\infty(0,T;D^{-1}(\mathbb{R}^n))} \leq K, \end{aligned} \tag{4.56}$$

where the constant K is independent of R. Estimates (4.56) applied imply that

$$\widetilde{u}_i^R \text{ is relatively compact in } C([0,T];L^2_\rho(\mathbb{R}^n)). \tag{4.57}$$

Next, using relations (4.56) and (4.57), the continuity of the embedding

$$C([0,T];L^2_\rho(\mathbb{R}^n)) \subset L^2([0,T];L^2_\rho(\mathbb{R}^n)),$$

and the continuity of f_i we may extract a subsequence of $\widetilde{u}_i^R$, denoted by $\widetilde{u}_i^{R_m}$, such that as $R_m \to \infty$ we get

$$\begin{aligned}
\widetilde{u}_i^{R_m} &\overset{*}{\longrightarrow} \widetilde{u}_i \quad \text{in } L^\infty(0,T;D(B_R)),\\
\frac{\partial \widetilde{u}_i^{R_m}}{\partial t} &\overset{*}{\longrightarrow} \frac{\partial u_i}{\partial t} \quad \text{in } L^\infty(0,T;L^l_\rho(B_R)),\\
\frac{\partial^2 \widetilde{u}_i^{R_m}}{\partial t^2} &\overset{*}{\longrightarrow} \frac{\partial^2 u_i}{\partial t^2} \quad \text{in } L^\infty(0,T;D^{-1}(B_R)),\\
f(\widetilde{u}_i^{R_m}) &\overset{*}{\longrightarrow} f(\widetilde{u}_i) \quad \text{in } L^\infty(0,T;D(B_R)).
\end{aligned} \tag{4.58}$$

For fixed $R = R_m$, let L_m denote the operator of restriction

$$L_m : [0,T] \times \mathbb{R}^n \to [0,T] \times B_R.$$

It is clear that the restricted subsequence $L_m \widetilde{u}_i^{R_m}$ satisfies the estimates obtained in Lemma 4.7. Therefore, there exists a subsequence $\widetilde{u}_i^{R_{m_j}} = \widetilde{u}_i^j$ for which it can be shown by following the procedure of Lemma 4.7, that $L_m \widetilde{u}_i^j$ converges weakly to solution $\widetilde{u}_i^m$. We have

$$\begin{cases}
\int_0^T (L_m \frac{\partial}{\partial t}(|\frac{\partial \widetilde{u}_1^j}{\partial t}|^{l-2}\frac{\partial \widetilde{u}_1^j}{\partial t}), v)_{L^l_\rho(B_R)} ds + \int_0^T (f_1(L_m \widetilde{u}_1^j, L_m \widetilde{u}_2^j), v)_{L^2_\rho(B_R)} ds\\
\quad + \int_0^T \int_{B_R} \nabla L_m \widetilde{u}_1^j \nabla v dx ds - \int_0^T \int_0^t \mu_1(t-s) \int_{B_R} \nabla \widetilde{u}_1^j \nabla v dx ds\\
\quad + \int_0^T \int_{B_R} \nabla L_m \widetilde{u}_1'^j \nabla v dx ds\\
\quad = \int_0^T (\frac{\partial}{\partial t}(|\frac{\partial \widetilde{u}_1^j}{\partial t}|^{l-2}\frac{\partial \widetilde{u}_1^j}{\partial t}), v)_{L^l_\rho(\mathbb{R}^n)} ds + \int_0^T (f_1(\widetilde{u}_1^j, \widetilde{u}_2^j), v)_{L^2_\rho(\mathbb{R}^n)}\\
\quad + \int_0^T \int_{\mathbb{R}^n} \nabla \widetilde{u}_1^j \nabla v dx ds - \int_0^T \int_0^t \mu_1(t-s) \int_{\mathbb{R}^n} \nabla \widetilde{u}_1^j \nabla v dx ds,\\
\int_0^T (L_m \frac{\partial}{\partial t}(|\frac{\partial \widetilde{u}_2^j}{\partial t}|^{l-2}\frac{\partial \widetilde{u}_2^j}{\partial t}), v)_{L^l_\rho(B_R)} ds + \int_0^T (f_2(L_m \widetilde{u}_1^j, L_m \widetilde{u}_2^j), v)_{L^2_\rho(B_R)} ds\\
\quad + \int_0^T \int_{B_R} \nabla L_m \widetilde{u}_2^j \nabla v dx ds - \int_0^T \int_0^t \mu_2(t-s) \int_{B_R} \nabla \widetilde{u}_2^j \nabla v dx ds\\
\quad + \int_0^T \int_{B_R} \nabla L_m \widetilde{u}_2'^j \nabla v dx ds\\
\quad = \int_0^T (\frac{\partial}{\partial t}(|\frac{\partial \widetilde{u}_2^j}{\partial t}|^{l-2}\frac{\partial \widetilde{u}_2^j}{\partial t}), v)_{L^l_\rho(\mathbb{R}^n)} ds + \int_0^T (f_2(\widetilde{u}_1^j, \widetilde{u}_2^j), v)_{L^2_\rho(\mathbb{R}^n)}\\
\quad + \int_0^T \int_{\mathbb{R}^n} \nabla \widetilde{u}_2^j \nabla v dx ds - \int_0^T \int_0^t \mu_2(t-s) \int_{\mathbb{R}^n} \nabla \widetilde{u}_2^j \nabla v dx ds,
\end{cases} \tag{4.59}$$

for every

$$v \in C_0^\infty([0,T] \times B_R).$$

Passing to the limit in (4.59) as $j \longrightarrow \infty$, we obtain that

$$L_m \tilde{u}_i = \tilde{u}_i^m.$$

The equality (4.59) holds for any

$$v \in C_0^\infty([0,T] \times \mathbb{R}^n),$$

since the radius R is arbitrarily chosen. Therefore, $\tilde{u}_i$ is a solution of the problem (4.55).

(b) *Uniqueness.* Let us assume that

$$(u_{11}, u_{21}), (u_{12}, u_{22}),$$

are two strong solutions of (4.32). Then

$$(z_1, z_2) = (u_{11} - u_{12}, u_{21} - u_{22}),$$

satisfies for all $w \in \mathscr{H}(\mathbb{R}^n)$,

$$\begin{cases} \int_{\mathbb{R}^n} (\rho(x)\frac{\partial}{\partial t}(|\frac{\partial z_1}{\partial t}|^{l-2}\frac{\partial z_1}{\partial t})w + \nabla z_1 \nabla w + \int_0^t \mu_1(s)\nabla z_1(s-t,x)\nabla w ds)dx \\ \quad + \int_{\mathbb{R}^n} \rho(x) f_1(z_1, z_2) w dx + \nabla \frac{\partial z_2}{\partial t} \nabla w = 0, \\ \int_{\mathbb{R}^n} (\rho(x)\frac{\partial}{\partial t}(|\frac{\partial z_2}{\partial t}|^{l-2}\frac{\partial z_2}{\partial t})w + \nabla z_2 \nabla w + \int_0^t \mu_2(s)\nabla z_2(s-t,x)\nabla w ds)dx \\ \quad + \int_{\mathbb{R}^n} \rho(x) f_2(z_1, z_2) w dx + \nabla \frac{\partial z_2}{\partial t} \nabla w = 0. \end{cases} \tag{4.60}$$

Substituting

$$w = \frac{\partial z_1}{\partial t},$$

in the first equation and

$$w = \frac{\partial z_2}{\partial t},$$

in the second equation, adding the resulting equations, integrating by parts and using (1.37) yield

$$\frac{d}{dt}\sum_{i=1}^{2}\left(\frac{l-1}{l}\left\|\frac{\partial z_i}{\partial t}\right\|_{L^l_\rho}^{l} + \frac{1}{2}\left(1 - \int_0^t \mu_i(s)ds\right)\|\nabla z_i\|_2^2 + \frac{1}{2}(g_i \circ \nabla z_i)\right)$$
$$\leq \int_{\mathbb{R}^n}\left(\left[f_1(u_{21}, u_{22}) + f_1(u_{11}, u_{12})\right]\frac{\partial z_1}{\partial t} + \left[f_2(u_{21}, u_{22}) + f_2(u_{11}, u_{12})\right]\frac{\partial z_2}{\partial t}\right)dx.$$

Making use of (4.41) and following similar arguments that used to obtain (4.48), we find

$$\begin{aligned}
&\int_{\mathbb{R}^n}\Big([f_1(u_{21},u_{22})+f_1(u_{11},u_{12})]\frac{\partial z_1}{\partial t}+[f_2(u_{21},u_{22})+f_2(u_{11},u_{12})]\frac{\partial z_2}{\partial t}\Big)dx\\
&\quad\le k\int_{\mathbb{R}^n}(1+|u_{11}|^{\beta_{11}-1}+|u_{12}|^{\beta_{11}-1}+|u_{21}|^{\beta_{12}-1}+|u_{22}|^{\beta_{12}-1})(|z_1|+|z_2|)\frac{\partial z_1}{\partial t}dx\\
&\qquad+k\int_{B_R}(1+|u_{11}|^{\beta_{21}-1}+|u_{12}|^{\beta_{21}-1}+|u_{21}|^{\beta_{22}-1}+|u_{22}|^{\beta_{22}-1})(|z_1|+|z_2|)\frac{\partial z_2}{\partial t}dx\\
&\quad\le c\sum_{i=1}^{2}\Big(\Big\|\frac{\partial z_i}{\partial t}\Big\|_{L^l_\rho}^{l}+\|\nabla z_i\|_2^2\Big).
\end{aligned}\tag{4.61}$$

Combining (4.60)–(4.61), integrating over $(0,t)$ and using Gronwall's lemma, then we deduce that

$$\sum_{i=1}^{2}\Big(\Big\|\frac{\partial z_i}{\partial t}\Big\|_{L^l_\rho}^{l}+\|z_i\|_2^2\Big)=0,\tag{4.62}$$

which means that $(u_{11},u_{21})=(u_{12},u_{22})$. This completes the proof. □

We can now state and prove the asymptotic behavior of the solution of (4.32).

4.2.2 Decay rate for linear cases

We show that our solution decays time asymptotically to zero and the rate of decay for the solution is similar to that of the memory terms, making some small perturbation in the associate energy, for this purpose, we introduce the next functional

$$\psi(t)=\sum_{i=1}^{2}\int_{\mathbb{R}^n}\rho(x)u_i\Big|\frac{\partial u_i}{\partial t}\Big|^{l-2}\frac{\partial u_i}{\partial t}dx.\tag{4.63}$$

The following lemma will be useful in the proof of our next result.

Lemma 4.8. *Under the assumptions* (1.37)–(1.38), *and* (1.58), *the functional* ψ *satisfies, along the solution of* (4.32),

$$\psi'(t)\le\sum_{i=1}^{2}\|u_i'\|_{L^l_\rho(\mathbb{R}^n)}^{l}-(k-1-\delta+|\alpha|c)\sum_{i=1}^{2}\|\nabla u_i\|_2^2+c\sum_{i=1}^{2}(\mu_i\circ\nabla u_i),\tag{4.64}$$

for positive constants c.

Proof. From (4.63), integrate by parts over $\mathbb{R}^n$, we have

$$\psi'(t) = \int_{\mathbb{R}^n} \rho(x)\frac{\partial u_1}{\partial t}^l dx + \int_{\mathbb{R}^n} \rho(x)u_1\left(\left|\frac{\partial u_1}{\partial t}\right|^{l-2}\frac{\partial u_1}{\partial t}\right)' dx$$

$$+ \int_{\mathbb{R}^n} \rho(x)\frac{\partial u_2}{\partial t}^l dx + \int_{\mathbb{R}^n} \rho(x)u_2\left(\left|\frac{\partial u_2}{\partial t}\right|^{l-2}\frac{\partial u_2}{\partial t}\right)' dx$$

$$= \int_{\mathbb{R}^n}\left(\rho(x)\frac{\partial u_1}{\partial t}^l - u_1\Delta u_1 - u_1\Delta\frac{\partial u}{\partial t}_1 - \alpha\rho(x)u_1u_2 + u_1\int_0^t \mu_1(t-s)\Delta u_1(s,x)ds\right)dx$$

$$+ \int_{\mathbb{R}^n}\left(\rho(x)\frac{\partial u_2}{\partial t}^l - u_2\Delta^2 u_2 - u_2\Delta\frac{\partial u}{\partial t}_2 - \alpha\rho(x)u_1u_2 + u_2\int_0^t \mu_2(t-s)\Delta u_2(s,x)ds\right)dx$$

$$= \sum_{i=1}^2\left\|\frac{\partial u_i}{\partial t}\right\|^l_{L^l_\rho(\mathbb{R}^n)} - \left(1-\int_0^t \mu_i(s)ds\right)\sum_{i=1}^2\|\nabla u_i\|_2^2$$

$$- \sum_{i=1}^2\left\|\nabla\frac{\partial u}{\partial t}_i\right\|_2^2 - 2\alpha\int_{\mathbb{R}^n}\rho(x)u_1u_2dx$$

$$+ \sum_{i=1}^2\int_{\mathbb{R}^n}\nabla u_i\int_0^t \mu_i(t-s)(\nabla u_i(s)-\nabla u_i(t))dsdx.$$

Recalling that

$$\int_0^t \mu_i(s)ds \le \int_0^\infty \mu_i(s)ds = 1-k_i,$$

using Young's inequality, Lemma 1.14 and Lemma 1.9, we obtain

$$\psi'(t) \le \sum_{i=1}^2\left\|\frac{\partial u_i}{\partial t}\right\|^l_{L^l_\rho(\mathbb{R}^n)} - \sum_{i=1}^2\left\|\nabla\frac{\partial u}{\partial t}_i\right\|_2^2 - (k_i - 1 + |\alpha|\|\rho\|^{-1}_{L^s(\mathbb{R}^n)})\sum_{i=1}^2\|\nabla u_i\|_2^2$$

$$+ \delta\sum_{i=1}^2\|\nabla u_i\|_2^2 + \frac{1}{4\delta}\sum_{i=1}^2\int_{\mathbb{R}^n}\left(\int_0^t \mu_i(t-s)|\nabla u_i(s)-\nabla u_i(t)|ds\right)^2 dx$$

$$\le \sum_{i=1}^2\left\|\frac{\partial u_i}{\partial t}\right\|^l_{L^l_\rho(\mathbb{R}^n)} - \sum_{i=1}^2\left\|\nabla\frac{\partial u}{\partial t}_i\right\|_2^2 - (k-1-\delta+|\alpha|c)\sum_{i=1}^2\|\nabla u_i\|_2^2$$

$$+ \frac{(1-k)}{4\delta}\sum_{i=1}^2(\mu_i \circ \nabla u_i).$$

For α small enough and

$$k = \min\{k_1, k_2\}.$$ □

Theorem 4.5. *Let*

$$(u_{10}, u_{11}), (u_{20}, u_{21}) \in \mathcal{H}(\mathbb{R}^n) \times L^l_\rho(\mathbb{R}^n),$$

and suppose that (1.37)–(1.38), *and* (1.58) *hold. Then there exist positive constants W, ω such that the energy of solution given by* (4.32) *satisfies*

$$E(t) \le WE(0) \exp\left(-\omega \int_0^t \xi(s)ds \right), \quad \forall t \ge 0. \tag{4.65}$$

In order to prove this theorem, let us define

$$L(t) = N_1 E(t) + \varepsilon\psi(t), \quad \forall \varepsilon > 0. \tag{4.66}$$

For $N_1 > 1$, we need the next lemma, which means that there is equivalent between the perturbed energy and energy functions.

Lemma 4.9. *For $N_1 > 1$, we have*

$$\beta_1 L(t) \le E(t) \le L(t)\beta_2, \quad \forall t \ge 0, \tag{4.67}$$

holds for two positive constants β_1 and β_2.

Proof. By (4.63) and (4.66), we have

$$\begin{aligned} \left|L(t) - N_1 E(t)\right| &\le \varepsilon\left|\psi_1(t)\right| \\ &\le \varepsilon \sum_{i=1}^{2} \int_{\mathbb{R}^n} \left|\rho(x) u_i \left|\frac{\partial u_i}{\partial t}\right|^{l-2} \frac{\partial u_i}{\partial t}\right| dx. \end{aligned}$$

Thanks to Hölder and Young's inequalities with exponents $\frac{l}{l-1}$, l, since

$$\frac{2n}{n+2} \ge l \ge 2,$$

we have by using Lemma 1.14,

$$\int_{\mathbb{R}^n} \left|\rho(x) u_i \left|\frac{\partial u_i}{\partial t}\right|^{l-2} \frac{\partial u_i}{\partial t}\right| dx \le \left(\int_{\mathbb{R}^n} \rho(x) |u_i|^l dx \right)^{1/l} \left(\int_{\mathbb{R}^n} \rho(x) \left|\frac{\partial u_i}{\partial t}\right|^l dx \right)^{(l-1)/l}$$

$$\leq \frac{1}{l}\Big(\int_{\mathbb{R}^n} \rho(x)|u_i|^l dx\Big) + \frac{l-1}{l}\Big(\int_{\mathbb{R}^n} \rho(x)\Big|\frac{\partial u_i}{\partial t}\Big|^l dx\Big)$$

$$\leq c\Big\|\frac{\partial u_i}{\partial t}\Big\|^l_{L^l_\rho(\mathbb{R}^n)} + c\|\rho\|^l_{L^s(\mathbb{R}^n)}\|\nabla u_i\|^l_2. \tag{4.68}$$

Then, since $l \geq 2$, we have by using (4.35),

$$\begin{aligned}|L(t) - N_1E(t)| &\leq \varepsilon c\sum_{i=1}^{2}\Big(\Big\|\frac{\partial u_i}{\partial t}\Big\|^l_{L^l_\rho(\mathbb{R}^n)} + \|\nabla u_i\|^l_2\Big)\\ &\leq \varepsilon c(E(t) + E^{l/2}(t))\\ &\leq \varepsilon cE(t)(1 + E^{[(l/2)-1]}(t))\\ &\leq \varepsilon cE(t)(1 + E^{[(l/2)-1]}(0))\\ &\leq \varepsilon cE(t).\end{aligned}$$

Consequently, (4.67) follows. □

Proof of Theorem 4.5. From (4.35), results of Lemma 4.8, we have

$$\begin{aligned}L'(t) &= N_1E'(t) + \varepsilon\psi'(t)\\ &\leq N_1\Big(\frac{1}{2}\sum_{i=1}^{2}(\mu_i' \circ \nabla u_i)(t) - \sum_{i=1}^{2}\Big\|\nabla\frac{\partial u}{\partial t}_i\Big\|^2_2\Big)\\ &\quad + \varepsilon\sum_{i=1}^{2}\Big(\Big\|\frac{\partial u_i}{\partial t}\Big\|^l_{L^l_\rho(\mathbb{R}^n)} - (k - 1 - \delta + |\alpha|c)\|\nabla u_i\|^2_2 + c(\mu_i \circ \nabla u_i)\Big).\end{aligned}$$

At this point, we choose N_1 large and ε so small such that

$$L'(t) \leq M_0\sum_{i=1}^{2}(\mu_i \circ \nabla u_i) - \varepsilon E(t), \quad \forall t \geq 0. \tag{4.69}$$

Multiplying (4.69) by $\xi(t)$ gives

$$\xi(t)L'(t) \leq -\varepsilon\xi(t)E(t) + M_0\xi(t)\sum_{i=1}^{2}(g_i \circ \nabla u_i). \tag{4.70}$$

The last term can be estimated, using (1.37) as follows:

$$\begin{aligned}\xi(t)\sum_{i=1}^{2}(g_i \circ \nabla u_i) &\leq \sum_{i=1}^{2}\xi_i(t)\int_{\mathbb{R}^n}\int_0^t \mu_i(t-s)|u_i(t) - u_i(s)|^2 dsdx\\ &\leq \sum_{i=1}^{2}\int_{\mathbb{R}^n}\int_0^t \xi_i(t-s)\mu_i(t-s)|u_i(t) - u_i(s)|^2 dsdx\end{aligned}$$

$$\leq -\sum_{i=1}^{2}\int_{\mathbb{R}^n}\int_0^t \mu'(t-s)|u_i(t)-u_i(s)|^2 ds dx$$

$$\leq -\sum_{i=1}^{2}(\mu' \circ \nabla u_i) \leq -E'(t). \tag{4.71}$$

Thus, (4.69) becomes

$$\xi(t)L'(t) + M_0E'(t) \leq -\varepsilon\xi(t)E(t) \quad \forall t \geq 0. \tag{4.72}$$

Using the fact that ξ is a nonincreasing continuous function as ξ_1 and ξ_2 are non-increasing and so ξ is differentiable, with $\xi'(t) \leq 0$ for a.e t, then

$$(\xi(t)L(t) + M_0E(t))' \leq \xi(t)L'(t) + M_0E'(t) \leq -\varepsilon\xi(t)E(t) \quad \forall t \geq 0. \tag{4.73}$$

Since, using (4.67),

$$F = \xi L + M_0E \sim E, \tag{4.74}$$

we obtain, for some positive constant ω,

$$F'(t) \leq -\omega\xi(t)F(t) \quad \forall t \geq 0. \tag{4.75}$$

Integration over $(0,t)$ leads to, for some constant $\omega > 0$ such that

$$F(t) \leq WF(0)\exp\left(-\omega\int_0^t \xi(s)ds\right), \quad \forall t \geq 0. \tag{4.76}$$

Recalling (4.74), estimate (4.76) yields the desired result (4.65). This completes the proof of Theorem (4.5). □

5 Wave equation of Kirchhoff type with density

5.1 Damped wave equation of Kirchhoff type

We consider the following equation:

$$\rho(x)\frac{\partial}{\partial t}\left(\left\|\frac{\partial u}{\partial t}\right\|^{q-2}\frac{\partial u}{\partial t}\right) - M(\|\nabla_x u\|_2^2)\Delta_x u + \int_0^t \mu(t-s)\Delta_x u(s)ds = 0, \quad x \in \mathbb{R}^n, t > 0, \tag{5.1}$$

where $q, n \geq 2$ and M is a positive C^1 function satisfying for

$$s \geq 0, \quad m_0 > 0, \quad m_1 \geq 0, \quad \gamma \geq 1,$$

and

$$M(s) = m_0 + m_1 s^\gamma,$$

and the scalar function $\mu(s)$ is assumed to satisfy (1.34). Equation (5.1) is a prototype for PDE of hyperbolic in Kirchhoff type with memory when it is equipped by the following initial data:

$$u(0,x) = u_0(x) \in \mathcal{H}(\mathbb{R}^n), \quad \frac{\partial u}{\partial t}(0,x) = u_1(x) \in L^q_\rho(\mathbb{R}^n), \tag{5.2}$$

where the weighted spaces $\mathcal{H}$ is given in Definition 1.3.

In this framework, it is well known that for any initial data

$$u_0 \in \mathcal{H}(\mathbb{R}^n), \quad u_1 \in L^q_\rho(\mathbb{R}^n),$$

the problems (5.1)–(5.2) have a unique solution

$$u \in C([0,T), \mathcal{H}(\mathbb{R}^n)), \quad \frac{\partial u}{\partial t} \in C([0,T), L^q_\rho(\mathbb{R}^n)),$$

for T small enough, under hypothesis (1.34), (1.44). The energy of u at time t is defined by

$$E(t) = \frac{(q-1)}{q}\left\|\frac{\partial u}{\partial t}\right\|^q_{L^q_\rho(\mathbb{R}^n)} + \frac{1}{2}\left(m_0 - \int_0^t \mu(s)ds\right)\|\nabla_x u\|_2^2 + \frac{1}{2}(\mu \circ \nabla_x u)$$
$$+ \frac{m_1}{2(\gamma+1)}\|\nabla_x u\|_2^{2(\gamma+1)}, \tag{5.3}$$

and the following energy functional law holds:

$$E'(t) = \frac{1}{2}(\mu' \circ \nabla_x u)(t) - \frac{1}{2}\mu(t)\|\nabla_x u(t)\|_2^2, \quad \text{for all } t \geq 0. \tag{5.4}$$

https://doi.org/10.1515/9783111633787-005

which means that, our energy is uniformly bounded and decreasing along the trajectories. To construct a Lyapunov functional L equivalent to E, we introduce the next functional

$$\psi_1(t) = \int_{\mathbb{R}^n} \rho(x)u\left\|\frac{\partial u}{\partial t}\right\|^{q-2}\frac{\partial u}{\partial t}dx, \tag{5.5}$$

$$\psi_2(t) = -\int_{\mathbb{R}^n} \rho(x)\left\|\frac{\partial u}{\partial t}\right\|^{q-2}\frac{\partial u}{\partial t}\int_0^t \mu(t-s)(u(t)-u(s))dsdx. \tag{5.6}$$

Lemma 5.1. *Under the assumptions* (1.34), (1.44), *the functional* ψ_1 *satisfies, along the solution of* (5.1), (5.2),

$$\psi_1'(t) \le \left\|\frac{\partial u}{\partial t}\right\|^q_{L^q_\rho(\mathbb{R}^n)} + c_1 m_1\|\nabla_x u\|_2^{2(\gamma+1)} + (\sigma-\mu_0)\|\nabla_x u\|_2^2 + \frac{(1-\mu_0)}{4\sigma}(\mu\circ\nabla_x u). \tag{5.7}$$

Proof. From (5.5), integrate over $\mathbb{R}^n$, we have

$$\begin{aligned}
\psi_1'(t) &= \int_{\mathbb{R}^n} \rho(x)\left\|\frac{\partial u}{\partial t}\right\|^q dx + \int_{\mathbb{R}^n} \rho(x)u\frac{\partial}{\partial t}\left(\left\|\frac{\partial u}{\partial t}\right\|^{q-2}\frac{\partial u}{\partial t}\right)dx \\
&= \int_{\mathbb{R}^n}\left(\rho(x)\left\|\frac{\partial u}{\partial t}\right\|^q + M(\|\nabla_x u\|_2^2)u\Delta_x u - u\int_0^t \mu(t-s)\Delta_x u(s,x)ds\right)dx \\
&\le \left\|\frac{\partial u}{\partial t}\right\|^q_{L^q_\rho(\mathbb{R}^n)} + m_1\|\nabla_x u\|_2^{2(\gamma+1)} - \mu_0\|\nabla_x u\|_2^2 \\
&\quad + \int_{\mathbb{R}^n} \nabla_x u\int_0^t \mu(t-s)(\nabla_x u(s)-\nabla_x u(t))dsdx.
\end{aligned}$$

Using Young's inequality and Lemma 1.9 for $\theta = 1/2$, we obtain

$$\begin{aligned}
\psi_1'(t) &\le \left\|\frac{\partial u}{\partial t}\right\|^q_{L^q_\rho(\mathbb{R}^n)} + m_1\|\nabla_x u\|_2^{2(\gamma+1)} - \mu_0\|\nabla_x u\|_2^2 \\
&\quad + \sigma\|\nabla_x u\|_2^2 + \frac{1}{4\sigma}\int_{\mathbb{R}^n}\left(\int_0^t \mu(t-s)|\nabla_x u(s)-\nabla_x u(t)|ds\right)^2 dx \\
&\le \left\|\frac{\partial u}{\partial t}\right\|^q_{L^q_\rho(\mathbb{R}^n)} + m_1\|\nabla_x u\|_2^{2(\gamma+1)} + (\sigma-\mu_0)\|\nabla_x u\|_2^2 + \frac{(1-\mu_0)}{4\sigma}(\mu\circ\nabla_x u).
\end{aligned} \tag{5.8}$$

□

Lemma 5.2. *Under the assumptions* (1.34), (1.44), *the functional* ψ_2 *satisfies, along the solution of* (5.1), (5.2), *for any* $\sigma \in (0, m_0)$,

$$\psi_2'(t) \le \left(\sigma - \int_0^t \mu(s)ds\right)\left\|\frac{\partial u}{\partial t}\right\|^q_{L^q_\rho(\mathbb{R}^n)} + m_1\|\nabla u\|_2^{2(\gamma+1)}$$
$$+ \sigma\|\nabla_x u\|_2^2 + \frac{c}{\sigma}(\mu\circ\nabla_x u) - c_\sigma\|\rho\|^q_{L^s(\mathbb{R}^n)}(g'\circ\nabla_x u)^{q/2}. \tag{5.9}$$

Proof. Exploiting (5.1), (5.6) to get

$$\begin{aligned}
\psi_2'(t) &= -\int_{\mathbb{R}^n}\rho(x)\frac{\partial}{\partial t}\left(\left\|\frac{\partial u}{\partial t}\right\|^{q-2}\frac{\partial u}{\partial t}\right)\int_0^t \mu(t-s)(u(t)-u(s))dsdx\\
&\quad - \int_{\mathbb{R}^n}\rho(x)\left\|\frac{\partial u}{\partial t}\right\|^{q-2}\frac{\partial u}{\partial t}\int_0^t \mu'(t-s)(u(t)-u(s))dsdx\\
&\quad - \int_0^t \mu(s)ds\left\|\frac{\partial u}{\partial t}\right\|^q_{L^q_\rho(\mathbb{R}^n)}\\
&= \int_{\mathbb{R}^n} M(\|\nabla u\|_2^2)\nabla_x u\int_0^t \mu(t-s)(\nabla_x u(t)-\nabla_x u(s))dsdx\\
&\quad - \int_{\mathbb{R}^n}\left(\int_0^t \mu(t-s)\nabla_x u(s,x)ds\right)\left(\int_0^t \mu(t-s)(\nabla_x u(t)-\nabla_x u(s))ds\right)dx\\
&\quad - \int_{\mathbb{R}^n}\rho(x)\left\|\frac{\partial u}{\partial t}\right\|^{q-2}\frac{\partial u}{\partial t}\int_0^t \mu'(t-s)(u(t)-u(s))dsdx\\
&\quad - \int_0^t \mu(s)ds\left\|\frac{\partial u}{\partial t}\right\|^q_{L^q_\rho(\mathbb{R}^n)}\\
&= \left(m_0 - \int_0^t \mu(s)ds\right)\int_{\mathbb{R}^n}\nabla_x u\int_0^t \mu(t-s)(\nabla_x u(t)-\nabla_x u(s))dsdx\\
&\quad + \int_{\mathbb{R}^n}\left(\int_0^t \mu(t-s)(\nabla_x u(t)-\nabla_x u(s))ds\right)^2 dx + c_1 m_1\|\nabla_x u\|_2^{2(\gamma+1)}\\
&\quad - \int_{\mathbb{R}^n}\rho(x)\left\|\frac{\partial u}{\partial t}\right\|^{q-2}\frac{\partial u}{\partial t}\int_0^t \mu'(t-s)(u(t)-u(s))dsdx\\
&\quad - \int_0^t \mu(s)ds\left\|\frac{\partial u}{\partial t}\right\|^q_{L^q_\rho(\mathbb{R}^n)} + c(\mu\circ\nabla_x u)(t).
\end{aligned}$$

By Hölder's and Young's inequalities and Lemma 1.14, we estimate

$$
\begin{aligned}
&-\int_{\mathbb{R}^n} \rho(x)\left\|\frac{\partial u}{\partial t}\right\|^{q-2}\frac{\partial u}{\partial t}\int_0^t \mu'(t-s)(u(t)-u(s))dsdx \\
&\leq \left(\int_{\mathbb{R}^n} \rho(x)\left\|\frac{\partial u}{\partial t}\right\|^q dx\right)^{(q-1)/q} \\
&\quad\times\left(\int_{\mathbb{R}^n} \rho(x)\left|\int_0^t \mu'(t-s)(u(t)-u(s))ds\right|^q\right)^{1/q} \\
&\leq \sigma\left\|\frac{\partial u}{\partial t}\right\|^q_{L^q_\rho(\mathbb{R}^n)} + c_\sigma\left\|\int_0^t -\mu'(t-s)(u(t)-u(s))ds\right\|^q_{L^q_\rho(\mathbb{R}^n)} \\
&\leq \sigma\left\|\frac{\partial u}{\partial t}\right\|^q_{L^q_\rho(\mathbb{R}^n)} - c_\sigma\|\rho\|^q_{L^s(\mathbb{R}^n)}(\mu'\circ\nabla_x u)^{q/2}(t).
\end{aligned}
$$

Using Young and Poincare's inequalities and Lemma 1.9 for $\theta = 1/2$, we obtain

$$
\begin{aligned}
\psi_2'(t) &\leq \sigma\|\nabla_x u\|_2^2 + \frac{c}{\sigma}(\mu\circ\nabla_x u) - c_\sigma\|\rho\|^q_{L^s(\mathbb{R}^n)}(\mu'\circ\nabla_x u)^{q/2} \\
&\quad+\left(\sigma - \int_0^t \mu(s)ds\right)\left\|\frac{\partial u}{\partial t}\right\|^{q/2}_{L^q_\rho(\mathbb{R}^n)} + c_1 m_1\|\nabla_x u\|_2^{2(\gamma+1)}.
\end{aligned}
$$

□

Theorem 5.1. *Let*

$$(u_0, u_1) \in \mathscr{H}(\mathbb{R}^n) \times L^q_\rho(\mathbb{R}^n),$$

and suppose that (1.34), (1.44) *hold. Then there exist constants*

$$\alpha_0, \alpha_1, \alpha_2, \alpha_3 > 0,$$

such that the energy of solution given by (5.1), (5.2) *satisfies*

$$E(t) \leq \alpha_3 H_1^{-1}(\alpha_1 t + \alpha_2), \quad \textit{for all } t \geq 0,$$

where

$$H_1(t) = \int_t^1 \frac{1}{sH_0'(\alpha_0 s)}ds \tag{5.10}$$

In order to prove this theorem, let us define

$$L(t) = \xi_1 E(t) + \psi_1(t) + \xi_2\psi_2(t) \tag{5.11}$$

for $\xi_1, \xi_2 > 1$. We need the next lemma, which means that there is equivalent between the Lyapunov and energy functions, i. e., for $\xi_1, \xi_2 > 1$, we have

$$\beta_1 L(t) \le E(t) \le \beta_2 L(t) \tag{5.12}$$

holds for two positive constants β_1 and β_2.

Lemma 5.3. *For $\xi_1, \xi_2 > 1$, we have*

$$L(t) \sim E(t). \tag{5.13}$$

Proof. By (5.11), we have

$$\begin{aligned}
|L(t) - \xi_1 E(t)| &\le |\psi_1(t)| + \xi_2 |\psi_2(t)| \\
&\le \int_{\mathbb{R}^n} \left| \rho(x) u \left\| \frac{\partial u}{\partial t} \right\|^{q-2} \frac{\partial u}{\partial t} \right| dx \\
&\quad + \xi_2 \int_{\mathbb{R}^n} \left| \rho(x) \left\| \frac{\partial u}{\partial t} \right\|^{q-2} \frac{\partial u}{\partial t} \int_0^t \mu(t-s)(u(t)-u(s))\,ds \right| dx.
\end{aligned}$$

Thanks to Hölder and Young's inequalities with exponents $\frac{q}{q-1}, q$, since

$$\frac{2n}{n+2} \ge q \ge 2,$$

we have by using Lemma 1.14,

$$\begin{aligned}
\int_{\mathbb{R}^n} \left| \rho(x) u \left\| \frac{\partial u}{\partial t} \right\|^{q-2} \frac{\partial u}{\partial t} \right| dx &\le \left(\int_{\mathbb{R}^n} \rho(x) |u|^q dx \right)^{1/q} \left(\int_{\mathbb{R}^n} \rho(x) \left\| \frac{\partial u}{\partial t} \right\|^q dx \right)^{(q-1)/q} \\
&< \frac{1}{q} \left(\int_{\mathbb{R}^n} \rho(x) |u|^q dx \right) + \frac{q-1}{q} \left(\int_{\mathbb{R}^n} \rho(x) \left\| \frac{\partial u}{\partial t} \right\|^q dx \right) \\
&\le c \left\| \frac{\partial u}{\partial t} \right\|^q_{L^q_\rho(\mathbb{R}^n)} + c \|\rho\|^q_{L^s(\mathbb{R}^n)} \|\nabla_x u\|_2^q
\end{aligned} \tag{5.14}$$

and

$$\begin{aligned}
&\int_{\mathbb{R}^n} \left| \left(\rho(x)^{\frac{q-1}{q}} \left\| \frac{\partial u}{\partial t} \right\|^{q-2} \frac{\partial u}{\partial t} \right) \left(\rho(x)^{\frac{1}{q}} \int_0^t \mu(t-s)(u(t)-u(s))\,ds \right) \right| dx \\
&\le \left(\int_{\mathbb{R}^n} \rho(x) \left\| \frac{\partial u}{\partial t} \right\|^q dx \right)^{(q-1)/q} \\
&\quad \times \left(\int_{\mathbb{R}^n} \rho(x) \left| \int_0^t \mu(t-s)(u(t)-u(s))\,ds \right|^q dx \right)^{1/q}
\end{aligned}$$

$$\leq \frac{q-1}{q}\left\|\frac{\partial u}{\partial t}\right\|^q_{L^q_\rho(\mathbb{R}^n)} + \frac{1}{q}\left\|\int_0^t \mu(t-s)(u(t)-u(s))ds\right\|^q_{L^q_\rho(\mathbb{R}^n)}$$
$$\leq \frac{q-1}{q}\left\|\frac{\partial u}{\partial t}\right\|^q_{L^q_\rho(\mathbb{R}^n)} + \frac{1}{q}\|\rho\|^q_{L^s(\mathbb{R}^n)}(\mu\circ\nabla_x u)^{q/2}(t).$$

Then, since $q \geq 2$, we have

$$\begin{aligned}|L(t)-\xi_1 E(t)| &\leq c(E(t)+E^{q/2}(t))\\ &\leq c(E(t)+E(t)E^{(q/2)-1}(t))\\ &\leq c(E(t)+E(t)E^{(q/2)-1}(0))\\ &\leq cE(t).\end{aligned}$$

Therefore, we can choose ξ_1 so that

$$L(t) \sim E(t). \tag{5.15}$$

□

Proof of Theorem 5.1. From (5.4), Lemma 5.1 and Lemma 5.2, we have

$$\begin{aligned}L'(t) &= \xi_1 E'(t) + \psi_1'(t) + \xi_2\psi_2'(t)\\ &\leq \left(\frac{1}{2}\xi_1 - c_\sigma\|\rho\|^q_{L^s(\mathbb{R}^n)}\xi_2\right)(\mu'\circ\nabla_x u)^{q/2} + M_0(\mu\circ\nabla_x u)\\ &\quad - M_1\left\|\frac{\partial u}{\partial t}\right\|^q_{L^q_\rho(\mathbb{R}^n)} - M_2\|\nabla_x u\|_2^2 + c_1 m_1\|\nabla_x u\|_2^{2(\gamma+1)}\end{aligned}$$

where

$$M_0 = \left(\frac{4\xi_2 c + (1-\mu_0)}{4\sigma}\right),$$
$$M_1 = \left(\xi_2\left(\int_0^{t_1}\mu(s)ds - \sigma\right) - 1\right),$$
$$M_2 = \left(-\xi_2\sigma + \frac{1}{2}\xi_1\mu(t_1) + (l-\sigma)\right),$$

and t_1 was introduced in Remark 1.1.

We choose σ so small that

$$\xi_1 > 2c_\sigma\|\rho\|^q_{L^s(\mathbb{R}^n)}\xi_2.$$

Whence σ is fixed, we can choose ξ_1, ξ_2 large enough so that $M_1, M_2 > 0$, which yields

$$L'(t) \leq M_0(\mu\circ\nabla_x u) + c_1 m_1\|\nabla_x u\|_2^{2(\gamma+1)} - cE(t), \quad \text{for all } t \geq t_1. \tag{5.16}$$

Now we set

$$F(t) = L(t) + cE(t),$$

which is equivalent to $E(t)$. Then by (5.16), we get for some

$$\begin{aligned} c &> 2c_1(\gamma+1), \\ F'(t) &= L'(t) + cE'(t) \\ &\leq -cE(t) + c\int_{\mathbb{R}^n}\int_{t_1}^{t} \mu(t-s)|\nabla_x u(t) - \nabla_x u(s)|^2 dsdx, \quad \text{for all } t \geq t_1. \end{aligned} \tag{5.17}$$

By (5.4) and Remark 1.1, we have for all $t \geq t_1$,

$$\begin{aligned} \int_{\mathbb{R}^n}\int_0^{t_1} \mu(t-s)|\nabla_x u(t) - \nabla_x u(s)|^2 dsdx &\leq -\frac{1}{k}\int_{\mathbb{R}^n}\int_0^{t_1} \mu(t-s)|\nabla_x u(t) - \nabla_x u(s)|^2 dsdx \\ &\leq -cE'(t). \end{aligned}$$

At this point, we define

$$I(t) = \int_{t_1}^{t} H_0(-\mu'(s))(\mu \circ \nabla_x u)(t)ds. \tag{5.18}$$

Since

$$\int_0^{+\infty} H_0(-\mu'(s))\mu(s)ds < +\infty,$$

from (5.4) we have

$$\begin{aligned} I(t) &= \int_{t_1}^{t} H_0(-\mu'(s)) \int_{\mathbb{R}^n} \mu(s)|\nabla_x u(t) - \nabla_x u(t-s)|^2 dxds \\ &\leq 2\int_{t_1}^{t} H_0(-\mu'(s))\mu(s) \int_{\mathbb{R}^n} |\nabla_x u(t)|^2 + |\nabla_x u(t-s)|^2 dxds \\ &\leq cE(0)\int_{t_1}^{t} H_0(-\mu'(s))\mu(s)ds < 1. \end{aligned} \tag{5.19}$$

Now, we define again a new functional $\lambda(t)$ related with $I(t)$ as

$$\lambda(t) = -\int_{t_1}^{t} H_0(-\mu'(s))\mu'(s) \int_{\mathbb{R}^n} \mu(s)|\nabla_x u(t) - \nabla_x u(t-s)|^2 dxds. \tag{5.20}$$

From (1.34), (1.44) and Remark 1.1, we get

$$H_0(-\mu'(s))\mu(s) \leq H_0(H(\mu(s)))\mu(s) = D(\mu(s))\mu(s) \leq k_0$$

for some positive constant k_0. Then, for all $t \geq t_1$,

$$\begin{aligned}
\lambda(t) &\leq -k_0 \int_{t_1}^{t} \mu'(s) \int_{\mathbb{R}^n} |\nabla_x u(t) - \nabla_x u(t-s)|^2 dxds \\
&\leq -k_0 \int_{t_1}^{t} \mu'(s) \int_{\mathbb{R}^n} |\nabla_x u(t)|^2 + |\nabla_x u(t-s)|^2 dxds \\
&\leq -cE(0) \int_{t_1}^{t} \mu'(s) ds \\
&\leq cE(0)\mu(t_1) \\
&< \min\{r, H(r), H_0(r)\}.
\end{aligned} \tag{5.21}$$

Using the properties of H_0 (strictly convex in $(0, r]$, $H_0(0) = 0$), then for $x \in (0, r]$, $\theta \in [0, 1]$,

$$H_0(\theta x) \leq \theta H_0(x).$$

Using Remark 1.1, (5.19), (5.21) and Jensen's inequality leads to

$$\begin{aligned}
\lambda(t) &= \frac{1}{I(t)} \int_{t_1}^{t} I(t) H_0[H_0^{-1}(-\mu'(s))] H_0(-\mu'(s))\mu'(s) \int_{\mathbb{R}^n} \mu(s)|\nabla_x u(t) - \nabla_x u(t-s)|^2 dxds \\
&\geq \frac{1}{I(t)} \int_{t_1}^{t} H_0[I(t) H_0^{-1}(-\mu'(s))] H_0(-\mu'(s))\mu'(s) \int_{\mathbb{R}^n} \mu(s)|\nabla_x u(t) - \nabla_x u(t-s)|^2 dxds \\
&\geq H_0\left(\frac{1}{I(t)} \int_{t_1}^{t} I(t) H_0^{-1}(-\mu'(s)) H_0(-\mu'(s))\mu'(s) \int_{\mathbb{R}^n} \mu(s)|\nabla_x u(t) - \nabla_x u(t-s)|^2 dxds\right) \\
&\geq H_0\left(\int_{t_1}^{t} \int_{\mathbb{R}^n} \mu(s)|\nabla_x u(t) - \nabla_x u(t-s)|^2 dxds\right),
\end{aligned}$$

which implies

$$\int_{t_1}^{t}\int_{\mathbb{R}^n} \mu(s)|\nabla_x u(t) - \nabla_x u(t-s)|^2 dxds \le H_0^{-1}(\lambda(t)).$$

Then

$$F'(t) \le -cE(t) + cH_0^{-1}(\lambda(t)), \quad \text{for all } t \ge t_1.$$

Now, using the fact that

$$E' \le 0, \quad 0 < H_0', \quad 0 < H_0'',$$

on $(0, r]$ to define the functional

$$F_1(t) = H_0'\left(\alpha_0 \frac{E(t)}{E(0)}\right)F(t) + cE(t), \quad \alpha_0 < r, 0 < c,$$

where

$$F_1(t) \sim E(t),$$

and

$$\begin{aligned} F_1'(t) &= \alpha_0 \frac{E'(t)}{E(0)} H_0''\left(\alpha_0 \frac{E(t)}{E(0)}\right)F(t) + H_0'\left(\alpha_0 \frac{E(t)}{E(0)}\right)F'(t) + cE'(t) \\ &\le -cE(t)H_0'\left(\alpha_0 \frac{E(t)}{E(0)}\right) + cH_0'\left(\alpha_0 \frac{E(t)}{E(0)}\right)H_0^{-1}(\lambda(t)) + cE'(t). \end{aligned}$$

Let H_0^* given in Remark 1.1 and using Young's inequality (1.47) with

$$A = H_0'\left(\alpha_0 \frac{E(t)}{E(0)}\right), \quad B = H_0^{-1}(\lambda(t)),$$

to get

$$\begin{aligned} F_1'(t) &\le -cE(t)H_0'\left(\alpha_0 \frac{E(t)}{E(0)}\right) + cH_0^*\left(H_0'\left(\alpha_0 \frac{E(t)}{E(0)}\right)\right) + c\lambda(t) + cE'(t) \\ &\le -cE(t)H_0'\left(\alpha_0 \frac{E(t)}{E(0)}\right) + c\alpha_0 \frac{E(t)}{E(0)} H_0'\left(\alpha_0 \frac{E(t)}{E(0)}\right) - c'E'(t) + cE'(t). \end{aligned}$$

Choosing α_0, c, c', such that for all $t \ge t_1$ we have

$$\begin{aligned} F_1'(t) &\le -k\frac{E(t)}{E(0)} H_0'\left(\alpha_0 \frac{E(t)}{E(0)}\right) \\ &= -kH_2\left(\frac{E(t)}{E(0)}\right), \end{aligned}$$

where

$$H_2(t) = tH_0'(\alpha_0 t).$$

Using the strict convexity of H_0 on $(0, r]$, to find that H_2', H_2 are strict positives on $(0, 1]$, then

$$R(t) = \tau \frac{k_1 F_1(t)}{E(0)} \sim E(t), \quad \tau \in (0,1) \tag{5.22}$$

and

$$R'(t) \leq -\tau k_0 H_2(R(t)), \quad k_0 \in (0, +\infty), t \geq t_1.$$

Then a simple integration and a suitable choice of τ yield

$$R(t) \leq H_1^{-1}(\alpha_1 t + \alpha_2), \quad \alpha_1, \alpha_2 \in (0, +\infty), t \geq t_1.$$

Here,

$$H_1(t) = \int_t^1 H_2^{-1}(s)ds.$$

From (5.22), for a positive constant α_3, we have

$$E(t) \leq \alpha_3 H_1^{-1}(\alpha_1 t + \alpha_2), \quad \alpha_1, \alpha_2 \in (0, +\infty), t \geq t_1.$$

The fact that H_1 is strictly decreasing function on $(0, 1]$ and due to properties of H_2, we have

$$\lim_{t \to 0} H_1(t) = +\infty.$$

Then

$$E(t) \leq \alpha_3 H_1^{-1}(\alpha_1 t + \alpha_2), \quad \text{for all } t \geq 0.$$

This completes the proof of Theorem 5.1. □

5.2 Wave equation of Kirchhoff type in viscoelasticity

We consider the following degenerate problem:

$$\rho(x)\frac{\partial}{\partial t}\left(\left|\frac{\partial u}{\partial t}\right|^{p-2}\frac{\partial u}{\partial t}\right) - M(\|\nabla_x u\|_2^2)\operatorname{div}[a(x)\nabla_x u] + \int_0^t \mu(t-s)\operatorname{div}[a(x)\nabla_x u(s)]ds = 0, \tag{5.23}$$

where $q, n \geq 2$, and M is a positive C^1 function satisfying for

$$s \geq 0, \quad m_0 > 0, \quad m_1 \geq 0, \quad \gamma \geq 1, \quad M(s) = m_0 + m_1 s^\gamma,$$

and the scalar function $\mu(s)$ is assumed to satisfy (1.34).

Equation (5.23) is equipped by the following initial data:

$$u(0,x) = u_0 \in \mathscr{H}(\mathbb{R}^n), \quad \frac{\partial u}{\partial t}(0,x) = u_1 \in L^p_\rho(\mathbb{R}^n), \tag{5.24}$$

where the weighted spaces $\mathscr{H}$ is given in Definition 1.3 and the density function satisfies (1.58).

The self-adjoint nonpositive operator of the form

$$Au = \operatorname{div}(a(x)\nabla_x u(t))$$

represent the class of degenerate term considered in (5.23), where $a \in C^1(\mathbb{R}^n)$ may vanish in a subset Ω_0 of $\mathbb{R}^n$.

The energy of u at time t is defined by

$$\begin{aligned} E(t) = {} & \frac{(p-1)}{p}\left|\frac{\partial u}{\partial t}\right|^p_{L^p_\rho(\mathbb{R}^n)} + \frac{1}{2}\left(m_0 - \int_0^t \mu(s)ds\right)\|\nabla_x u\|^2_{V_a} \\ & + \frac{1}{2}(\mu \circ \nabla_x u) + \frac{m_1}{2(\gamma+1)}\|\nabla_x u\|_2^{2(\gamma+1)} \end{aligned} \tag{5.25}$$

and the following energy functional law holds:

$$E'(t) = \frac{1}{2}(\mu' \circ \nabla_x u)(t) - \frac{1}{2}\mu(t)\|\nabla_x u(t)\|^2_{V_a}, \quad \text{for all } t \geq 0, \tag{5.26}$$

which means that, our energy is uniformly bounded and decreasing along the trajectories.

First, we recall and make use the following assumptions on the functions a and follows.

Let $a \in C^1(\mathbb{R}^n)$ be a positive function such that

$$a(x) \geq A^2 > 0,$$

and

$$V_a = \left\{u \in L^2(\mathbb{R}^n) : \int_{\mathbb{R}^n} a(x)|\nabla u|^2 dx < \infty\right\} \tag{5.27}$$

is a Hilbert space endowed with the norm

$$\|u\|_{V_a}^2 = \int_{\mathbb{R}^n} a|\nabla u|^2 dx.$$

Note that

$$\|\nabla_x u\|_2 \le \frac{1}{|A|}\|u\|_{V_a}$$

for any $u \in V_a$. Really, we have

$$\begin{aligned}\|\nabla_x u\|_2^2 &= \int_{\mathbb{R}^n} |\nabla_x u|^2 dx\\ &= \int_{\mathbb{R}^n} \frac{a(x)}{a(x)}|\nabla_x u|^2 dx\\ &\le \frac{1}{A^2}\int_{\mathbb{R}^n} a(x)|\nabla_x u|^2 dx\\ &= \frac{1}{A^2}\|u\|_{V_a}^2,\end{aligned}$$

whereupon we get the desired estimate.

We will start with the following useful lemma.

Lemma 5.4. *Let ρ satisfies* (1.58), *then for any $u \in \mathscr{H}(\mathbb{R}^n)$ there exists $\alpha > 0$ such that*

$$\|u\|_{L^p_\rho(\mathbb{R}^n)} \le \alpha^{\frac{1}{p}} \|\rho\|_{L^{s_1}(\mathbb{R}^n)}^{\frac{1}{p}} \|\nabla_x u\|_{L^2(\mathbb{R}^n)} \tag{5.28}$$

with $s_1 = \frac{2n(q-1)}{nq-2n+2q}$, $2 \le q$.

Proof. Let $u \in \mathscr{H}(\mathbb{R}^n)$ be arbitrarily chosen and fixed. Since $\mathscr{H}(\mathbb{R}^n)$ is embedded continuously in $L^{\frac{2n}{n-2}}(\mathbb{R}^n)$, there exists a positive constant α such that

$$\|u\|_{L^{\frac{2n}{n-2}}(\mathbb{R}^n)} \le \alpha^{\frac{1}{p}} \|\nabla_x u\|_{L^2(\mathbb{R}^n)}. \tag{5.29}$$

Now we apply Hölder's inequality for

$$s_1 = \frac{2n(q-1)}{nq-2n+2q} \quad \text{and} \quad s_2 = \frac{2n}{(n-2)p}$$

and we get

$$\begin{aligned}\|u\|_{L^p_\rho(\mathbb{R}^n)}^p &= \int_{\mathbb{R}^n} \rho|u|^p dx\\ &\le \Big(\int_{\mathbb{R}^n} \rho^{s_1} dx\Big)^{\frac{1}{s_1}} \Big(\int_{\mathbb{R}^n} |u|^{\frac{2n}{n-2}} dx\Big)^{\frac{p(n-2)}{2n}}\end{aligned}$$

$$= \|\rho\|_{L^{s_1}(\mathbb{R}^n)} \|u\|^p_{L^{\frac{2n}{n-2}}(\mathbb{R}^n)}.$$

Now we apply (5.29) and we get

$$\|u\|^p_{L^p_\rho(\mathbb{R}^n)} \le \alpha \|\rho\|_{L^{s_1}(\mathbb{R}^n)} \|\nabla_x u\|^p_{L^2(\mathbb{R}^n)},$$

which completes the proof. □

To construct a Lyapunov functional L, equivalent to E, we introduce the following functional:

$$\psi_1(t) = \int_{\mathbb{R}^n} \rho(x) u \left|\frac{\partial u}{\partial t}\right|^{p-2} \frac{\partial u}{\partial t} dx, \tag{5.30}$$

$$\psi_2(t) = -\int_{\mathbb{R}^n} \rho(x) \left|\frac{\partial u}{\partial t}\right|^{p-2} \frac{\partial u}{\partial t} \int_0^t \mu(t-s)(u(t)-u(s)) ds dx. \tag{5.31}$$

Lemma 5.5. *Under the assumptions* (1.34), (1.44), *the functional* ψ_1 *satisfies, along the solution of* (5.23), (5.24), *the following inequality:*

$$\psi_1'(t) \le \left|\frac{\partial u}{\partial t}\right|^p_{L^p_\rho(\mathbb{R}^n)} + \frac{m_1}{A^{2\gamma}} \|u\|^{2+2\gamma}_{V_a} + (\sigma - \mu_0)\|w\|^2_{V_a} + \frac{1}{4\sigma}(m_0 - \mu_0)(\mu \circ \nabla_x u)$$

for any $\sigma > 0$.

Proof. We differentiate (5.30) and using (5.23), we get

$$\begin{aligned}
\psi_1'(t) &= \int_{\mathbb{R}^n} \rho(x) \left|\frac{\partial u}{\partial t}\right|^p dx + \int_{\mathbb{R}^n} \rho(x) u \frac{\partial}{\partial t}\left(\left|\frac{\partial u}{\partial t}\right|^{p-2} \frac{\partial u}{\partial t}\right) dx \\
&= \left\|\frac{\partial u}{\partial t}\right\|^p_{L^p_\rho(\mathbb{R}^n)} + \int_{\mathbb{R}^n} u \left(M(\|\nabla_x u\|_2^2) \operatorname{div}[a(x)\nabla_x u] - \int_0^t \mu(t-s) \operatorname{div}[a(x)\nabla_x u(s)] ds \right) dx \\
&= \left\|\frac{\partial u}{\partial t}\right\|^p_{L^p_\rho(\mathbb{R}^n)} \\
&\quad + \int_{\mathbb{R}^n} u \left((m_0 + m_1 \|\nabla_x u\|_2^{2\gamma}) \operatorname{div}[a(x)\nabla_x u] - \int_0^t \mu(t-s) \operatorname{div}[a(x)\nabla_x u(s)] ds \right) dx \\
&= \left\|\frac{\partial u}{\partial t}\right\|^p_{L^p_\rho(\mathbb{R}^n)} + (m_0 + m_1 \|\nabla_x u\|_2^{2\gamma}) \int_{\mathbb{R}^n} u \operatorname{div}[a(x)\nabla_x u] dx \\
&\quad - \int_{\mathbb{R}^n} u \int_0^t \mu(t-s) \operatorname{div}[a(x)\nabla_x u(s)] ds dx
\end{aligned}$$

$$
\begin{aligned}
&= \left\|\frac{\partial u}{\partial t}\right\|_{L^p_\rho(\mathbb{R}^n)}^p - (m_0 + m_1\|\nabla_x u\|_2^{2\gamma}) \int_{\mathbb{R}^n} a(x)|\nabla_x u|^2 dx \\
&\quad - \int_{\mathbb{R}^n} u \int_0^t \mu(t-s)\operatorname{div}[a(x)\nabla_x u(s)]dsdx \\
&= \left\|\frac{\partial u}{\partial t}\right\|_{L^p_\rho(\mathbb{R}^n)}^p - (m_0 + m_1\|\nabla_x\|_2^{2\gamma})\|u\|_{V_a}^2 \\
&\quad + \int_{\mathbb{R}^n} \nabla_x u \int_0^t \mu(t-s)a(x)\nabla_x u(s)dsdx \\
&= \left\|\frac{\partial u}{\partial t}\right\|_{L^p_\rho(\mathbb{R}^n)}^p - (m_0 + m_1\|\nabla_x u\|_2^{2\gamma})\|u\|_{V_a}^2 \\
&\quad + \int_{\mathbb{R}^n} \nabla_x u \int_0^t \mu(t-s)a(x)(\nabla_x u(s) - \nabla_x u(t))dsdx \\
&\quad + \left(\int_{\mathbb{R}^n} a(x)|\nabla_x u|^2 dx\right)\int_0^t \mu(s)ds \\
&\le \left\|\frac{\partial u}{\partial t}\right\|_{L^p_\rho(\mathbb{R}^n)} - (m_0 + m_1\|\nabla_x u\|_2^{2\gamma})\|u\|_{V_a}^2 \\
&\quad + \|u\|_{V_a}^2 \int_0^\infty \mu(s)ds + \int_{\mathbb{R}^n} \nabla_x u \int_0^t \mu(t-s)a(x)(\nabla_x u(s) - \nabla_x u(t))dsdx \\
&= \left\|\frac{\partial u}{\partial t}\right\|_{L^p_\rho(\mathbb{R}^n)}^p - (m_0 + m_1\|\nabla_x u\|_2^{2\gamma})\|u\|_{V_a}^2 + (m_0 - \mu_0)\|u\|_{V_a}^2 \\
&\quad + \int_{\mathbb{R}^n} \nabla_x u \int_0^t \mu(t-s)a(x)(\nabla_x u(s) - \nabla_x u(t))dsdx \\
&= \left\|\frac{\partial u}{\partial t}\right\|_{L^p_\rho(\mathbb{R}^n)}^p - m_1\|\nabla_x u\|_2^{2\gamma}\|u\|_{V_a}^2 - \mu_0\|u\|_{V_a}^2 \\
&\quad + \int_{\mathbb{R}^n} \nabla_x u \int_0^t \mu(t-s)a(x)(\nabla_x u(s) - \nabla_x u(t))dsdx \\
&\le \left\|\frac{\partial u}{\partial t}\right\|_{L^p_\rho(\mathbb{R}^n)}^p + m_1\|\nabla_x u\|_2^{2\gamma}\|u\|_{V_a}^2 - \mu_0\|u\|_{V_a}^2 \\
&\quad + \int_{\mathbb{R}^n} \nabla_x u \int_0^t \mu(t-s)a(x)(\nabla_x u(s) - \nabla_x u(t))dsdx
\end{aligned}
$$

$$\leq \left\| \frac{\partial u}{\partial t} \right\|^p_{L^p_\rho(\mathbb{R}^n)} + \frac{m_1}{A^{2\gamma}} \|u\|^{2+2\gamma}_{V_a} - \mu_0 \|u\|^2_{V_a}$$
$$+ \int_{\mathbb{R}^n} \nabla_x u \int_0^t \mu(t-s) a(x) (\nabla_x u(s) - \nabla_x u(t)) ds dx,$$

i. e.,

$$\psi_1'(t) \leq \left\| \frac{\partial u}{\partial t} \right\|^p_{L^p_\rho(\mathbb{R}^n)} + \frac{m_1}{A^{2\gamma}} \|u\|^{2+2\gamma}_{V_a} - \mu_0 \|u\|^2_{V_a}$$
$$+ \int_{\mathbb{R}^n} \nabla_x u \int_0^t \mu(t-s) a(x) (\nabla_x u(s) - \nabla_x u(t)) ds dx.$$

Let $\sigma > 0$ be arbitrarily chosen. Now we apply Young's inequality with σ and we get

$$\psi_1'(t) \leq \left\| \frac{\partial u}{\partial t} \right\|^p_{L^p_\rho(\mathbb{R}^n)} + \frac{m_1}{A^{2\gamma}} \|u\|^{2+2\gamma}_{V_a} - \mu_0 \|u\|^2_{V_a} + \sigma \int_{\mathbb{R}^n} a(x) |\nabla_x u|^2 dx$$
$$+ \frac{1}{4\sigma} \int_{\mathbb{R}^n} \left(\int_0^t \mu(t-s) (a(x))^{\frac{1}{2}} (\nabla_x u(s) - \nabla_x u(t)) ds \right)^2 dx$$
$$= \left\| \frac{\partial u}{\partial t} \right\|^p_{L^p_\rho(\mathbb{R}^n)} + \frac{m_1}{A^{2\gamma}} \|u\|^{2+2\gamma}_{V_a} + (\sigma - \mu_0) \|u\|^2_{V_a}$$
$$+ \frac{1}{4\sigma} \int_{\mathbb{R}^n} \left(\int_0^t \mu(t-s) (a(x))^{\frac{1}{2}} (\nabla_x u(s) - \nabla_x u(t)) ds \right)^2 dx.$$

Now we apply Lemma 5.4 for $\theta = \frac{1}{2} \mu$ and

$$(a(x))^{\frac{1}{2}} (\nabla_x u(s) - \nabla_x u(t)),$$

and we obtain

$$\psi_1'(t) \leq \left\| \frac{\partial u}{\partial t} \right\|^p_{L^p_\rho(\mathbb{R}^n)} + \frac{m_1}{A^{2\gamma}} \|u\|^{2+2\gamma}_{V_a} + (\sigma - \mu_0) \|u\|^2_{V_a}$$
$$+ \frac{1}{4\sigma} \int_{\mathbb{R}^n} \left(\int_0^t \mu(s) ds \right) \int_0^t \mu(t-s) a(x) \left| \nabla_x w(s) - \nabla_x u(t) \right|^2 ds dx$$
$$\leq \left\| \frac{\partial u}{\partial t} \right\|^p_{L^p_\rho(\mathbb{R}^n)} + \frac{m_1}{A^{2\gamma}} \|u\|^{2+2\gamma}_{V_a} + (\sigma - \mu_0) \|u\|^2_{V_a}$$
$$+ \frac{1}{4\sigma} \int_0^\infty \mu(s) ds \int_{\mathbb{R}^n} \int_0^t \mu(t-s) a(x) \left| \nabla_x u(s) - \nabla_x u(t) \right|^2 ds dx$$

$$= \left\|\frac{\partial u}{\partial t}\right\|^p_{L^p_\rho(\mathbb{R}^n)} + \frac{m_1}{A^{2\gamma}}\|u\|^{2+2\gamma}_{V_a} + (\sigma - \mu_0)\|u\|^2_{V_a} + \frac{1}{4\sigma}(m_0 - \mu_0)(\mu \circ \nabla_x u),$$

which completes the proof. □

Remark 5.1. By the proof of Lemma 5.5, it follows

$$\int_{\mathbb{R}^n} \nabla_x u \int_0^t \mu(t-s)a(x)(\nabla_x u(s) - \nabla_x u(t))dsdx \le \sigma\left\|\frac{\partial u}{\partial t}\right\|^2_{V_a} + \frac{1}{4\sigma}(m_0 - \mu_0)(\mu \circ \nabla_x u) \tag{5.32}$$

for every $\sigma > 0$.

Lemma 5.6. *Under the assumptions* (1.34), (1.44), *the functional* ψ_2 *satisfies, along the solution of* (5.23), (5.24), *the following inequality:*

$$\begin{aligned}\psi_2'(t) \le{}& \left(m_0 + m_1\|\nabla_x u\|_2^{2\gamma} - \int_0^t \mu(s)ds\right)\left(\sigma\left\|\frac{\partial u}{\partial t}\right\|^2_{V_a} + \frac{m_0 - l}{4\sigma}(\mu \circ \nabla_x u)\right)\\ &+ \left(\sigma - \int_0^t \mu(s)ds\right)\left\|\frac{\partial u}{\partial t}\right\|^p_{L^p_\rho(\mathbb{R}^n)} + (\mu \circ \nabla_x u)\int_0^t \mu(s)ds\\ &+ \frac{aQ^{\frac{p}{2}}C(\sigma)}{A_2^p}\|\rho\|_{L^{s_1}(\mathbb{R}^n)}(|\mu'| \circ \nabla_x u)^{\frac{p}{2}}\end{aligned}$$

for any $\sigma > 0$ *and for some* $a > 0$. *Here,*

$$Q = \max_{t\in[0,\infty)} \int_0^t |\mu'(s)|ds, \quad C(\sigma) = \frac{1}{(\sigma p)^{\frac{q}{p}} q}.$$

Proof. By the definition of $\psi_2(t)$, we get

$$\begin{aligned}\psi_2'(t) ={}& -\int_{\mathbb{R}^n} \rho(x)\frac{\partial}{\partial t}\left(\left|\frac{\partial u}{\partial t}\right|^{p-2}\frac{\partial u}{\partial t}\right)\int_0^t \mu(t-s)(u(t) - u(s))dsdx\\ &- \int_{\mathbb{R}^n} \rho(x)\left|\frac{\partial u}{\partial t}\right|^{p-2}\frac{\partial u}{\partial t}\int_0^t \mu'(t-s)(u(t) - u(s))dsdx\\ &- \int_{\mathbb{R}^n} \rho(x)\left|\frac{\partial u}{\partial t}\right|^p dx \int_0^t \mu(s)ds\end{aligned}$$

$$
\begin{aligned}
&= -\int_{\mathbb{R}^n}\left(M(\|\nabla_x u\|_2^2)\operatorname{div}[a(x)\nabla_x u] - \int_0^t \mu(t-s)\operatorname{div}[a(x)\nabla_x u(s)]ds\right)\\
&\quad\times\left(\int_0^t \mu(t-s)(u(t)-u(s))ds\right)dx\\
&\quad-\int_{\mathbb{R}^n}\rho(x)\left|\frac{\partial u}{\partial t}\right|^{p-2}\frac{\partial u}{\partial t}\int_0^t \mu'(t-s)(u(t)-u(s))dsdx - \left\|\frac{\partial u}{\partial t}\right\|_{L^p_\rho(\mathbb{R}^n)}\int_0^t \mu(s)ds\\
&= -\int_{\mathbb{R}^n} M(\|\nabla_x u\|_2^2)\operatorname{div}[a(x)\nabla_x u]\int_0^t \mu(t-s)(u(t)-u(s))dsdx\\
&\quad+\int_{\mathbb{R}^n}\left(\int_0^t \mu(t-s)\operatorname{div}[a(x)\nabla_x u(s)]ds\right)\left(\int_0^t \mu(t-s)(u(t)-u(s))ds\right)dx\\
&\quad-\int_{\mathbb{R}^n}\rho(x)\left|\frac{\partial u}{\partial t}\right|^{p-2}\frac{\partial u}{\partial t}\int_0^t \mu'(t-s)(u(t)-u(s))dsdx - \left\|\frac{\partial u}{\partial t}\right\|^p_{L^p_\rho(\mathbb{R}^n)}\int_0^t \mu(s)ds\\
&= \int_{\mathbb{R}^n} M(\|\nabla_x u\|_2^2)a(x)\nabla_x u\int_0^t \mu(t-s)(\nabla_x u(t)-\nabla_x u(s))dsdx\\
&\quad-\int_{\mathbb{R}^n}\left(\int_0^t \mu(t-s)a(x)\nabla_x u(s)ds\right)\left(\int_0^t \mu(t-s)(\nabla_x u(t)-\nabla_x u(s))ds\right)dx\\
&\quad-\int_{\mathbb{R}^n}\rho(x)\left|\frac{\partial u}{\partial t}\right|^{p-2}\frac{\partial u}{\partial t}\int_0^t \mu'(t-s)(u(t)-u(s))dsdx - \left\|\frac{\partial u}{\partial t}\right\|^p_{L^p_\rho(\mathbb{R}^n)}\int_0^t \mu(s)ds\\
&= \int_{\mathbb{R}^n}(m_0 + m_1\|\nabla_x\|_2^{2\gamma})a(x)\nabla_x u\int_0^t \mu(t-s)(\nabla_x u(t)-\nabla_x u(s))dsdx\\
&\quad+\int_{\mathbb{R}^n}\left(\int_0^t \mu(t-s)(a(x))^{\frac{1}{2}}(\nabla_x u(t)-\nabla_x u(s))ds\right)^2dx\\
&\quad-\int_{\mathbb{R}^n}\rho(x)\left|\frac{\partial u}{\partial t}\right|^{p-2}\frac{\partial u}{\partial t}\int_0^t \mu'(t-s)(u(t)-u(s))dsdx - \left\|\frac{\partial u}{\partial t}\right\|_{L^p_\rho(\mathbb{R}^n)}\int_0^t \mu(s)ds\\
&\quad-\int_{\mathbb{R}^n}\nabla_x u(t)a(x)\int_0^t \mu(t-s)(\nabla_x u(t)-\nabla_x u(s))dsdx\int_0^t \mu(s)ds\\
&= \left(m_0 + m_1\|\nabla_x u\|_2^{2\gamma} - \int_0^t \mu(s)ds\right)\int_{\mathbb{R}^n} a(x)\nabla_x u(t)\int_0^t \mu(t-s)(\nabla_x u(t)-\nabla_x u(s))dsdx
\end{aligned}
$$

$$+ \int_{\mathbb{R}^n} \left(\int_0^t \mu(t-s)(a(x))^{\frac{1}{2}} (\nabla_x u(t) - \nabla_x u(s)) ds \right)^2 dx$$
$$- \int_{\mathbb{R}^n} \rho(x) \left| \frac{\partial u}{\partial t} \right|^{p-2} \frac{\partial u}{\partial t} \int_0^t \mu'(t-s)(u(t)-u(s)) ds dx - \left\| \frac{\partial u}{\partial t} \right\|^p_{L^p_\rho(\mathbb{R}^n)} \int_0^t \mu(s) ds,$$

i. e.,

$$\begin{aligned} \psi_2'(t) \le & \left(m_0 + m_1 \|\nabla_x u\|_2^{2\gamma} - \int_0^t \mu(s) ds \right) \int_{\mathbb{R}^n} a(x) \nabla_x u(t) \int_0^t \mu(t-s)(\nabla_x u(t) - \nabla_x u(s)) ds dx \\ & + \int_{\mathbb{R}^n} \left(\int_0^t \mu(t-s)(a(x))^{\frac{1}{2}} (\nabla_x u(t) - \nabla_x u(s)) ds \right)^2 dx \\ & - \int_{\mathbb{R}^n} \rho(x) \left| \frac{\partial u}{\partial t} \right|^{p-2} \frac{\partial u}{\partial t} \int_0^t \mu'(t-s)(u(t)-u(s)) ds dx - \left\| \frac{\partial u}{\partial t} \right\|^p_{L^p_\rho(\mathbb{R}^n)} \int_0^t \mu(s) ds. \end{aligned} \tag{5.33}$$

Let $\sigma > 0$ be arbitrarily chosen. Then using (5.32), we get that

$$\begin{aligned} \psi_2'(t) \le & \left(m_0 + m_1 \|\nabla_x u\|_2^{2\gamma} - \int_0^t \mu(s) ds \right) \left(\sigma \|u\|^2_{V_a} + \frac{1}{4\sigma} (m_0 - \mu_0)(\mu \circ \nabla_x u) \right) \\ & + \int_{\mathbb{R}^n} \left(\int_0^t \mu(t-s)(a(x))^{\frac{1}{2}} (\nabla_x u(t) - \nabla_x u(s)) ds \right)^2 dx \\ & - \left\| \frac{\partial u}{\partial t} \right\|^p_{L^p_\rho(\mathbb{R}^n)} \int_0^t \mu(s) ds - \int_{\mathbb{R}^n} \rho(x) \left| \frac{\partial u}{\partial t} \right|^{p-2} \frac{\partial u}{\partial t} \int_0^t \mu'(t-s)(u(t)-u(s)) ds dx. \end{aligned} \tag{5.34}$$

We apply Lemma 5.4 for $\theta = \frac{1}{2}$, μ and

$$(a(x))^{\frac{1}{2}} (\nabla_x u(t) - \nabla_x u(s)),$$

and we obtain

$$\begin{aligned} & \int_{\mathbb{R}^n} \left(\int_0^t \mu(t-s)(a(x))^{\frac{1}{2}} (\nabla_x u(t) - \nabla_x u(s)) ds \right)^2 dx \\ & \quad \le \int_{\mathbb{R}^n} \left(\int_0^t \mu(s) ds \right) \left(\int_0^t \mu(t-s) a(x) (\nabla_x u(t) - \nabla_x u(s))^2 ds \right) dx \end{aligned}$$

$$= \left(\int_{\mathbb{R}^n} a(x) \int_0^t \mu(t-s)(\nabla_x u(t) - \nabla_x u(s))^2 ds dx \right) \int_0^t \mu(s) ds$$
$$= (\mu \circ \nabla_x u)(t) \int_0^t \mu(s) ds.$$

Hence by (5.34), we get

$$\begin{aligned} \psi_2'(t) \le & \left(m_0 + m_1 \|\nabla_x u\|_2^{2\gamma} - \int_0^t \mu(s) ds \right) \left(\sigma \|u\|_{V_a}^2 + \frac{1}{4\sigma} (m_0 - \mu_0)(\mu \circ \nabla_x u)(t) \right) \\ & + (\mu \circ \nabla_x u)(t) \int_0^t \mu(s) ds - \left\| \frac{\partial u}{\partial t} \right\|_{L_\rho^p(\mathbb{R}^n)}^p \int_0^t \mu(s) ds \qquad (5.35) \\ & - \int_{\mathbb{R}^n} \rho(x) \left| \frac{\partial u}{\partial t} \right|^{p-2} \frac{\partial u}{\partial t} \int_0^t \mu'(t-s)(u(t) - u(s)) ds dx. \end{aligned}$$

Now we consider

$$- \int_{\mathbb{R}^n} \rho(x) \left| \frac{\partial u}{\partial t} \right|^{p-2} \frac{\partial u}{\partial t} \int_0^t \mu'(t-s)(u(t) - u(s)) ds dx.$$

Applying Hölder and Young's inequalities with σ, we get

$$\begin{aligned} & - \int_{\mathbb{R}^n} \rho(x) \left| \frac{\partial u}{\partial t} \right|^{p-2} \frac{\partial u}{\partial t} \int_0^t \mu'(t-s)(u(t) - u(s)) ds dx \\ & \le \int_{\mathbb{R}^n} \rho(x) \left| \frac{\partial u}{\partial t} \right|^{p-1} \int_0^t |\mu'(t-s)| |u(t) - u(s)| ds dx \\ & = \int_{\mathbb{R}^n} (\rho(x))^{\frac{1}{q}} \left| \frac{\partial u}{\partial t} \right|^{p-1} (\rho(x))^{\frac{1}{p}} \int_0^t |\mu'(t-s)| |u(t) - u(s)| ds dx \\ & \le \left(\int_{\mathbb{R}^n} \rho(x) \left| \frac{\partial u}{\partial t} \right|^p dx \right)^{\frac{1}{q}} \left(\int_{\mathbb{R}^n} \rho(x) \left| \int_0^t |\mu'(t-s)| |u(t) - u(s)| ds \right|^p dx \right)^{\frac{1}{p}} \\ & \le \sigma \left(\int_{\mathbb{R}^n} \rho(x) \left| \frac{\partial u}{\partial t} \right|^p dx \right) + C(\sigma) \left(\int_{\mathbb{R}^n} \rho(x) \left| \int_0^t |\mu'(t-s)| |u(t) - u(s)| ds \right|^p dx \right)^{\frac{1}{p}} \\ & = \sigma \left\| \frac{\partial u}{\partial t} \right\|_{L_\rho^p(\mathbb{R}^n)}^p + C(\sigma) \left\| \int_0^t |\mu'(t-s)| |u(t) - u(s)| ds \right\|_{L_\rho^p(\mathbb{R}^n)}^p . \end{aligned}$$

Hence by (5.35), we get

$$\psi_2'(t) \le \left(m_0 + m_1 \|\nabla_x u\|_2^{2\gamma} - \int_0^t \mu(s)ds \right)\left(\sigma \|u\|_{V_a}^2 + \frac{1}{4\sigma}(m_0 - \mu_0)(\mu \circ \nabla_x u)(t) \right)$$
$$+ (\mu \circ \nabla_x u)(t) \int_0^t \mu(s)ds - \left\| \frac{\partial u}{\partial t} \right\|_{L_\rho^p(\mathbb{R}^n)}^p \int_0^t \mu(s)ds \tag{5.36}$$
$$+ \sigma \left\| \frac{\partial u}{\partial t} \right\|_{L_\rho^p(\mathbb{R}^n)}^p + C(\sigma) \left\| \int_0^t |\mu'(t-s)||u(t) - u(s)|ds \right\|_{L_\rho^p(\mathbb{R}^n)}^p .$$

Now we consider

$$\left\| \int_0^t |\mu'(t-s)||u(t) - u(s)|ds \right\|_{L_\rho^p(\mathbb{R}^n)}^p .$$

Applying Lemma 5.4, we obtain

$$\begin{aligned}
&\left\| \int_0^t |\mu'(t-s)||u(t) - u(s)|ds \right\|_{L_\rho^p(\mathbb{R}^n)}^p \\
&\le \alpha \|\rho\|_{L^{s_1}(\mathbb{R}^n)} \left\| \int_0^t |\mu'(t-s)||\nabla_x u(t) - \nabla_x u(s)|ds \right\|_{L^2(\mathbb{R}^n)}^p \\
&= \alpha \|\rho\|_{L^{s_1}(\mathbb{R}^n)} \left(\int_{\mathbb{R}^n} \left| \int_0^t |\mu'(t-s)||\nabla_x u(t) - \nabla_x u(s)|ds \right|^2 dx \right)^{\frac{p}{2}} \\
&= \alpha \|\rho\|_{L^{s_1}(\mathbb{R}^n)} \left(\int_{\mathbb{R}^n} \frac{a(x)}{a(x)} \left(\int_0^t |\mu'(t-s)||\nabla_x u(t) - \nabla_x u(s)|ds \right)^2 dx \right)^{\frac{p}{2}} \\
&\le \frac{\alpha}{A_2^p} \|\rho\|_{L^{s_1}(\mathbb{R}^n)} \left(\int_{\mathbb{R}^n} \left(\int_0^t |\mu'(t-s)|(a(x))^{\frac{1}{2}} |\nabla_x u(t) - \nabla_x u(s)|ds \right)^2 dx \right)^{\frac{p}{2}} \\
&\le \frac{\alpha}{A_2^p} \|\rho\|_{L^{s_1}(\mathbb{R}^n)} \left(\int_{\mathbb{R}^n} \left(\int_0^t |\mu'(t-s)|ds \right)\left(\int_0^t |\mu'(t-s)|a(x)|\nabla_x u(t) - \nabla_x u(s)|^2 ds \right) dx \right)^{\frac{p}{2}} \\
&\le \frac{\alpha Q^{\frac{p}{2}}}{A_2^p} \|\rho\|_{L^{s_1}(\mathbb{R}^n)} \left(\int_{\mathbb{R}^n} a(x) \int_0^t |\mu'(t-s)||\nabla_x u(t) - \nabla_x u(s)|^2 dsdx \right)^{\frac{1}{2}} \\
&= \frac{\alpha Q^{\frac{p}{2}}}{A_2^p} \|\rho\|_{L^{s_1}(\mathbb{R}^n)} ((|\mu'| \circ \nabla_x u)(t))^{\frac{p}{2}} .
\end{aligned}$$

Hence by (5.36), we get

$$\psi_2'(t) \le \left(m_0 + m_1\|\nabla_x u\|_2^{2\gamma} - \int_0^t \mu(s)ds\right)\left(\sigma\left\|\frac{\partial u}{\partial t}\right\|_{V_a}^2 + \frac{1}{4\sigma}(m_0 - \mu_0)(\mu \circ \nabla_x u)(t)\right)$$
$$+ (\mu \circ \nabla_x u)(t)\int_0^t \mu(s)ds - \left\|\frac{\partial u}{\partial t}\right\|_{L_\rho^p(\mathbb{R}^n)}^p \int_0^t \mu(s)ds$$
$$+ \sigma\left\|\frac{\partial u}{\partial t}\right\|_{L_\rho^p(\mathbb{R}^n)}^p + C(\sigma)\frac{\alpha Q^{\frac{p}{2}}}{A_2^p}\|\rho\|_{L^{s_1}(\mathbb{R}^n)}((|\mu'| \circ \nabla_x u)(t))^{\frac{p}{2}},$$

which completes the proof. □

Remark 5.2. By the proof of Lemma 5.6, it follows that

$$\left\|\int_0^t \mu'(t-s)(u(t)-u(s))ds\right\|_{L_\rho^p(\mathbb{R}^n)}^p \le \frac{\alpha Q^{\frac{p}{2}}}{A_2^p}\|\rho\|_{L^{s_1}(\mathbb{R}^n)}((|\mu'| \circ \nabla_x u)(t))^{\frac{p}{2}}. \tag{5.37}$$

We define the Lyapunov function as

$$L(t) = \xi_1 E(t) + \psi_1(t) + \xi_2\psi_2(t) \tag{5.38}$$

for $\xi_1, \xi_2 > 1$. We need the next lemma, which means that the Lyapunov function and energy function are equivalent, i. e., for $\xi_1, \xi_2 > 1$, we have

$$\beta_1 L(t) \le E(t) \le \beta_2 L(t) \tag{5.39}$$

holds for some positive constants β_1 and β_2.

Lemma 5.7. *For $\xi_1, \xi_2 > 1$, we have*

$$L(t) \sim E(t). \tag{5.40}$$

Proof. By (5.38), we have

$$|L(t) - \xi_1 E(t)| \le |\psi_1(t)| + \xi_2|\psi_2(t)|$$
$$\le \int_{\mathbb{R}^n}\left|\rho(x)u\left|\frac{\partial u}{\partial t}\right|^{p-2}\frac{\partial u}{\partial t}\right|dx$$
$$+ \xi_2\int_{\mathbb{R}^n}\left|\rho(x)\left|\frac{\partial u}{\partial t}\right|^{p-2}\frac{\partial u}{\partial t}\int_0^t \mu(t-s)(u(t)-u(s))ds\right|dx.$$

Using Hölder and Young's inequalities, Lemma 5.4 and Lemma 5.6, we have

$$\begin{aligned}\int_{\mathbb{R}^n}\left|\rho(x)u\left|\frac{\partial u}{\partial t}\right|^{p-2}\frac{\partial u}{\partial t}\right|dx &\le \left(\int_{\mathbb{R}^n}\rho(x)|u|^p dx\right)^{1/p}\left(\int_{\mathbb{R}^n}\rho(x)\left|\frac{\partial u}{\partial t}\right|^p dx\right)^{(p-1)/p}\\ &\le \frac{1}{p}\left(\int_{\mathbb{R}^n}\rho(x)|u|^p dx\right)+\frac{p-1}{p}\left(\int_{\mathbb{R}^n}\rho(x)\left|\frac{\partial u}{\partial t}\right|^p dx\right)\\ &= \frac{1}{p}\|u\|^p_{L^p_\rho(\mathbb{R}^n)^p}+\frac{p-1}{p}\left\|\frac{\partial u}{\partial t}\right\|^p_{L^p_\rho(\mathbb{R}^n)}\\ &\le \frac{\alpha}{p}\|\rho\|_{L^{s_1}(\mathbb{R}^n)}\|\nabla_x u\|^p_{L^2(\mathbf{R}^n)}+\frac{p-1}{p}\left\|\frac{\partial u}{\partial t}\right\|^p_{L^p_\rho(\mathbb{R}^n)},\end{aligned}$$

and

$$\begin{aligned}&\int_{\mathbb{R}^n}\left|\rho(x)\left|\frac{\partial u}{\partial t}\right|^{p-2}\frac{\partial u}{\partial t}\int_0^t \mu(t-s)(u(t)-u(s))ds\right|dx\\ &\quad\le \sigma\left\|\frac{\partial u}{\partial t}\right\|^p_{L^p_\rho(\mathbb{R}^n)}+C(\sigma)\frac{\alpha Q^{\frac{p}{2}}}{A_2^p}\|\rho\|_{L^{s_1}(\mathbb{R}^n)}((\mu\circ\nabla_x u)(t))^{\frac{p}{2}}.\end{aligned}$$

Therefore,

$$\begin{aligned}|L(t)-\xi_1 E(t)| &\le \frac{\alpha}{p}\|\rho\|_{L^{s_1}(\mathbb{R}^n)}\|\nabla_x u\|^p_{L^2(\mathbf{R}^n)}+\frac{p-1}{p}\left\|\frac{\partial u}{\partial t}\right\|^p_{L^p_\rho(\mathbb{R}^n)}\\ &\quad+\xi_2\left(\sigma\left\|\frac{\partial u}{\partial t}\right\|^p_{L^p_\rho(\mathbb{R}^n)}+C(\sigma)\frac{\alpha Q^{\frac{p}{2}}}{A_2^p}\|\rho\|_{L^{s_1}(\mathbb{R}^n)}((\mu\circ\nabla_x u)(t))^{\frac{p}{2}}\right)\\ &= \frac{\alpha}{p}\|\rho\|_{L^{s_1}(\mathbb{R}^n)}\|\nabla_x u\|^p_{L^2(\mathbb{R}^n)}\\ &\quad+\left(\frac{p-1}{p}+\sigma\xi_2\right)\left\|\frac{\partial u}{\partial t}\right\|^p_{L^p_\rho(\mathbb{R}^n)}\\ &\quad+\xi_2 C(\sigma)\frac{\alpha Q^{\frac{p}{2}}}{A_2^p}\|\rho\|_{L^{s_1}(\mathbb{R}^n)}((\mu\circ\nabla_x u)(t))^{\frac{p}{2}}\\ &\le BE(t)\end{aligned}$$

for some positive constant B. If there is a need, we enlarge ξ_1 so that $\xi_1 \ge B$. Therefore,

$$-BE(t) \le L(t)-\xi_1 E(t) \le BE(t)$$

or

$$(\xi_1 - B)E(t) \le L(t) \le (\xi_1 + B)E(t),$$

which completes the proof. □

Theorem 5.2. *Let*

$$(u_0, u_1) \in \mathscr{H}(\mathbb{R}^n) \times L^q_\rho(\mathbb{R}^n),$$

and suppose that (1.34), (1.44) *hold. Then there exist positive constants*

$$\beta_0, \beta_1, \beta_2, \beta_3,$$

such that the energy of solution given by (5.23), (5.24) *satisfies*

$$E(t) \le \beta_3 h^{-1}(\beta_1 t + \beta_2), \quad \textit{for all } t \ge 0,$$

where

$$h(t) = \int_t^1 \frac{1}{sH_0'(\beta_0 s)} ds. \tag{5.41}$$

Proof. Note that

$$L'(t) = \xi_1 E'(t) + \psi_1'(t) + \xi_2 \psi_2'(t).$$

By Lemma 5.5 and Lemma 5.6, it follows that we can choose $\xi_1 > 1$ and $\xi_2 > 1$ large enough and $\sigma > 0$ small enough so that

$$L'(t) \le N_0(\mu \circ \nabla_x u) + c_1 m_1 \|\nabla_x u\|_{V_a}^{2(\gamma+1)} - cE(t), \quad \text{for all } t \ge 0, \tag{5.42}$$

for some positive constants c_1 and c, where

$$N_0 = \frac{4\xi_2 c + m_0 - l}{4\sigma}.$$

Now we set

$$F(t) = L(t) + cE(t).$$

Because

$$L(t) \sim E(t),$$

we have that

$$F(t) \sim E(t).$$

Then by (5.42), if there is a need we enlarge c, we get

$$F'(t) = L'(t) + cE'(t)$$
$$\leq -cE(t) + c \int_{\mathbb{R}^n} \int_{t_1}^{t} \mu(t-s)a(x)|\nabla_x u(t) - \nabla_x u(s)|^2 dsdx, \quad \text{for all } t \geq t_1. \tag{5.43}$$

By (5.26) and Remark 1.2, we have for any $t \geq t_1$,

$$\int_{\mathbb{R}^n} \int_0^{t_1} \mu(t-s)a(x)|\nabla_x u(t) - \nabla_x u(s)|^2 dsdx$$
$$\leq -\frac{1}{\mu_0} \int_{\mathbb{R}^n} \int_0^{t_1} \mu'(t-s)a(x)|\nabla_x u(t) - \nabla_x u(s)|^2 dsdx$$
$$\leq -cE'(t).$$

We define

$$J(t) = \int_{t_1}^{t} H_0(-\mu'(s))(\mu \circ \nabla_x u)(t) ds. \tag{5.44}$$

Since

$$\int_0^{+\infty} H_0(-\mu'(s))\mu(s) ds < +\infty,$$

if there is a need we enlarge t_1, from (5.26) we have

$$J(t) = \int_{t_1}^{t} H_0(-\mu'(s)) \int_{\mathbb{R}^n} \mu(s)a(x)|\nabla_x u(t) - \nabla_x u(t-s)|^2 dxds$$
$$\leq 2 \int_{t_1}^{t} H_0(-\mu'(s))\mu(s) \int_{\mathbb{R}^n} (a(x)|\nabla_x u(t)|^2 + a(x)|\nabla_x u(t-s)|^2) dxds \tag{5.45}$$
$$\leq cE(0) \int_{t_1}^{t} H_0(-\mu'(s))\mu(s) ds$$
$$< 1.$$

We define the following functional:

$$\Psi(t) = -\int_{t_1}^{t} H_0(-\mu'(s))\mu'(s) \int_{\mathbb{R}^n} \mu(s)a(x)|\nabla_x u(t) - \nabla_x u(t-s)|^2 dxds. \tag{5.46}$$

Using Remark 1.2 and (1.34), (1.44), we get that there exists a positive constant k_0 such that

$$H_0(-\mu'(s)) \le k_0 \quad \text{for any } s \ge t_1.$$

Hence, for any $t \ge t_1$, we have

$$\begin{aligned}
\Psi(t) &\le -k_0 \int_{t_1}^{t} \mu'(s) \int_{\mathbb{R}^n} a(x)|\nabla_x u(t) - \nabla_x u(t-s)|^2 dxds \\
&\le -2k_0 \int_{t_1}^{t} \mu'(s) \int_{\mathbb{R}^n} (a(x)|\nabla_x u(t)|^2 + a(x)|\nabla_x u(t-s)|^2) dxds \\
&\le -cE(0) \int_{t_1}^{t} \mu'(s) ds \\
&\le cE(0)\mu(t_1) \\
&< \min\{r, H(r), H_0(r)\}.
\end{aligned} \tag{5.47}$$

Using the properties of H_0 (strictly convex in $(0,r]$, $H_0(0) = 0$), then for $x \in (0,r]$, $\theta \in [0,1]$, we have

$$H_0(\theta x) \le \theta H_0(x).$$

Using Remark 1.2, (5.45), (5.47) and Jensen's inequality, we go to

$$\begin{aligned}
\Psi(t) &- \frac{1}{J(t)} \int_{t_1}^{t} J(t) H_0[H_0^{-1}(-\mu'(s))] H_0(-\mu'(s)) \int_{\mathbb{R}^n} \mu(s) a(x)|\nabla_x u(t) - \nabla_x u(t-s)|^2 dxds \\
&\ge \frac{1}{J(t)} \int_{t_1}^{t} H_0[J(t) H_0^{-1}(-\mu'(s))] H_0(-\mu'(s)) \int_{\mathbb{R}^n} \mu(s) a(x)|\nabla_x u(t) - \nabla_x u(t-s)|^2 dxds \\
&\ge H_0\left(\frac{1}{J(t)} \int_{t_1}^{t} J(t) H_0^{-1}(-\mu'(s)) H_0(-\mu'(s)) \int_{\mathbb{R}^n} \mu(s) a(x)|\nabla_x u(t) - \nabla_x u(t-s)|^2 dxds\right) \\
&\ge H_0\left(\int_{t_1}^{t} \int_{\mathbb{R}^n} \mu(s) a(x)|\nabla_x u(t) - \nabla_x u(t-s)|^2 dxds\right),
\end{aligned}$$

which implies

$$\int_{t_1}^{t} \int_{\mathbb{R}^n} \mu(s) a(x)|\nabla_x u(t) - \nabla_x u(t-s)|^2 dxds \le H_0^{-1}(\Psi(t)). \tag{5.48}$$

We have

$$F'(t) \le -cE(t) + cH_0^{-1}(\Psi(t)), \quad \text{for all } t \ge t_1.$$

Now we define the functional

$$F_1(t) = H_0'\left(\beta_0 \frac{E(t)}{E(0)}\right)F(t) + cE(t), \quad \beta_0 < r, 0 < c.$$

Since

$$F(t) \sim E(t),$$

we have that

$$F_1(t) \sim E(t).$$

Also, using that $E'(t) \le 0$ for any $t \ge 0$ and $H_0' > 0$, $H_0'' >$ on $(0, r]$, we get

$$\begin{aligned} F_1'(t) &= \beta_0 \frac{E'(t)}{E(0)} H_0''\left(\beta_0 \frac{E(t)}{E(0)}\right)F(t) + H_0'\left(\beta_0 \frac{E(t)}{E(0)}\right)F'(t) + cE'(t) \\ &\le -cE(t)H_0'\left(\beta_0 \frac{E(t)}{E(0)}\right) + cH_0'\left(\beta_0 \frac{E(t)}{E(0)}\right)H_0^{-1}(\Psi(t)) + cE'(t). \end{aligned}$$

Let H_0^* be given as in Remark 1.2. Hence, using Young's inequality (1.47) with

$$A = H_0'\left(\beta_0 \frac{E(t)}{E(0)}\right), \quad B = H_0^{-1}(\Psi(t)),$$

we obtain

$$\begin{aligned} F_1'(t) &\le -cE(t)H_0'\left(\beta_0 \frac{E(t)}{E(0)}\right) + cH_0^*\left(H_0'\left(\beta_0 \frac{E(t)}{E(0)}\right)\right) + c\Psi(t) + cE'(t) \\ &\le -cE(t)H_0'\left(\beta_0 \frac{E(t)}{E(0)}\right) + c\beta_0 \frac{E(t)}{E(0)} H_0'\left(\beta_0 \frac{E(t)}{E(0)}\right) - c'E'(t) + cE'(t). \end{aligned}$$

We can choose β_0, c and c' such that

$$\begin{aligned} F_1'(t) &\le -k \frac{E(t)}{E(0)} H_0'\left(\beta_0 \frac{E(t)}{E(0)}\right) \\ &= -kH_2\left(\frac{E(t)}{E(0)}\right) \quad \text{for any } t \ge t_1, \end{aligned}$$

where

$$H_2(t) = tH_0'(\beta_0 t).$$

Using the strict convexity of H_0 on $(0, r]$, we conclude that H_2' and H_2 are strict positive on $(0, 1]$. Therefore,

$$R(t) = \tau \frac{k_1 F_1(t)}{E(0)} \sim E(t), \quad \tau \in (0,1), \tag{5.49}$$

for some positive constant k_1. Hence,

$$R'(t) \leq -\tau k_2 H_2(R(t)) \quad \text{for any } t \geq t_1$$

and for some positive constant k_2. Then a simple integration and a suitable choice of τ yield

$$R(t) \leq h^{-1}(\beta_1 t + \beta_2), \quad \beta_1, \beta_2 \in (0, +\infty), t \geq t_1.$$

Here,

$$h(t) = \int_t^1 H_2^{-1}(s) ds.$$

From (5.49), for a positive constant β_3, we have

$$E(t) \leq \beta_3 h^{-1}(\beta_1 t + \beta_2), \quad \beta_1, \beta_2 \in (0, +\infty), t \geq t_1.$$

The fact that h is strictly decreasing function on $(0, 1]$ and due to properties of H_2, we have

$$\lim_{t \to 0} h(t) = +\infty.$$

Then

$$E(t) \leq \beta_3 h^{-1}(\beta_1 t + \beta_2) \quad \text{for all } t \geq 0.$$

This completes the proof of Theorem 5.2. □

6 Wave equation with logarithmic nonlinearities in Kirchhoff type

We study a viscoelastic wave equations of the Kirchhoff type

$$\frac{\partial^2 u}{\partial t^2} - \phi(x)\left(M(\|\nabla_x u\|_2^2)\Delta_x u - \int_0^t \mu(t-s)\Delta_x u(s)ds \right) = au \ln |u|^k \tag{6.1}$$

defined in any spaces dimension. We use weighted spaces to establish the long-time behavior of solution of (6.1).

We consider (6.1), with $n \geq 2, k, a > 0$ and M is a positive C^1 function satisfying for

$$s \geq 0, \quad m_0 > 0, \quad m_1 \geq 0, \quad \gamma \geq 1,$$

and

$$M(s) = m_0 + m_1 s^\gamma,$$

and the scalar function $\mu(s)$ is assumed to satisfy (1.34).

Equation (6.1) is equipped by the following initial data:

$$u(0,x) = u_0(x) \in \mathcal{H}(\mathbb{R}^n), \quad \frac{\partial u}{\partial t}(0,x) = u_1(x) \in L_\rho^2(\mathbb{R}^n), \tag{6.2}$$

where the weighted spaces $\mathcal{H}$ is given in Definition 1.3 and the density function $\phi(x) > 0$, $\forall x \in \mathbb{R}^n$, $(\phi(x))^{-1} = \rho(x)$ satisfy (1.58). Now, using Lemma 1.14, we give the following lemma concerning logarithmic Sobolev inequality.

Lemma 6.1. [1] *Let $u \in \mathcal{H}(\mathbb{R}^n)$ be any function and $c_1, c_2 > 0$ be any numbers. Then*

$$\begin{aligned} 2\int_{\mathbb{R}^n} \rho(x)|u|^2 \ln\left(\frac{|u|}{\|u\|_{L_\rho^2}^2}\right)dx + n(1+c_1)\|u\|_{L_\rho^2}^2 \\ \leq c_2 \frac{\|\rho\|_{L^2}^2}{\pi}\|\nabla_x u\|_2^2 \end{aligned}$$

Definition 6.1. By the weak solution of (6.1) over $[0,T]$, we mean a function

$$u \in C([0,T], \mathcal{H}(\mathbb{R}^n)) \cap C^1([0,T], L_\rho^2(\mathbb{R}^n)) \cap C^2([0,T], \mathcal{H}^{-1}(\mathbb{R}^n))$$

with

---

1 H. Chen, P. Luo and G. Liu. Global solution and blow-up of a semilinear heat equation with logarithmic nonlinearity. J. Math. Anal. Appl. 422 (2015), 84–98.

https://doi.org/10.1515/9783111633787-006

$$\frac{\partial u}{\partial t} \in L^2([0,T], \mathscr{H}(\mathbb{R}^n)),$$

such that

$$u(0) = u_0, \quad \frac{\partial u}{\partial t}(0) = u_1,$$

and for all $v \in \mathscr{H}$, $t \in [0,T]$,

$$\begin{aligned}
&\int_{\mathbb{R}^n} \rho(x) u \ln |u|^k v dx \\
&\quad = \int_{\mathbb{R}^n} \rho(x) \frac{\partial^2 u}{\partial t^2} v dx + M(\|\nabla_x u\|_2^2) \int_{\mathbb{R}^n} \nabla_x u \nabla_x v dx \\
&\quad - \int_{\mathbb{R}^n} \int_0^t \mu(t-s) \nabla_x u(s) ds \nabla_x v dx.
\end{aligned}$$

Multiplying the equation (6.1) by $\rho(x)\frac{\partial u}{\partial t}$, and integrating by parts over $\mathbb{R}^n$, we have the energy of u at time t is given by

$$\begin{aligned}
E(t) &= \frac{1}{2}\left(\left\| \frac{\partial u}{\partial t} \right\|_{L_\rho^2}^2 + \left(m_0 - \int_0^t \mu(s) ds \right) \|\nabla_x u\|_2^2 \right. \\
&\quad \left. + (\mu \circ \nabla_x u) - \int_{\mathbb{R}^n} \rho(x) u^2 \ln |u|^k dx \right) \\
&\quad + \frac{k}{4} \|u\|_{L_\rho^2}^2 + \frac{m_1}{2(\gamma+1)} \|\nabla_x u\|_2^{2(\gamma+1)}
\end{aligned} \tag{6.3}$$

and the following energy functional law holds:

$$E'(t) = \frac{1}{2}(\mu' \circ \nabla_x u)(t) - \frac{1}{2}\mu(t) \|\nabla_x u(t)\|_2^2, \quad \forall t \geq 0. \tag{6.4}$$

which means that, our energy is uniformly bounded and decreasing along the trajectories for $u \in \mathscr{H}(\mathbb{R}^n)$, $t \geq 0$.

6.1 Global existence in time

Theorem 6.1 (Local existence). *Let*

$$u_0(x) \in \mathscr{H}(\mathbb{R}^n), \quad u_1(x) \in L_\rho^2(\mathbb{R}^n),$$

be given. Then, under hypothesis (1.34), (1.44) *and* (1.58), *problem* (6.1) *has a unique local solution*

$$u \in C([0,T], \mathscr{H}(\mathbb{R}^n)) \cap C^1([0,T], L_\rho^2(\mathbb{R}^n)).$$

Now, we introduce two functionals:

$$J(t) = \frac{1}{2}\left(\left(m_0 - \int_0^t \mu(s)ds\right)\|\nabla_x u\|_2^2 + (\mu \circ \nabla_x u) - \int_{\mathbb{R}^n} \rho(x)u^2 \ln|u|^k dx\right) + \frac{k}{4}\|u\|_{L^2_\rho}^2 + \frac{m_1}{2(\gamma+1)}\|\nabla_x u\|_2^{2(\gamma+1)} \tag{6.5}$$

and

$$I(t) = \left(m_0 - \int_0^t \mu(s)ds\right)\|\nabla_x u\|_2^2 + (\mu \circ \nabla_x u) - \int_{\mathbb{R}^n} \rho(x)u^2 \ln|u|^k dx + \frac{m_1}{2(\gamma+1)}\|\nabla_x u\|_2^{2(\gamma+1)}. \tag{6.6}$$

Then

$$J(t) = \frac{1}{2}I(t) + \frac{k}{4}\|u\|_{L^2_\rho}^2. \tag{6.7}$$

To establish the corresponding method of potential wells. which is related to the logarithmic nonlinear term, we introduce the stable set as follows:

$$W = \{u \in \mathcal{H}(\mathbb{R}^n) : I(t) > 0, J(t) < d\} \cup \{0\}. \tag{6.8}$$

Remark 6.1. We notice that the mountain pass level d given in (6.8) defined by

$$d = \inf\Big\{\sup_{u\in\mathcal{H}(\mathbb{R}^n)\setminus\{0\}\mu\geq 0} J(\mu u)\Big\}. \tag{6.9}$$

Also, by introducing the so-called “Nehari manifold”

$$\mathcal{N} = \{u \in \mathcal{H}(\mathbb{R}^n)\setminus\{0\} : I(t) = 0\}.$$

It is readily seen that the potential depth d is also characterized by

$$d = \inf_{u\in\mathcal{N}} J(t). \tag{6.10}$$

This characterization of d shows that

$$\operatorname{dist}(0, \mathcal{N}) = \min_{u\in\mathcal{N}} \|u\|_{\mathcal{H}(\mathbb{R}^n)}. \tag{6.11}$$

By the fact that (6.4), we will prove the invariance of the set W. That is, if for some $t_0 > 0$ if $u(t_0) \in W$, then $u(t) \in W, \forall t \geq t_0$. Let us begin by giving the existence lemma of the potential depth.

Lemma 6.2. *The potential depth d is positive constant.*

Lemma 6.3. *Let*

$$u \in \mathscr{H}(\mathbb{R}^n),$$

and

$$\beta = e^{\frac{1}{2}n(1+c_1)}.$$

1. *If*

$$0 < \|u\|_{L_\rho^2}^2 < \beta,$$

then $I(t) > 0$.

2. *If*

$$I(t) = 0, \quad \|u\|_2^2 \neq 0,$$

then $\|u\|_{L_\rho^2}^2 > \beta$.

Proof. By (1.33), (6.6) and Lemma 6.1, we have

$$\begin{aligned} I(t) &= \left(m_0 - \int_0^t \mu(s)ds\right)\|\nabla_x u\|_2^2 + (\mu \circ \nabla_x u) \\ &\quad - \int_{\mathbb{R}^n} \rho(x)u^2 \ln |u|^k dx + \frac{m_1}{2(\gamma+1)}\|\nabla_x u\|_2^{2(\gamma+1)} \\ &\geq \mu_0\|\nabla_x u\|_2^2 - k\int_{\mathbb{R}^n} \rho(x)u^2\left(\ln \frac{|u|}{\|u\|_{L_\rho^2}^2} + \ln \|u\|_{L_\rho^2}^2\right)dx \\ &\geq \left(\mu_0 - \frac{kc_2}{2\pi}\|\rho\|_{L_\rho^2}^2\right)\|\nabla_x u\|_2^2 + \frac{1}{2}kn(1+c_1)\|u\|_{L_\rho^2}^2 \\ &\quad - k\|u\|_{L_\rho^2}^2 \ln \|u\|_{L_\rho^2}^2. \end{aligned}$$

Choosing c_2 such that

$$\mu_0 > \frac{kc_2}{2\pi}\|\rho\|_{L_\rho^2}^2,$$

then

$$I(t) \geq k\left(\frac{1}{2}n(1 + c_1) - \ln \|u\|^2_{L^2_\rho}\right)\|u\|^2_{L^2_\rho}.$$

Therefore:

1. If
$$0 < \|u\|^2_{L^2_\rho} < \beta,$$
then $I(t) > 0$.
2. If $I(t) = 0$, $\|u\|_2^2 \neq 0$, we have
$$\beta < \|u\|^2_{L^2_\rho},$$
then
$$\|u\|^2_{L^2_\rho} > \beta.$$
□

Theorem 6.2 (Global existence). *Let*

$$u_0(x) \in \mathcal{H}(\mathbb{R}^n), \quad u_1(x) \in L^2_\rho(\mathbb{R}^n),$$

and

$$0 < E(0) < d, \quad I(0) > 0.$$

Then, under hypothesis (1.34), (1.44) *and conditions* (1.58), *problem* (6.1) *has a global solution in time.*

Proof. From the definition of energy for solution and by (6.4), we have

$$\frac{1}{2}\left\|\frac{\partial u}{\partial t}\right\|^2_{L^2_\rho} + J(t) \leq \frac{1}{2}\|u_1\|^2_{L^2_\rho} + J(0), \quad \forall t \in [0, T_{\max}) \tag{6.12}$$

where $T_{\max}$ is the maximal existence time of solution of u. Then, by the definition of the stable set and using Lemma 6.3, we have $u \in W$, $\forall t \in [0, T_{\max})$. □

6.2 General decay estimates

We apply the multiplier techniques to obtain useful estimates and prepare some functional associated with the nature of our problem to introduce appropriate Lyapunov functions. For this purpose, we introduce the functional

$$\psi_1(t) = \int_{\mathbb{R}^n} \rho(x)u\frac{\partial u}{\partial t}dx. \tag{6.13}$$

Lemma 6.4. *Under the hypothesis* (1.34), (1.44), *the functional* ψ_1 *satisfies, along the solution of* (6.1), (6.2),

$$\psi_1'(t) \le \left\|\frac{\partial u}{\partial t}\right\|_{L^2_\rho}^2 + m_1\|\nabla_x u\|_2^{2(\gamma+1)} + \frac{(1-\mu_0)}{4\sigma}(\mu \circ \nabla_x u) + \left[\left(\sigma + \frac{kc_2}{2\pi}\|\rho\|_{L^2}^2 - \mu_0\right) + k\|\rho\|_{L^2}^2\left(\ln\|u\|_{L^2_\rho}^2 - \frac{1}{2}n(1+c_1)\right)\right]\|\nabla u\|_2^2.$$

Proof. From (6.13), integrate over $\mathbb{R}^n$, we have

$$\begin{aligned}
\psi_1'(t) &= \int_{\mathbb{R}^n} \rho(x)\left\|\frac{\partial u}{\partial t}\right\|^2 dx + \int_{\mathbb{R}^n} \rho(x) u \frac{\partial^2 u}{\partial t^2} dx \\
&= \int_{\mathbb{R}^n}\left(\rho(x)\left\|\frac{\partial u}{\partial t}\right\|^2 + M(\|\nabla_x u\|_2^2) u\Delta_x u - u\int_0^t \mu(t-s)\Delta_x u(s,x)ds\right)dx \\
&\quad + \int_{\mathbb{R}^n} \rho(x) u^2 \ln|u|^k dx \\
&\le \left\|\frac{\partial u}{\partial t}\right\|_{L^2_\rho(\mathbb{R}^n)}^2 + m_1\|\nabla_x u\|_2^{2(\gamma+1)} - \mu_0\|\nabla_x u\|_2^2 \\
&\quad + k\int_{\mathbb{R}^n} \rho(x) u^2\left(\ln\left(\frac{|u|}{\|u\|_{L^2_\rho}^2}\right) + \ln\|u\|_{L^2_\rho}^2\right)dx \\
&\quad + \int_{\mathbb{R}^n} \nabla_x u \int_0^t \mu(t-s)(\nabla_x u(s) - \nabla_x u(t))dsdx.
\end{aligned}$$

By using the logarithmic Sobolev inequality in Lemma 6.1 and generalized version of Poincare's inequality in Lemma 1.14 along with with Young's inequality and Lemma 1.9 for $\theta = 1/2$, we obtain

$$\begin{aligned}
\psi_1'(t) &\le \left\|\frac{\partial u}{\partial t}\right\|_{L^2_\rho}^2 + m_1\|\nabla_x u\|_2^{2(\gamma+1)} + \left(\frac{kc_2}{2\pi}\|\rho\|_{L^2}^2 - \mu_0\right)\|\nabla_x u\|_2^2 \\
&\quad + k\|u\|_{L^2_\rho}^2 \ln\|u\|_{L^2_\rho}^2 \\
&\quad + \sigma\|\nabla_x u\|_2^2 + \frac{1}{4\sigma}\int_{\mathbb{R}^n}\left(\int_0^t \mu(t-s)|\nabla_x u(s) - \nabla_x u(t)|ds\right)^2 dx \\
&\quad - \frac{1}{2}kn(1+c_1)\|u\|_{L^2_\rho}^2 \\
&\le \left\|\frac{\partial u}{\partial t}\right\|_{L^2_\rho}^2 + m_1\|\nabla_x u\|_2^{2(\gamma+1)} + \left(\sigma + \frac{kc_2}{2\pi}\|\rho\|_{L^2}^2 - \mu_0\right)\|\nabla_x u\|_2^2
\end{aligned}$$

$$+\frac{(1-\mu_0)}{4\sigma}(\mu\circ\nabla_x u)+k\left(\ln\|u\|^2_{L^2_\rho}-\frac{1}{2}n(1+c_1)\right)\|u\|^2_{L^2_\rho}.$$

Then

$$\psi_1'(t)\le\left\|\frac{\partial u}{\partial t}\right\|^2_{L^2_\rho}+m_1\|\nabla_x u\|_2^{2(\gamma+1)}+\frac{(1-\mu_0)}{4\sigma}(\mu\circ\nabla_x u)$$
$$+\left[\left(\sigma+\frac{kc_2}{2\pi}\|\rho\|^2_{L^2}-\mu_0\right)+k\|\rho\|^2_{L^2}\left(\ln\|u\|^2_{L^2_\rho}-\frac{1}{2}n(1+c_1)\right)\right]\|\nabla u\|_2^2. \qquad \square$$

The existence of the memory term forces us to make second modification of the associate energy functional. Set

$$\psi_2(t)=-\int_{\mathbb{R}^n}\rho(x)\frac{\partial u}{\partial t}\int_0^t\mu(t-s)(u(t)-u(s))dsdx. \tag{6.14}$$

Lemma 6.5. *Under the hypothesis* (1.34), (1.44), *the functional* ψ_2 *satisfies, along the solution of* (6.1), (6.2), *for any* $\sigma\in(0,m_0)$,

$$\psi_2'(t)\le\left[\sigma+k\left(\sigma\frac{c_2}{2\pi}+\ln\|u\|^2_{L^2_\rho}-\frac{n(1+c_1)}{2}\right)\right]\|\nabla_x u\|_2^2$$
$$+cm_1\|\nabla_x u\|_2^{2(\gamma+1)}+c_\sigma\left(1+\left(k\frac{c_2}{2\pi}+1\right)\|\rho\|^2_{L^2}\right)(\mu\circ\nabla_x u)$$
$$-c_\sigma\|\rho\|^2_{L^2}(\mu'\circ\nabla_x u)+\left(\sigma-\int_0^t\mu(s)ds\right)\left\|\frac{\partial u}{\partial t}\right\|^2_{L^2_\rho}.$$

Proof. Exploiting (6.1), (6.14) to get

$$\psi_2'(t)=-\int_{\mathbb{R}^n}\rho(x)\frac{\partial^2 u}{\partial t^2}\int_0^t\mu(t-s)(u(t)-u(s))dsdx$$
$$-\int_{\mathbb{R}^n}\rho(x)\frac{\partial u}{\partial t}\int_0^t\mu'(t-s)(u(t)-u(s))dsdx$$
$$-\int_0^t\mu(s)ds\left\|\frac{\partial u}{\partial t}\right\|^2_{L^2_\rho}$$
$$=\int_{\mathbb{R}^n}M(\|\nabla u\|_2^2)\nabla_x u\int_0^t\mu(t-s)(\nabla_x u(t)-\nabla_x u(s))dsdx$$
$$-\int_{\mathbb{R}^n}\rho(x)u\ln|u|^k\int_0^t\mu(t-s)(u(t)-u(s))dsdx$$

$$
\begin{aligned}
&-\int_{\mathbb{R}^n}\left(\int_0^t \mu(t-s)\nabla_x u(s,x)ds\right)\\
&\times\left(\int_0^t \mu(t-s)(\nabla_x u(t)-\nabla_x u(s))ds\right)dx\\
&-\int_{\mathbb{R}^n}\rho(x)\frac{\partial u}{\partial t}\int_0^t \mu'(t-s)(u(t)-u(s))dsdx\\
&-\int_0^t \mu(s)ds\left\|\frac{\partial u}{\partial t}\right\|_{L^2_\rho}^2.
\end{aligned}
$$

By (1.34), we have

$$
\begin{aligned}
\psi_2'(t) = {}&\left(m_0-\int_0^t \mu(s)ds\right)\\
&\times\int_{\mathbb{R}^n}\nabla_x u\int_0^t \mu(t-s)(\nabla_x u(t)-\nabla_x u(s))dsdx\\
&+\int_{\mathbb{R}^n}\left(\int_0^t \mu(t-s)(\nabla_x u(t)-\nabla_x u(s))ds\right)^2 dx\\
&+cm_1\|\nabla_x u\|_2^{2(\gamma+1)}\\
&-\int_{\mathbb{R}^n}\rho(x)u\ln|u|^k\int_0^t \mu(t-s)(u(t)-u(s))dsdx\\
&-\int_{\mathbb{R}^n}\rho(x)\frac{\partial u}{\partial t}\int_0^t \mu'(t-s)(u(t)-u(s))dsdx\\
&-\int_0^t \mu(s)ds\left\|\frac{\partial u}{\partial t}\right\|_{L^2_\rho}^2+c(\mu\circ\nabla_x u)(t).
\end{aligned}
$$

By Hölder and Young's inequalities and Lemma 1.14, we estimate

$$
\begin{aligned}
&-\int_{\mathbb{R}^n}\rho(x)\frac{\partial u}{\partial t}\int_0^t \mu'(t-s)(u(t)-u(s))dsdx\\
&\leq\left(\int_{\mathbb{R}^n}\rho(x)\left\|\frac{\partial u}{\partial t}\right\|^2 dx\right)^{1/2}
\end{aligned}
$$

$$\times\left(\int_{\mathbb{R}^n}\rho(x)\left|\int_0^t \mu'(t-s)(u(t)-u(s))ds\right|^2\right)^{1/2}$$
$$\leq \sigma\left\|\frac{\partial u}{\partial t}\right\|_{L^2_\rho}^2 + c_\sigma\left\|\int_0^t -\mu'(t-s)(u(t)-u(s))ds\right\|_{L^2_\rho}^2$$
$$\leq \sigma\left\|\frac{\partial u}{\partial t}\right\|_{L^2_\rho}^2 - c_\sigma\|\rho\|_{L^2}^2(\mu'\circ\nabla_x u)(t),$$

and

$$\int_{\mathbb{R}^n}\rho(x)\frac{\partial u}{\partial t}\int_0^t \mu(t-s)(u(t)-u(s))dsdx$$
$$\leq \sigma\left\|\frac{\partial u}{\partial t}\right\|_{L^2_\rho}^2 + c_\sigma\|\rho\|_{L^2}^2(\mu\circ\nabla_x u)(t),$$

and by Lemma 1.14 and Lemma 6.1 and conditions in Lemma 6.3, we have

$$-\int_{\mathbb{R}^n}\rho(x)\ln|u|^k u\int_0^t \mu(t-s)(u(t)-u(s))dsdx$$
$$\leq k\int_{\mathbb{R}^n}\rho(x)\left(\ln\left(\frac{|u|}{\|u\|_{L^2_\rho}^2}\right)+\ln\|u\|_{L^2_\rho}^2\right)u$$
$$\times\int_0^t \mu(t-s)(u(t)-u(s))dsdx$$
$$\leq k\left(\ln\|u\|_{L^2_\rho}^2 - \frac{n(1+c_1)}{2}\right)\|u\|_{L^2_\rho}^2$$
$$+ k\frac{c_2}{2\pi}\left\|u\int_0^t \mu(t-s)(u(t)-u(s))ds\right\|_{L^2_\rho}^2$$
$$\leq k\left(\ln\|u\|_{L^2_\rho}^2 - \frac{n(1+c_1)}{2}\right)\|\rho\|_{L^2}^2\|\nabla_x u\|_2^2$$
$$+ k\frac{c_2}{2\pi}\|\rho\|_{L^2}^2\left\|\nabla u\int_0^t \mu(t-s)(\nabla u(t)-\nabla u(s))ds\right\|_{L^2_\rho}^2$$
$$\leq k\left(\sigma\frac{c_2}{2\pi}+\ln\|u\|_{L^2_\rho}^2 - \frac{n(1+c_1)}{2}\right)\|\rho\|_{L^2}^2\|\nabla_x u\|_2^2$$
$$+ c_\sigma k\frac{c_2}{2\pi}\|\rho\|_{L^2}^2(\mu\circ\nabla_x u).$$

Using Young and Poincare's inequalities and Lemma 1.9 for $\theta = 1/2$, we obtain

$$\begin{aligned}\psi_2'(t) \le & \left[\sigma + k\left(\sigma\frac{c_2}{2\pi} + \ln\|u\|_{L^2_\rho}^2 - \frac{n(1+c_1)}{2}\right)\right]\|\nabla_x u\|_2^2 \\ & + cm_1\|\nabla_x u\|_2^{2(\gamma+1)} \\ & + c_\sigma\left(1 + \left(k\frac{c_2}{2\pi} + 1\right)\|\rho\|_{L^2}^2\right)(\mu \circ \nabla_x u) - c_\sigma\|\rho\|_{L^2}^2(\mu' \circ \nabla_x u) \\ & + \left(\sigma - \int_0^t \mu(s)ds\right)\left\|\frac{\partial u}{\partial t}\right\|_{L^2_\rho}^2 .\end{aligned}$$

□

Now, let us define

$$L(t) = \xi_1 E(t) + \psi_1(t) + \xi_2\psi_2(t) \tag{6.15}$$

for $\xi_1, \xi_2 > 1$. We need the next lemma, which means that there is equivalent between the Lyapunov and energy functions, i. e., for $\xi_1, \xi_2 > 1$, we have

$$\beta_1 L(t) \le E(t) \le \beta_2 L(t)$$

that holds for two positive constants β_1 and β_2.

Lemma 6.6. *For $\xi_1, \xi_2 > 1$, we have*

$$L(t) \sim E(t).$$

Proof. By (6.15), we have

$$\begin{aligned}|L(t) - \xi_1 E(t)| & \le |\psi_1(t)| + \xi_2|\psi_2(t)| \\ & \le \int_{\mathbb{R}^n}\left|\rho(x)u\frac{\partial u}{\partial t}\right|dx \\ & \quad + \xi_2\int_{\mathbb{R}^n}\left|\rho(x)\frac{\partial u}{\partial t}\int_0^t \mu(t-s)(u(t)-u(s))ds\right|dx.\end{aligned}$$

Thanks to Hölder and Young's inequalities, we have by using Lemma 1.14,

$$\begin{aligned}&\int_{\mathbb{R}^n}\left|\rho(x)u\frac{\partial u}{\partial t}\right|dx \\ &\quad \le \left(\int_{\mathbb{R}^n}\rho(x)|u|^2dx\right)^{1/2}\left(\int_{\mathbb{R}^n}\rho(x)\left\|\frac{\partial u}{\partial t}\right\|^2dx\right)^{1/2}\end{aligned}$$

$$\leq \frac{1}{2}\Big(\int_{\mathbb{R}^n} \rho(x)|u|^2 dx\Big) + \frac{1}{2}\Big(\int_{\mathbb{R}^n} \rho(x)\Big\|\frac{\partial u}{\partial t}\Big\|^2 dx\Big)$$

$$\leq c\Big\|\frac{\partial u}{\partial t}\Big\|^2_{L^2_\rho} + c\|\rho\|^2_{L^2}\|\nabla_x u\|_2^2$$

and

$$\int_{\mathbb{R}^n}\Big|\Big(\rho(x)^{\frac{1}{2}}\frac{\partial u}{\partial t}\Big)\Big(\rho(x)^{\frac{1}{2}}\int_0^t \mu(t-s)(u(t)-u(s))ds\Big)\Big|dx$$

$$\leq \Big(\int_{\mathbb{R}^n} \rho(x)\Big\|\frac{\partial u}{\partial t}\Big\|^2 dx\Big)^{1/2}$$

$$\times \Big(\int_{\mathbb{R}^n} \rho(x)\Big|\int_0^t \mu(t-s)(u(t)-u(s))ds\Big|^2 dx\Big)^{1/2}$$

$$\leq \frac{1}{2}\Big\|\frac{\partial u}{\partial t}\Big\|^2_{L^2_\rho} + \frac{1}{2}\Big\|\int_0^t \mu(t-s)(u(t)-u(s))ds\Big\|^2_{L^2_\rho}$$

$$\leq \frac{1}{2}\Big\|\frac{\partial u}{\partial t}\Big\|^2_{L^2_\rho} + \frac{1}{2}\|\rho\|^2_{L^2}(\mu\circ\nabla_x u).$$

Then

$$|L(t) - \xi_1 E(t)| \leq cE(t).$$

Therefore, we can choose ξ_1 so that

$$L(t) \sim E(t). \tag{6.16}$$

□

Lemma 6.7. *For all $t \geq t_1 > 0$, we have*

$$\int_{t_1}^t (\mu\circ\nabla_x u)(s)ds \leq H_0^{-1}\Big(-\int_{t_1}^t H_0(-\mu'(s))\mu'(s) \times \int_{\mathbb{R}^n} \mu(s)|\nabla_x u(t) - \nabla_x u(t-s)|^2 dxds\Big),$$

where H_0 is introduced in Remark 1.1.

Proof. By (6.4) and Remark 1.1, we have for all $t \geq t_1$,

$$\begin{aligned}
&\int_{\mathbb{R}^n}\int_0^{t_1} \mu(t-s)\left|\nabla_x u(t) - \nabla_x u(s)\right|^2 ds dx \\
&\leq -\frac{1}{k}\int_{\mathbb{R}^n}\int_0^{t_1} \mu(t-s)\left|\nabla_x u(t) - \nabla_x u(s)\right|^2 ds dx \\
&\leq -cE'(t).
\end{aligned}$$

Now, we define

$$I(t) = \int_{t_1}^{t} H_0(-\mu'(s))(\mu \circ \nabla_x u)(t) ds. \tag{6.17}$$

Since

$$\int_0^{+\infty} H_0(-\mu'(s))\mu(s) ds < +\infty,$$

from (6.4) we have

$$\begin{aligned}
I(t) &= \int_{t_1}^{t} H_0(-\mu'(s)) \int_{\mathbb{R}^n} \mu(s)\left|\nabla_x u(t) - \nabla_x u(t-s)\right|^2 dx ds \\
&\leq 2\int_{t_1}^{t} H_0(-\mu'(s))\mu(s) \int_{\mathbb{R}^n} \left|\nabla_x u(t)\right|^2 + \left|\nabla_x u(t-s)\right|^2 dx ds \\
&\leq cE(0)\int_{t_1}^{t} H_0(-\mu'(s))\mu(s) ds < 1.
\end{aligned} \tag{6.18}$$

We define again a new functional $\lambda(t)$ related with $I(t)$ as

$$\lambda(t) = -\int_{t_1}^{t} H_0(-\mu'(s))\mu'(s) \int_{\mathbb{R}^n} \mu(s)\left|\nabla_x u(t) - \nabla_x u(t-s)\right|^2 dx ds.$$

From (1.34), (1.44) and Remark 1.1, we get

$$H_0(-\mu'(s))\mu(s) \leq H_0(H(\mu(s)))\mu(s) = D(\mu(s))\mu(s) \leq k_0,$$

for some positive constant k_0. Then, for all $t \geq t_1$,

$$\begin{aligned}\lambda(t) &\leq -k_0 \int_{t_1}^{t} \mu'(s) \int_{\mathbb{R}^n} |\nabla_x u(t) - \nabla_x u(t-s)|^2 dxds \\ &\leq -k_0 \int_{t_1}^{t} \mu'(s) \int_{\mathbb{R}^n} |\nabla_x u(t)|^2 + |\nabla_x u(t-s)|^2 dxds \\ &\leq -cE(0) \int_{t_1}^{t} \mu'(s) ds \\ &\leq cE(0)\mu(t_1) \\ &< \min\{r, H(r), H_0(r)\}. \end{aligned} \tag{6.19}$$

Using the properties of H_0 (strictly convex in $(0, r]$, $H_0(0) = 0$), then for $x \in (0, r]$, $\theta \in [0, 1]$,

$$H_0(\theta x) \leq \theta H_0(x).$$

Using the hypothesis in Remark 1.1, (6.18), (6.19) and Jensen's inequality leads to

$$\begin{aligned}\lambda(t) &= \frac{1}{I(t)} \int_{t_1}^{t} I(t) H_0[H_0^{-1}(-\mu'(s))] H_0(-\mu'(s))\mu'(s) \\ &\quad \times \int_{\mathbb{R}^n} \mu(s) |\nabla_x u(t) - \nabla_x u(t-s)|^2 dxds \\ &\geq \frac{1}{I(t)} \int_{t_1}^{t} H_0[I(t) H_0^{-1}(-\mu'(s))] H_0(-\mu'(s))\mu'(s) \\ &\quad \times \int_{\mathbb{R}^n} \mu(s) |\nabla_x u(t) - \nabla_x u(t-s)|^2 dxds \\ &\geq H_0\Bigg(\frac{1}{I(t)} \int_{t_1}^{t} I(t) H_0^{-1}(-\mu'(s)) H_0(-\mu'(s))\mu'(s) \\ &\quad \times \int_{\mathbb{R}^n} \mu(s) |\nabla_x u(t) - \nabla_x u(t-s)|^2 dxds \Bigg) \\ &\geq H_0\Bigg(\int_{t_1}^{t} \int_{\mathbb{R}^n} \mu(s) |\nabla_x u(t) - \nabla_x u(t-s)|^2 dxds \Bigg), \end{aligned}$$

which implies

$$\int_{t_1}^{t}\int_{\mathbb{R}^n} \mu(s)\left|\nabla_x u(t) - \nabla_x u(t-s)\right|^2 dxds \le H_0^{-1}(\lambda(t)).$$ □

Theorem 6.3. *Let*

$$(u_0, u_1) \in \mathscr{H}(\mathbb{R}^n) \times L^2_\rho(\mathbb{R}^n),$$

and suppose that (1.34), (1.44) *hold. Then there exist positive constants*

$$\alpha_0, \alpha_1, \alpha_2, \alpha_3,$$

such that the energy of solution given by (6.1), (6.2) *satisfies*

$$E(t) \le \alpha_3 H_1^{-1}(\alpha_1 t + \alpha_2), \quad \textit{for all } t \ge 0,$$

where

$$H_1(t) = \int_t^1 \left(sH_0'(\alpha_0 s)\right)^{-1} ds.$$

Proof. From (6.4), Lemma 6.4 and Lemma 6.5, we have

$$\begin{aligned} L'(t) &= \xi_1 E'(t) + \psi_1'(t) + \xi_2\psi_2'(t) \\ &\le \left(\frac{1}{2}\xi_1 - c_\sigma\|\rho\|^2_{L^2}\xi_2\right)(\mu' \circ \nabla_x u) + M_0(\mu \circ \nabla_x u) \\ &\quad - M_1\left\|\frac{\partial u}{\partial t}\right\|^2_{L^2_\rho} - M_2\|\nabla_x u\|_2^2 + (c\xi_2 + 1)m_1\|\nabla_x u\|_2^{2(\gamma+1)} \end{aligned}$$

where

$$\begin{aligned} M_0 &= \left(\xi_2 c_\sigma\left(1 + \left(k\frac{c_2}{2\pi} + 1\right)\|\rho\|^2_{L^2}\right) + \frac{(1-\mu_0)}{4\sigma}\right) > 0, \\ M_1 &= \left(\xi_2\left(\int_0^{t_1}\mu(s)ds - \sigma\right) - 1\right), \\ M_2 &= \frac{1}{2}\xi_1\mu(t_1) - \left[\left(\sigma + \frac{kc_2}{2\pi}\|\rho\|^2_{L^2} - \mu_0\right)\right. \\ &\quad \left. + k\|\rho\|^2_{L^2}\left(\ln\|u\|^2_{L^2_\rho} - \frac{1}{2}n(1+c_1)\right)\right] \\ &\quad - \xi_2\left[\sigma + k\left(\sigma\frac{c_2}{2\pi} + \ln\|u\|^2_{L^2_\rho} - \frac{n(1+c_1)}{2}\right)\right] \end{aligned}$$

and t_1 was introduced in Remark 1.1. We choose σ so small that

$$\xi_1 > 2c_\sigma \|\rho\|_{L^2}^2 \xi_2.$$

Whence σ is fixed, we can choose

$$\xi_2 > \left(\int_0^{t_1} \mu(s)ds - \sigma \right)^{-1}$$

and ξ_1 large enough so that $M_2 > 0$, which yields

$$L'(t) \leq M_0(\mu \circ \nabla_x u) + (c\xi_2 + 1)m_1 \|\nabla_x u\|_2^{2(\gamma+1)} - cE'(t), \quad \forall t \geq t_1.$$

Now we set

$$F(t) = L(t) + cE(t),$$

which is equivalent to $E(t)$. Then we get for some

$$\begin{aligned} c &> 2(c\xi_2 + 1)(\gamma + 1), \\ F'(t) &= L'(t) + cE'(t) \\ &\leq -cE(t) + c \int_{\mathbb{R}^n} \int_{t_1}^{t} \mu(t-s) |\nabla_x u(t) - \nabla_x u(s)|^2 dsdx, \\ &\quad \text{for all } t \geq t_1. \end{aligned} \tag{6.20}$$

Using Lemma 6.7, we obtain

$$F'(t) \leq -cE(t) + cH_0^{-1}(\lambda(t)), \quad \text{for all } t \geq t_1.$$

Now, we will use the fact that

$$E' \leq 0, \quad 0 < H_0', \quad 0 < H_0'',$$

on $(0, r]$ to define the functional

$$F_1(t) = H_0'\left(\alpha_0 \frac{E(t)}{E(0)} \right) F(t) + cE(t), \quad \alpha_0 < r, 0 < c,$$

where

$$F_1(t) \sim E(t),$$

and

$$\begin{aligned}F_1'(t) &= \alpha_0\frac{E'(t)}{E(0)}H_0''\left(\alpha_0\frac{E(t)}{E(0)}\right)F(t)\\&\quad+H_0'\left(\alpha_0\frac{E(t)}{E(0)}\right)F'(t)+cE'(t)\\&\le -cE(t)H_0'\left(\alpha_0\frac{E(t)}{E(0)}\right)\\&\quad+cH_0'\left(\alpha_0\frac{E(t)}{E(0)}\right)H_0^{-1}(\lambda(t))+cE'(t).\end{aligned}$$

Let H_0^* given in Remark 1.1 and using Young's inequality (1.47) with

$$A=H_0'\left(\alpha_0\frac{E(t)}{E(0)}\right),\quad B=H_0^{-1}(\lambda(t)),$$

to get

$$\begin{aligned}F_1'(t) &\le -cE(t)H_0'\left(\alpha_0\frac{E(t)}{E(0)}\right)+cH_0^*\left(H_0'\left(\alpha_0\frac{E(t)}{E(0)}\right)\right)\\&\quad+c\lambda(t)+cE'(t)\\&\le -cE(t)H_0'\left(\alpha_0\frac{E(t)}{E(0)}\right)+c\alpha_0\frac{E(t)}{E(0)}H_0'\left(\alpha_0\frac{E(t)}{E(0)}\right)\\&\quad-c'E'(t)+cE'(t).\end{aligned}$$

Choosing α_0, c, c', such that for all $t\ge t_1$ we have

$$\begin{aligned}F_1'(t) &\le -k\frac{E(t)}{E(0)}H_0'\left(\alpha_0\frac{E(t)}{E(0)}\right)\\&= -kH_2\left(\frac{E(t)}{E(0)}\right),\end{aligned}$$

where

$$H_2(t)=tH_0'(\alpha_0 t).$$

Using the strict convexity of H_0 on $(0,r]$, to find that H_2', H_2 are strict positives on $(0,1]$, then

$$R(t)=\tau\frac{k_1F_1(t)}{E(0)}\sim E(t),\quad \tau\in(0,1)\tag{6.21}$$

and

$$R'(t)\le -\tau k_0H_2(R(t)),\quad k_0\in(0,+\infty), t\ge t_1.$$

Then a simple integration and a suitable choice of τ yield

$$R(t) \leq H_1^{-1}(\alpha_1 t + \alpha_2), \quad \alpha_1, \alpha_2 \in (0, +\infty), t \geq t_1.$$

Here,

$$H_1(t) = \int_t^1 H_2^{-1}(s)ds.$$

From (6.21), for a positive constant α_3, we have

$$E(t) \leq \alpha_3 H_1^{-1}(\alpha_1 t + \alpha_2), \quad \alpha_1, \alpha_2 \in (0, +\infty), t \geq t_1.$$

The fact that H_1 is a strictly decreasing function on $(0, 1]$ and due to properties of H_2, we have

$$\lim_{t \to 0} H_1(t) = +\infty.$$

Then

$$E(t) \leq \alpha_3 H_1^{-1}(\alpha_1 t + \alpha_2), \quad \text{for all } t \geq 0.$$

This completes the proof of Theorem 6.3. □

7 Petrowsky–Petrowsky system in $\mathbb{R}^n$

We consider the following Petrowsky–Petrowsky system:

$$\begin{cases} \frac{\partial^2 u}{\partial t^2} + \phi(x)(\Delta^2 u - \int_{-\infty}^{t} \mu(t-s)\Delta^2 u(s)ds) + \alpha v = 0, \\ \frac{\partial^2 v}{\partial t^2} + \phi(x)\Delta^2 v + \alpha u = 0, \\ (u_0, v_0) \in \mathscr{D}^{2,2}(\mathbb{R}^n), \quad (u_1, v_1) \in L^2_\rho(\mathbb{R}^n) \end{cases} \tag{7.1}$$

where $\phi(x) > 0$, $\alpha \neq 0$, $(\phi(x))^{-1} = \rho(x)$, where the density function $\rho(x)$ is defined in (1.58), the function μ is as in (1.33) and the spaces $\mathscr{D}^{2,2}(\mathbb{R}^n)$, $L^2_\rho(\mathbb{R}^n)$ defined as the following.

Definition 7.1. [1] We define the function spaces of our problem and its norm as follows:

$$\mathscr{D}^{2,2}(\mathbb{R}^n) = \{f \in L^{2n/(n-4)}(\mathbb{R}^n) : \Delta_x f \in L^2(\mathbb{R}^n)\}$$

and the spaces $L^2_\rho(\mathbb{R}^n)$ to be the closure of $C_0^\infty(\mathbb{R}^n)$ functions with respect to the inner product

$$(f, h)_{L^2_\rho(\mathbb{R}^n)} = \int_{\mathbb{R}^n} gfh dx,$$

adding a new variable η to the system, which corresponds to the relative displacement history. Let us define the auxiliary variable

$$\eta = \eta^t(x,s) = u(x,t) - u(x,t-s), \quad (x,s) \in \mathbb{R}^n \times \mathbb{R}^+, \ t \geq 0. \tag{7.2}$$

By differentiation, we have

$$\eta^t_t(x,s) = -\eta^t_s(x,s) + \frac{\partial u}{\partial t}(x,t), \quad (x,s) \in \mathbb{R}^n \times \mathbb{R}^+, \ t \geq 0, \tag{7.3}$$

and we can take as the initial condition ($t = 0$),

$$\eta^0(x,s) = u_0(x,0) - u_0(x,-s), \quad (x,s) \in \mathbb{R}^n \times \mathbb{R}^+. \tag{7.4}$$

Thus, the original memory term can be rewritten as

$$\int_{-\infty}^{t} \mu(t-s)\Delta^2 u(s)ds = \int_{0}^{\infty} \mu(s)\Delta^2 u(t-s)ds$$

1 R. C. A. M. Van Der Vorst. Best constant for the embedding of the space $H^2 \cap H^1_0(\Omega)$ into $L^{2n/n-4}(\Omega)$. Diff. Int. Eq. 6 (2) (1993), 259–276.

https://doi.org/10.1515/9783111633787-007

$$= \left(\int_0^\infty \mu(s) ds \right) \Delta^2 u - \int_0^\infty \mu(s) \Delta^2 \eta^t(s) ds,$$

Problem (7.1) is transformed into the system

$$\begin{cases} \frac{\partial^2 u}{\partial t^2} + \mu_0 \phi(x) \Delta^2 u + \int_0^\infty \mu(s) \phi(x) \Delta^2 \eta(s) ds + \alpha v = 0, \\ \frac{\partial^2 v}{\partial t^2} + \phi(x) \Delta^2 v + \alpha u = 0, \\ \eta_t + \eta_s - \frac{\partial u}{\partial t} = 0, \end{cases} \tag{7.5}$$

and initial conditions

$$u(x,0) = u_0(x), \quad \frac{\partial u}{\partial t}(x,0) = u_1(x), \quad \eta^0(x,s) = \eta_0(x,s), \tag{7.6}$$

where

$$\eta_0(x,s) = u_0(x,0) - u_0(x,-s), \quad (x,s) \in \mathbb{R}^n \times \mathbb{R}^+.$$

The energy of u at times t is given by

$$E(t) = \frac{1}{2} \left(\left\| \frac{\partial u}{\partial t} \right\|_{L^2_\rho}^2 + \left\| \frac{\partial v}{\partial t} \right\|_{L^2_\rho}^2 + \mu_0 \|\Delta u\|_{L^2}^2 + \|\Delta v\|_{L^2}^2 + \|\Delta \eta^t\|_{L^2_\rho}^2 + 2\alpha \int_{\mathbb{R}^n} \rho u v dx \right). \tag{7.7}$$

For α small enough, we deduce that

$$E(t) \geq \frac{1}{2}(1 - |\alpha|\gamma) \left(\left\| \frac{\partial u}{\partial t} \right\|_{L^2_\rho}^2 + \left\| \frac{\partial v}{\partial t} \right\|_{L^2_\rho}^2 + \mu_0 \|\Delta u\|_{L^2}^2 + \|\Delta v\|_{L^2}^2 + \|\Delta \eta^t\|_{L^2_\rho}^2 \right).$$

The function spaces $\mathscr{D}^{2,2}(\mathbb{R}^n)$ can be embedded continuously in $L^{2n/(n-4)}(\mathbb{R}^n)$, i. e., there exists $k > 0$ such that

$$\|u\|_{L^{2n/(n-4)}} \leq k \|u\|_{\mathscr{D}^{2,2}}. \tag{7.8}$$

We shall frequently use the following version of the generalized Poincaré's inequality:

$$\int_{\mathbb{R}^n} |\Delta u|^2 dx \geq \gamma \int_{\mathbb{R}^n} \rho u^2 dx \tag{7.9}$$

for all $u \in C_0^\infty$ and $\rho \in L^{\frac{n}{2}}$, where

$$\gamma =: k^{-2} \|\rho\|_{\frac{n}{4}}^{-1}.$$

In order to consider the relative displacement η as a new variable, we introduce the weighted L^2-spaces

$$L^2_\mu(\mathbb{R}^+; H^2(\mathbb{R}^n)) = \left\{ \xi : \mathbb{R}^+ \to H^2(\mathbb{R}^n) \;\middle|\; \int_0^\infty \mu(s)\|\xi(s)\|_V^2 ds < \infty \right\},$$

where

$$L^2_\mu(\mathbb{R}^+; H^2(\mathbb{R}^n)),$$

is the Hilbert space of $H^2(\mathbb{R}^n)$-valued functions on $\mathbb{R}^+$ endowed with the inner product

$$(\xi, \xi)_\mu = \int_0^\infty \mu(s)(\xi(s), \xi(s))_V ds, \quad \|\xi\|^2_\mu = \int_0^\infty \mu(s)\|\xi\|^2 ds.$$

Moreover, we have the following compact embedding.

Lemma 7.1. *Let*

$$\rho \in L^{n/2}(\mathbb{R}^n) \cap L^\infty(\mathbb{R}^n).$$

Then the embedding

$$D^{2,2} \subset L^2_\rho,$$

is compact.

We give the definition of the weak solution for this problem.

Definition 7.2. A weak solution of the problem (7.1) is a function (u, v, η) such that:
(i) $u, v \in L^2[0, T; D^{2,2}(\mathbb{R}^n)]$, $\frac{\partial u}{\partial t}, \frac{\partial v}{\partial t} \in L^2[0, T; L^2_\rho(\mathbb{R}^n)]$, $\frac{\partial^2 u}{\partial t^2}, \frac{\partial^2 v}{\partial t^2} \in L^2[0, T; D^{-2,2}(\mathbb{R}^n)]$,
(ii) For all

$$\varphi, \phi \in C_0^\infty([0, T] \times \mathbb{R}^n), (u, v, \eta),$$

it satisfies the generalized formula

$$\begin{aligned}&\int_0^T \left(\frac{\partial^2 u}{\partial t^2}(\tau), \varphi(\tau)\right)_{L^2_\rho} d\tau + \alpha \int_0^T (v(\tau), \varphi(\tau))_{L^2_\rho} d\tau + \int_0^T \left(\frac{\partial u}{\partial t}(\tau), \varphi(\tau)\right)_{L^2_\rho} d\tau \\ &+ (1 - \mu_0) \int_0^T \int_{\mathbb{R}^n} \Delta u(\tau) \Delta\varphi(\tau) dx d\tau + \int_0^T \int_{\mathbb{R}^n} \int_0^{+\infty} \mu(s) \Delta\eta^\tau(s) \Delta\varphi(\tau) ds dx d\tau = 0,\end{aligned} \tag{7.10}$$

and

$$\int_0^T \left(\frac{\partial^2 v}{\partial t^2}(\tau), \phi(\tau)\right)_{L^2_\rho} d\tau + \alpha \int_0^T (u(\tau), \phi(\tau))_{L^2_\rho} d\tau + \int_0^T \int_{\mathbb{R}^n} \Delta v(\tau) \Delta\phi(\tau) dx d\tau = 0, \tag{7.11}$$

(iii) (u, v) satisfies the initial conditions

$$(u_0(x), v_0(x)) \in (D^{2,2}(\mathbb{R}^n))^2, \quad (u_1(x), v_1(x)) \in (L^2_\rho(\mathbb{R}^n))^2, \quad \eta_0^t \in L^2_\rho(\mathbb{R}^+, H^1(\mathbb{R}^n)).$$

We denote by $\mathscr{A}$ the unbounded operator in the energy space:

$$\mathscr{H} = H^1(\mathbb{R}^n) \times H^1(\mathbb{R}^n) \times L^2(\mathbb{R}^n) \times L^2(\mathbb{R}^n),$$

defined by

$$\mathscr{D}(\mathscr{A}) = (H^2(\mathbb{R}^n) \cap H^1(\mathbb{R}^n))^2 \times (H^1(\mathbb{R}^n))^2,$$

and

$$\mathscr{A}\mathscr{U} = \left(\tilde{u}, (1 - \mu_0)\phi(x)\Delta^2 u + \int_0^\infty \mu(s)\phi(x)\Delta^2\eta(s)ds - \alpha v - \tilde{u}, \tilde{v}, \phi(x)\Delta^2 v - \alpha u \right)^T,$$

and

$$U = (u, \tilde{u}, v, \tilde{v})^T.$$

Problem (7.1) can then be reformulated under the abstract from

$$\mathscr{U}' + \mathscr{A}\mathscr{U} = 0. \tag{7.12}$$

Theorem 7.1. *Assume that* (1.33), (1.36) *hold. Let*

$$u_0, v_0 \in D^{2,2}(\mathbb{R}^n),$$

and

$$u_1, v_1 \in L^2_\rho(\mathbb{R}^n),$$

and

$$\eta_0^t \in L^2_\rho(\mathbb{R}^+, H^1(\mathbb{R}^n)).$$

Then there exists positive constant δ_1 and δ_2 such that the solution of (7.5) *satisfies*

$$E(t) \le \frac{C}{t}(E(0) + E_2(0) + E_3(0)), \quad \forall t \in \mathbb{R}^+. \tag{7.13}$$

7.1 Lack of exponential stability

We use Theorem 1.1 to prove the lack of exponential stability, i. e., we show that there exists a sequence of values μ_m such that

$$\|(i\mu_m I - \mathscr{A})^{-1}\|_{\mathscr{L}(\mathscr{H})} \to \infty. \tag{7.14}$$

It is equivalent to prove that there exist a sequence of data $F_m \in \mathscr{H}$ and a sequence of real numbers $\mu_m \in \mathbb{R}$, with $\|F_m\|_{\mathscr{H}} \leq 1$ such that

$$\|(i\mu_m I - \mathscr{A})^{-1} F_m\|_{\mathscr{H}} \to \infty. \tag{7.15}$$

The eigenvalue problem

$$\phi(x)\Delta^2 u = \mu u, \tag{7.16}$$

has a complete system of eigensolutions $\{w_j, \mu_j\}$ satisfying the following properties:

$$\begin{cases} \phi(x)\Delta^2 w_j = \mu_j w_j, \quad j = 1, \dots, \quad w_j \in \mathscr{D}^{2,2}(\mathbb{R}^n), \\ 0 < \mu_1 \leq \mu_2 \leq \cdots, \quad \mu_j \to +\infty \quad \text{as } j \to +\infty. \end{cases} \tag{7.17}$$

Theorem 7.2. *Assume that the kernel is of the form*

$$\mu(s) = e^{-\mu s}, \quad s \in \mathbb{R}^+,$$

with $\mu > 1$. The semigroup $S(t)$ on $\mathscr{H}$ is not exponentially stable.

Proof. We will find a sequence of bounded functions

$$F_m = (f_{1,m}, f_{2,m}, f_{3,m}, f_{4,m}, f_{5,m}) \in \mathscr{H},$$

for which the corresponding solutions of the resolvent equations is not bounded. This will prove that the resolvant operator is not uniformly bounded. Let us consider the equation

$$i\mu U_m - \mathscr{A} U_m = F_m.$$

The equation reads

$$\begin{cases} i\mu u - \varphi = f_1, \\ i\mu\varphi + \mu_0 \phi(x)\Delta^2 u + \alpha v + \int_0^\infty \mu(s)\Delta^2 \eta(x,s)\, ds = f_2, \\ i\mu v - \psi = f_3, \\ i\mu\psi + \Delta^2 v + \alpha v = f_4, \\ i\mu\eta - \varphi + \eta_s = f_5. \end{cases} \tag{7.18}$$

Let us consider

$$f_{1m} = f_{3m} = f_{5m} = 0,$$

and

$$f_{2m} = f_{4m} = w_m$$

to obtain

$$\varphi = i\mu u$$

and

$$\psi = i\mu v.$$

Then system (7.18) becomes

$$\begin{cases} -\mu^2 u + \mu_0 \phi(x)\Delta^2 u + \alpha v + \int_0^\infty \mu(s)\Delta^2 \eta(x,s)\, ds = w_m, \\ -\mu^2 v + \Delta^2 v + \alpha v = w_m, \\ i\mu\eta - i\mu u + \eta_s = 0. \end{cases} \tag{7.19}$$

We look for solutions of the form

$$u = aw_m, \quad v = bw_m, \quad \varphi = cw_m, \quad \psi = dw_m, \quad \eta(x,s) = \gamma(s)w_m$$

with

$$a, b, c, d \in \mathbb{C}$$

and $\gamma(s)$ depends on μ and will be determined explicitly in what follows. From (7.19), we get a and b that satisfy

$$\begin{cases} -\mu^2 a + \mu_0 a \mu_m + \alpha b + \mu_m \int_0^\infty \mu(s)\gamma(s)\, ds = 1, \\ -\mu^2 b + \mu_m b + \alpha a = 1, \\ \gamma_s + i\mu\gamma - i\mu a = 0. \end{cases} \tag{7.20}$$

Solving $(7.20)_3$, we get

$$\gamma(s) = a - ae^{-i\mu s}. \tag{7.21}$$

Then, from (7.21), we have

$$\int_0^\infty \mu(s)\gamma(s)\,ds = a(1-\mu_0) - a\int_0^\infty \mu(s)e^{-i\mu s}\,ds. \tag{7.22}$$

Now, choosing $\mu = \sqrt{\mu_m}$, using equations $(7.20)_1$ and $(7.20)_2$ we obtain

$$\begin{aligned}
a &= \frac{1}{\alpha},\\
b &= \frac{\mu_m(1-\mu_0)}{\alpha^2} - \frac{\mu_m}{\alpha}\int_0^\infty \mu(s)\gamma(s)\,ds + \frac{1}{\alpha},\\
c &= i\frac{\sqrt{\mu_m}}{\alpha},\\
d &= i\sqrt{\mu_m}\left(\frac{\mu_m(1-\mu_0)}{\alpha^2} - \frac{\mu_m}{\alpha}\int_0^\infty \mu(s)\gamma(s)\,ds + \frac{1}{\alpha}\right).
\end{aligned}$$

Recalling that

$$\varphi = cw_m = i\frac{\sqrt{\mu_m}}{\alpha},$$

we get

$$\|\varphi\|^2_{L^2_\rho} = \frac{\mu_m}{\alpha}.$$

Therefore, we have

$$\lim_{m\to\infty}\|U_m\|^2_{\mathcal{H}} \geq \lim_{m\to\infty}\|\varphi\|^2_{L^2_\rho} = \lim_{m\to\infty}\frac{\mu_m}{\alpha} = +\infty,$$

which completes the proof. □

7.2 Polynomial stability

Let us introduce the second- and third-order energy

$$E_2(t) = E\left(\frac{\partial u}{\partial t}, \frac{\partial v}{\partial t}, \eta_t\right),$$

and

$$E_3(t) = E\left(\frac{\partial^2 u}{\partial t^2}, \frac{\partial^2 v}{\partial t^2}, \eta_{tt}\right).$$

Lemma 7.2. *The energies $E(t)$, $E_2(t)$ and $E_3(t)$ associated to solutions of problem* (7.5) *satisfies*

$$\frac{d}{dt}E(t) = -\frac{k_1}{2}\int_0^{+\infty}\mu(s)\|\Delta\eta^t\|_2^2 ds \le 0, \quad \forall t \ge 0, \tag{7.23}$$

$$\frac{d}{dt}E_2(t) = -\frac{k_1}{2}\int_0^{+\infty}\mu(s)\|\Delta\eta_t^t\|_2^2 ds \le 0, \quad \forall t \ge 0, \tag{7.24}$$

$$\frac{d}{dt}E_3(t) = -\frac{k_1}{2}\int_0^{+\infty}\mu(s)\|\Delta\eta_{tt}^t\|_2^2 ds \le 0, \quad \forall t \ge 0. \tag{7.25}$$

Proof. Multiplying $(7.5)_1$ by $g\frac{\partial u}{\partial t}$ and $(7.5)_1$ by $g\frac{\partial v}{\partial t}$, summing the results obtained follows the conclusion of inequality (7.23). The inequalities (7.24) and (7.25) are obtained using the same procedure as in (7.23). □

Lemma 7.3. *Let (u, v, η) be the solution of* (7.1). *Let the functional F_1 defined by*

$$F_1(t) = -\int_{\mathbb{R}^n}\rho(x)\frac{\partial u}{\partial t}\left(\int_0^{+\infty}\mu(s)\eta^t(s)ds\right)dx. \tag{7.26}$$

For any $\varepsilon_1 > 0$, there exists a positive constant

$$C_{\varepsilon_1} = \frac{\mu_0\mu_1}{2\varepsilon_1} + \mu_1 + \frac{\alpha\mu_1}{2\varepsilon_1},$$

such that

$$\begin{aligned}\frac{d}{dt}F_1(t) \le &-\frac{\mu_1}{2}\int_{\mathbb{R}^n}\rho(x)\frac{\partial u}{\partial t}^2 dx + \frac{\mu_0\varepsilon_1}{2}\int_{\mathbb{R}^n}|\Delta u|^2 dx \\ &+ \frac{\alpha\varepsilon_1}{2}\int_{\mathbb{R}^n}\rho(x)|v|^2 dx + C_\varepsilon\int_0^{+\infty}\mu(s)\left\|\Delta\eta^t(s)\right\|^2 ds,\end{aligned} \tag{7.27}$$

where

$$\mu_1 = \int_0^{\infty}\mu(s)ds.$$

Proof. Using (7.1) and integrating by parts, we obtain

$$\begin{aligned}\frac{d}{dt}F_1(t) = {}&\mu_0\int_{\mathbb{R}^n}\Delta u\left(\int_0^{+\infty}\mu(s)\Delta\eta^t(s)ds\right)dx + \int_{\mathbb{R}^n}\left|\int_0^{+\infty}\mu(s)\Delta\eta^t(s)ds\right|^2 dx \\ &+ \alpha\int_{\mathbb{R}^n}\rho(x)v\left(\int_0^{+\infty}\mu(s)\eta^t(s)ds\right)dx - \mu_1\int_{\mathbb{R}^n}\rho(x)\frac{\partial u}{\partial t}^2 dx\end{aligned}$$

$$+\int_{\mathbb{R}^n} \rho(x)\frac{\partial u}{\partial t}\left(\int_0^{+\infty} \mu(s)\eta_s^t(s)ds\right)dx.$$

Since

$$\int_{\mathbb{R}^n} \rho(x)\frac{\partial u}{\partial t}\left(\int_0^{+\infty} \mu(s)\eta_s^t(s)ds\right)dx = -\int_{\mathbb{R}^n} \rho(x)\frac{\partial u}{\partial t}\left(\int_0^{+\infty} \mu'(s)\eta^t(s)ds\right)dx, \tag{7.28}$$

and

$$\int_{\mathbb{R}^n}\left|\int_0^{+\infty} \mu(s)\Delta\eta^t(s)ds\right|^2 dx \le \mu_1 \int_{\mathbb{R}^n}\left(\int_0^{+\infty} \mu(s)|\Delta\eta^t(s)|^2 ds\right)dx. \tag{7.29}$$

Then using Young's inequality and hypotheses (1.33), our conclusion follows. □

Let us consider the following functional:

$$F_1(t) = F_2\left(\frac{\partial u}{\partial t}, \frac{\partial v}{\partial t}, \eta_t^t\right). \tag{7.30}$$

Then using Lemma 7.3, we obtain

$$\frac{d}{dt}F_2(t) \le -\frac{\mu_1}{2}\int_{\mathbb{R}^n} \rho(x)\frac{\partial^2 u}{\partial t^2}^2 dx + \frac{\mu_0\varepsilon_1}{2}\int_{\mathbb{R}^n}\left|\Delta\frac{\partial u}{\partial t}\right|^2 dx \tag{7.31}$$

$$+\frac{\alpha\varepsilon_1}{2}\int_{\mathbb{R}^n} \rho(x)\left|\frac{\partial v}{\partial t}\right|^2 dx + C_\varepsilon \int_0^{+\infty} \mu(s)\|\Delta\eta_t^t(s)\|_2^2 ds.$$

Lemma 7.4. *Let (u, v, η) be the solution of* (7.1). *Then the functional Z_1 defined*

$$Z_1(t) = \int_{\mathbb{R}^n} \rho(x)\frac{\partial u}{\partial t}.udx, \tag{7.32}$$

satisfies, along the solution, the estimate

$$\frac{d}{dt}Z_1(t) \le \int_{\mathbb{R}^n} \rho(x)\frac{\partial u}{\partial t}^2 dx - \frac{\mu_0}{2}\int_{\mathbb{R}^n} |\Delta u|^2 dx - \alpha\int_{\mathbb{R}^n} \rho(x)uvdx + \frac{\mu_1}{2\mu_0}\int_0^{+\infty} \mu(s)\|\Delta\eta^t(s)\|^2 ds, \tag{7.33}$$

where

$$\mu_1 = \int_0^{\infty} \mu(s)ds.$$

Proof. Using $(7.5)_1$ and integrating by parts, we get

$$\frac{d}{dt}Z_1(t) = -\mu_0 \int_{\mathbb{R}^n} |\Delta u|^2 - \int_{\mathbb{R}^n} \int_0^{+\infty} \mu(s)\Delta\eta^t \Delta u ds dx - \alpha \int_{\mathbb{R}^n} \rho(x) uv dx + \int_{\mathbb{R}^n} \rho(x) \frac{\partial u}{\partial t}^2 dx. \tag{7.34}$$

Then using Young's inequality, our conclusion follows. □

Let us denote by $Z_2(t)$ the following functional:

$$Z_1(t) = Z_2\left(\frac{\partial u}{\partial t}, \frac{\partial v}{\partial t}, \eta_t^t\right). \tag{7.35}$$

Then, from (7.33), we have

$$\begin{aligned} \frac{d}{dt}Z_2(t) \le &\int_{\mathbb{R}^n} \rho(x) \frac{\partial^2 u}{\partial t^2}^2 dx - \frac{\mu_0}{2} \int_{\mathbb{R}^n} \left|\Delta \frac{\partial u}{\partial t}\right|^2 dx \\ &- \alpha \int_{\mathbb{R}^n} \rho(x) \frac{\partial u}{\partial t}\frac{\partial v}{\partial t} dx + \frac{\mu_1}{2\mu_0} \int_0^{+\infty} \mu(s) \|\Delta\eta_t^t(s)\|^2 ds. \end{aligned} \tag{7.36}$$

The following lemma shows the dissipative property of v. For this, let us consider the following functional:

$$\Phi(t) = \int_{\mathbb{R}^n} \frac{\partial^2 u}{\partial t^2}\frac{\partial v}{\partial t} dx + \mu_0 \int_{\mathbb{R}^n} \Delta \frac{\partial u}{\partial t}.\Delta v dx + \int_{\mathbb{R}^n} \left(\int_0^{+\infty} \mu(s)\Delta\eta_t^t(s)\Delta v ds\right) dx. \tag{7.37}$$

Lemma 7.5. *For any $\varepsilon_2 > 0$, we have*

$$\begin{aligned} \frac{d}{dt}\Phi(t) \le &-\alpha \left\|\frac{\partial v}{\partial t}\right\|_{L_\rho^2}^2 + \left(\frac{1}{2\varepsilon_2} + \frac{\mu_0}{2\varepsilon_2}\right) \int_{\mathbb{R}^n} \left(\int_0^{\infty} \mu(s)|\Delta\eta_{tt}^t|^2 ds\right) dx \\ &+ \left(\frac{k_2}{2\varepsilon_2} + \frac{\mu_1}{2\varepsilon_2} + \frac{k_2\mu_0}{2\varepsilon_2}\right) \int_{\mathbb{R}^n} \left(\int_0^{\infty} \mu(s)|\Delta\eta^t|^2 ds\right) dx + \frac{\mu(0)}{2\varepsilon_2} \left\|\frac{\partial u}{\partial t}\right\|^2 dx \\ &+ \left(\frac{1}{\mu_1} + \frac{\mu(0)+1}{2} + \frac{\mu_1}{\mu_0}\right)\varepsilon_2 \|\Delta v\|_2^2 + \frac{\alpha}{2}\left\|\frac{\partial^2 u}{\partial t^2}\right\|_{L_\rho^2}^2 dx + \frac{\alpha C_0}{2}\|\Delta u\|_2^2. \end{aligned}$$

Proof. Differentiating $\Phi(t)$ with respect to t and using $(7.1)_1$, we obtain

$$\begin{aligned} \frac{d}{dt}\Phi(t) = &-\alpha \int_{\mathbb{R}^n} \rho(x) \left|\frac{\partial v}{\partial t}\right|^2 dx + \int_{\mathbb{R}^n} \Delta \frac{\partial^2 u}{\partial t^2}.\Delta v dx - \alpha \int_{\mathbb{R}^n} \rho(x) \frac{\partial^2 u}{\partial t^2} u dx \\ &+ \int_{\mathbb{R}^n} \left(\int_0^{\infty} \Delta\eta_{tt}^t ds\right).\Delta v dx + \mu_0 \int_{\mathbb{R}^n} \Delta \frac{\partial^2 u}{\partial t^2} \Delta v dx. \end{aligned} \tag{7.38}$$

On the other hand, we have

$$\begin{aligned} &\eta_t^t(0) = 0, \quad \eta_s^t(0) = \frac{\partial u}{\partial t}(t), \\ &\frac{\partial^2 u}{\partial t^2} = \eta_{tt}^t - \eta_{ss}^t. \end{aligned} \tag{7.39}$$

Recalling that

$$\mu_1 = \int_0^\infty \mu(s)ds,$$

and using hypotheses (1.33) and (7.39), we obtain

$$\begin{aligned} &\mu_1 \int_{\mathbb{R}^n} \Delta \frac{\partial^2 u}{\partial t^2}.\Delta v dx \\ &= \int_{\mathbb{R}^n} \left(\int_0^\infty \mu(s) \Delta \frac{\partial^2 u}{\partial t^2} ds \right).\Delta v dx \\ &= \int_{\mathbb{R}^n} \left(\int_0^\infty \mu(s)(\Delta \eta_{tt}^t - \Delta \eta_{ss}^t) ds \right).\Delta v dx \\ &= \int_{\mathbb{R}^n} \left(\int_0^\infty \mu(s) \Delta \eta_{tt}^t \right).\Delta v dx - \int_{\mathbb{R}^n} \left(\int_0^\infty \mu(s) \Delta \eta_{ss}^t \right) ds).\Delta v dx \\ &= \int_{\mathbb{R}^n} \left(\int_0^\infty \mu(s) \Delta \eta_{tt}^t \right).\Delta v dx + \mu(0) \int_{\mathbb{R}^n} \Delta \frac{\partial u}{\partial t} \Delta v dx - \int_{\mathbb{R}^n} \left(\int_0^\infty \mu''(s) \Delta \eta^t \right).\Delta v dx \\ &\leq \int_{\mathbb{R}^n} \left(\int_0^\infty \mu(s) \Delta \eta_{tt}^t \right).\Delta v dx + \mu(0) \int_{\mathbb{R}^n} \Delta \frac{\partial u}{\partial t} \Delta v dx - k_2 \int_{\mathbb{R}^n} \left(\int_0^\infty \mu(s) \Delta \eta^t \right).\Delta v dx \end{aligned} \tag{7.40}$$

Substituting (7.40) into (7.38) and using Poincaré and Young inequalities, we arrive at

$$\begin{aligned} \frac{d}{dt}\Phi(t) \leq &-\alpha \left\| \frac{\partial v}{\partial t} \right\|_{L_\rho^2}^2 + \frac{1}{2\varepsilon_2 \mu_1} \int_{\mathbb{R}^n} \left(\int_0^\infty \mu(s) \Delta \eta_{tt}^t ds \right)^2 dx \\ &+ \frac{\varepsilon_2}{2\mu_1} \int_{\mathbb{R}^n} |\Delta v|^2 dx + \frac{\alpha}{2} \int_{\mathbb{R}^n} \rho(x) \left| \frac{\partial^2 u}{\partial t^2} \right|^2 dx + \frac{\alpha C_0}{2} \int_{\mathbb{R}^n} |\Delta u|^2 dx \\ &+ \frac{k_2}{2\varepsilon_2 \mu_1} \int_{\mathbb{R}^n} \left(\int_0^\infty \mu(s) |\Delta \eta^t| ds \right)^2 dx + \frac{\varepsilon_2}{2\mu_1} \int_{\mathbb{R}^n} |\Delta v|^2 dx \end{aligned}$$

$$+\frac{k_2}{2\varepsilon_2}\int_{\mathbb{R}^n}\left(\int_0^\infty \mu(s)|\Delta\eta^t|ds\right)^2 dx+\frac{\mu(0)+1}{2}\varepsilon_2\int_{\mathbb{R}^n}|\Delta v|^2dx+\frac{\mu(0)}{2\varepsilon_2}\int_{\mathbb{R}^n}\left|\frac{\partial u}{\partial t}\right|^2 dx$$
$$+\frac{\mu_0}{2\varepsilon_2\mu_1}\int_{\mathbb{R}^n}\left(\int_0^\infty \mu(s)\Delta\eta^t_{tt}ds\right)^2 dx+\frac{\varepsilon_2\mu_0}{2\mu_1}\int_{\mathbb{R}^n}|\Delta v|^2dx$$
$$+\frac{k_2\mu_0}{2\varepsilon_2\mu_1}\int_{\mathbb{R}^n}\left(\int_0^\infty \mu(s)|\Delta\eta^t|ds\right)^2 dx+\frac{\varepsilon_2\mu_0}{2\mu_1}\int_{\mathbb{R}^n}|\Delta v|^2dx. \tag{7.41}$$

Noting that

$$\int_{\mathbb{R}^n}\left(\int_0^\infty \mu(s)\Delta\eta^t_{tt}\right)^2 dx\le\mu_1\int_{\mathbb{R}^n}\left(\int_0^\infty \mu(s)|\Delta\eta^t_{tt}|^2ds\right)dx \tag{7.42}$$

and

$$\int_{\mathbb{R}^n}\left(\int_0^\infty \mu(s)\Delta\eta^t_{t}\right)^2 dx\le\mu_1\int_{\mathbb{R}^n}\left(\int_0^\infty \mu(s)|\Delta\eta^t_{t}|^2ds\right)dx, \tag{7.43}$$

our conclusion follows. □

Finally, we define the functional

$$\Psi(t)=\int_{\mathbb{R}^n}\rho(x)\frac{\partial v}{\partial t}.vdx. \tag{7.44}$$

From $(7.1)_2$,

$$\frac{d}{dt}\Psi(t)=-\int_{\mathbb{R}^n}|\Delta v|^2dx+\int_{\mathbb{R}^n}\rho(x)\left|\frac{\partial v}{\partial t}\right|^2dx-\alpha\int_{\mathbb{R}^n}\rho(x)u.vdx. \tag{7.45}$$

Proof of Theorem 7.1. We define $\mathscr{L}$ as

$$\mathscr{L}(t)=N_1(E(t)+E_2(t)+E_3(t))+N_2\sum_{i=1}^{2}F_i(t)+N_3\sum_{i=1}^{2}Z_i(t)+N_4\Phi(t)+\Psi(t), \tag{7.46}$$

where N_1, N_2, N_3 and N_4 are positive constants. Then from Lemma 7.2–Lemma 7.5, we get

$$
\begin{aligned}
\frac{d}{dt}\mathscr{L}(t) \le & -\Lambda_1 \int_{\mathbb{R}^n}\left(\int_0^\infty \mu(s)|\Delta\eta^t|^2 ds\right)dx - \Lambda_1 \int_{\mathbb{R}^n}\left(\int_0^\infty \mu(s)|\Delta\eta_t^t|^2 ds\right)dx \\
& - \Lambda_1 \int_{\mathbb{R}^n}\left(\int_0^\infty \mu(s)|\Delta\eta_{tt}^t|^2 ds\right)dx - \Lambda_2 \int_{\mathbb{R}^n} |\Delta v|^2 dx \\
& - \left(\frac{\mu_1}{2}N_2 - N_3 - \frac{\alpha}{2}N_4\right)\int_{\mathbb{R}^n} \left|\frac{\partial u}{\partial t}\right|^2 dx - \left(\frac{\mu_1}{2}N_2 - N_3 - \frac{\alpha}{2}N_4\right)\int_{\mathbb{R}^n} \left|\frac{\partial^2 u}{\partial t^2}\right|^2 dx \\
& - \left(\alpha N_4 - \frac{\alpha\varepsilon_1}{2}N_2 - 1\right)\int_{\mathbb{R}^n} \left|\frac{\partial v}{\partial t}\right|^2 dx - \left(\frac{\mu_0}{2}N_3 - \frac{\mu_0\varepsilon_1}{2}N_2 - \frac{\alpha}{2}N_4\right)\int_{\mathbb{R}^n} |\Delta u|^2 dx \\
& - \left(\frac{\mu_0}{2}N_3 - \frac{\mu_0\varepsilon_1}{2}N_2 - \frac{\mu(0)}{2\varepsilon_2}N_4\right)\int_{\mathbb{R}^n} \left|\Delta\frac{\partial u}{\partial t}\right|^2 dx,
\end{aligned}
\tag{7.47}
$$

where

$$
\Lambda_1 = -\left(N_1\frac{k_1}{2} - \left[\left(\frac{\mu_0\mu_1}{2\varepsilon_1} + \mu_1 + \frac{\alpha\mu_1}{2\varepsilon_1} + \frac{k_0 C}{2}\right)N_2 + \frac{\mu_1}{2\mu_0}N_3 + \left(\frac{k_2}{2\varepsilon_2} + \frac{\mu_1}{2\varepsilon_2} + \frac{k_2\mu_0}{\varepsilon_2}\right)N_4\right]\right)
$$

and

$$
\Lambda_2 = \left(1 - \frac{\alpha C\varepsilon_1}{2}N_2 - \left(\frac{1}{\mu_1} + \frac{\mu(0)+1}{2} + \frac{\mu_0}{\mu_1}\right)\varepsilon_2 N_4\right).
$$

Now, choosing ε_1, ε_2 small enough such that

$$
0 < \varepsilon_1 < \varepsilon_2 < \frac{\mu_1}{2},
$$

and $N_4 > 0$ large enough, we get

$$
\alpha > \frac{\mu_0\mu_1(\mu_1 - 2\varepsilon_1)}{N_4(\mu_0\mu_1(\mu_1 - 2\varepsilon_1) - 2\varepsilon_1(\varepsilon_1(\mu_0 + \mu_1 C)))} > 0
$$

is small enough. Then taking

$$
\begin{aligned}
N_1 > & \frac{2}{k_1}\left[\left(\frac{\mu_0\mu_1 + \alpha\mu_1}{2\varepsilon_1} + \mu_1 + \frac{k_0 C}{2}\right) + \left(2\frac{(\varepsilon_1+\alpha)\mu_0\varepsilon_2 + \mu(0)\mu_1}{\mu_0\mu_1\varepsilon_2(\mu_1 - 2\varepsilon_1)} + \alpha\right)\right. \\
& \left. + \frac{\mu_1^2(\varepsilon_1\varepsilon_2\alpha + \mu(0)\mu_1)}{2\mu_0(\mu_1 - 2\varepsilon_1)} + \frac{k_2 + \mu_1 + 2k_2\mu_0}{2\varepsilon_2}\right]N_4, \\
N_2 > & \max\left\{\left(\frac{2(\varepsilon_1\mu_0 + \mu_1 C\varepsilon_1)}{\mu_0\mu_1(\mu_1 - 2\varepsilon_1)} + \alpha\right)N_4, \left(\frac{(\varepsilon_1+\alpha)\mu_0\varepsilon_2 + \mu(0)\mu_1}{\mu_0\mu_1(\mu_1 - 2\varepsilon_1)} + \alpha\right)N_4\right\}
\end{aligned}
$$

and

$$
N_3 > \max\left\{\left(\frac{(\varepsilon_1\mu_0 + \mu_1 C\varepsilon_1)}{\mu_0\mu_1(\mu_1 - 2\varepsilon_1)}\right)N_4, \left(\frac{\varepsilon_1\varepsilon_2\alpha\mu_0 + \mu(0)\mu_1}{\mu_0\mu_1(\mu_1 - 2\varepsilon_1)}\right)N_4\right\}
$$

we get that there is a positive constant $\omega > 0$ such that

$$\frac{d}{dt}\mathscr{L}(t) \leq -wE(t). \tag{7.48}$$

Therefore,

$$w\int_0^t E(s)ds \leq \mathscr{L}(0) - \mathscr{L}(t), \quad \forall t > 0. \tag{7.49}$$

On the other hand, it is not difficult to prove that there exists a constant $\zeta > 0$ such that

$$\mathscr{L}(0) - \mathscr{L}(t) \leq \frac{\zeta}{w}(E(0) + E_2(0) + E_3(0)), \quad \forall t > 0. \tag{7.50}$$

From (7.41) and (7.42), we obtain

$$\int_{\mathbb{R}^n} E(s)ds \leq \frac{\zeta}{w}(E(0) + E_2(0) + E_3(0)). \tag{7.51}$$

Finally, since

$$\frac{d}{dt}tE(t) = E(t) + t\frac{d}{dt}E(t) \leq E(t). \tag{7.52}$$

By (7.43), we have

$$E(t) \leq \frac{C}{t}(E(0) + E_2(0) + E_3(0)), \tag{7.53}$$

where

$$C = \frac{\zeta}{w}.$$

Finally, we obtain (7.13), which completes the proof. □

8 System of three wave equations

8.1 System of three wave equations with infinite memories

We consider the following system:

$$\begin{cases} \frac{\partial^2 u_1}{\partial t^2} - \phi(x)(\Delta u_1 + \int_0^\infty \mu_1(s)\Delta u_1(t-s)ds) + \alpha(u_2+u_3) = 0, \\ \frac{\partial^2 u_2}{\partial t^2} - \phi(x)(\Delta u_2 + \int_0^\infty \mu_2(s)\Delta u_2(t-s)ds) + \alpha(u_1+u_3) = 0, \\ \frac{\partial^2 u_3}{\partial t^2} - \phi(x)\Delta u_3 + \alpha(u_1+u_2) = 0, \end{cases} \tag{8.1}$$

where the weighted function $(\phi(x))^{-1} = \rho(x)$ satisfy (1.58) and the relaxation kernels μ_1, μ_2 satisfy (1.33), (1.36). The initial conditions are given by

$$\begin{cases} u_i(-t) = u_0(t), \quad \forall t \in \mathbb{R}^+, \quad i = \overline{1,2}, \\ u_i(0) = u_i^0 \in \mathcal{H}(\mathbb{R}^n), \quad \frac{\partial u_i}{\partial t}(0) = u_i^1 \in L_\rho^2(\mathbb{R}^n), \quad i = \overline{1,3}, \end{cases} \tag{8.2}$$

where the spaces $\mathcal{H}$ is defined as in Definition 1.3. We add a new variable η to the system, which corresponds to the relative displacement history. Let us define

$$\begin{aligned} \eta^1 &= \eta^1(x,s) = u_1(x,t) - u_1(x,t-s), \quad (x,s) \in \mathbb{R}^n \times \mathbb{R}^+, t \geq 0, \\ \eta^2 &= \eta^2(x,s) = u_2(x,t) - u_2(x,t-s), \quad (x,s) \in \mathbb{R}^n \times \mathbb{R}^+, t \geq 0, \end{aligned} \tag{8.3}$$

By differentiation, we have

$$\begin{aligned} \eta_t^1(x,s) &= -\eta_s^1(x,s) + \frac{\partial u_1}{\partial t}(x,t), \\ \eta_t^2(x,s) &= -\eta_s^2(x,s) + \frac{\partial u_2}{\partial t}(x,t), \end{aligned} \tag{8.4}$$

and we can take $(t = 0)$,

$$\eta_0^i(x,s) = u_i^0(x,0) - u_i^0(x,-s), \quad (x,s) \in \mathbb{R}^n \times \mathbb{R}^+, \ i = 1,2. \tag{8.5}$$

Thus, the original memories terms can be rewritten as

$$\int_0^\infty \mu_i(s)\phi(x)\Delta u_i(t-s)ds = \left(\int_0^\infty \mu_i(s)ds\right)\phi(x)\Delta u_i - \int_0^\infty \mu_i(s)\phi(x)\Delta\eta^i(s)ds.$$

The problem (8.1) is can be transformed into the next system

https://doi.org/10.1515/9783111633787-008

$$\begin{cases} \frac{\partial^2 u_1}{\partial t^2} - \mu_1^0 \phi(x)\Delta u_1 - \int_0^\infty \mu_1(s)\phi(x)\Delta\eta^1(s)ds + \alpha(u_2+u_3) = 0, & \mathbb{R}^n \times \mathbb{R}^+, \\ \frac{\partial^2 u_2}{\partial t^2} - \mu_2^0 \phi(x)\Delta u_2 - \int_0^\infty \mu_2(s)\phi(x)\Delta\eta^2(s)ds + \alpha(u_1+u_3) = 0, & \mathbb{R}^n \times \mathbb{R}^+, \\ \frac{\partial^2 u_3}{\partial t^2} - \phi(x)\Delta u_3 + \alpha(u_1+u_2) = 0, & \mathbb{R}^n \times \mathbb{R}^+, \\ \eta_t^1 + \eta_s^1 - \frac{\partial u_1}{\partial t} = 0, & \mathbb{R}^n \times \mathbb{R}^+, \\ \eta_t^2 + \eta_s^2 - \frac{\partial u_2}{\partial t} = 0, & \mathbb{R}^n \times \mathbb{R}^+, \end{cases} \tag{8.6}$$

and initial conditions

$$u_i(x,0) = u_0(x), \quad \frac{\partial u}{\partial t}_i(x,0) = u_{i1}(x), \quad \eta_i^0(x,s) = \eta_0^i(x,s), \tag{8.7}$$

where

$$\eta_0^i(x,s) = u_i^0(x,0) - u_i^0(x,-s), \quad (x,s) \in \mathbb{R}^n \times \mathbb{R}^+, \ i = 1,2.$$

The energy of the problem is defined as

$$\begin{aligned} E(t) = \frac{1}{2}\left(\sum_{i=1}^{3} \left\| \frac{\partial u_i}{\partial t} \right\|_{L_\rho^2}^2 + \sum_{i=1}^{2} (\mu_i^0 \|\nabla u_i\|_{L^2}^2 + \|\nabla \eta^i\|_{L_\mu^2}^2) + \|\nabla u_3\|_{L^2}^2 \right) \\ + \alpha \int_{\mathbb{R}^n} (u_1 u_2 + u_1.u_3 + u_2.u_3)dx. \end{aligned} \tag{8.8}$$

In order to deal with (8.1), we need to know more information for the operator $-\phi\Delta$. We consider

$$-\phi(x)\Delta u(x) = \eta(x), \tag{8.9}$$

without boundary conditions. For every u, v in $C_0^\infty(\mathbb{R}^n)$,

$$(-\phi\Delta u, v)_{L_\rho^2} = \int_{\mathbb{R}^n} \nabla u \nabla v dx, \tag{8.10}$$

and $L_\rho^2(\mathbb{R}^n)$ is defined as the closure of $C_0^\infty(\mathbb{R}^n)$. We may consider equation (8.9) as an operator equation, for any $\eta \in L_\rho^2(\mathbb{R}^n)$,

$$A_0 u = \eta, \quad A_0 : D(A_0) \subseteq L_\rho^2(\mathbb{R}^n) \to L_\rho^2(\mathbb{R}^n).$$

By (8.10), we have the operator $A_0 = -\phi\Delta$ is symmetric with domain of definition

$$D(A_0) = C_0^\infty(\mathbb{R}^n).$$

Because of the appearance of $\phi(x)$, the operator A_0 is not symmetric in the standard Lebesgue space $L^2(\mathbb{R}^n)$ [5, pp. 185–187]. We have

$$(A_0 u, u)_{L^2_\rho} \geq c\|u\|_{L^2_\rho}, \quad \text{for all } u \in D(A_0). \tag{8.11}$$

From (8.11), we conclude that A_0 is a symmetric, strongly monotone operator on $L^2_\rho(\mathbb{R}^n)$. Hence, the Friedrichs extension theorem[1] is applicable.

The energy extension

$$A_E = -\phi\Delta \text{ of } A_0,$$

namely

$$-\phi\Delta : \mathscr{H}(\mathbb{R}^n) \to \mathscr{H}^{-1}(\mathbb{R}^n), \tag{8.12}$$

is defined to be the duality mapping of $\mathscr{H}(\mathbb{R}^n)$. Define $D(A)$ to be the set of all solutions of equation (8.10), for arbitrary $\eta \in L^2_\rho(\mathbb{R}^n)$. The Friedrichs extension A of A_0 is the restriction of the energy extension A_E to the set $D(A)$. Its domain $D(A)$ is a Hilbert space with respect to the scalar product

$$(u, v)_{D(A)} = (u, v)_{L^2_\rho} + (Au, Av)_{L^2_\rho}, \quad \text{for all } u, v \in D(A).$$

The norm induced is

$$\|u\|_{D(A)} = \left\{\int_{\mathbb{R}^n} \rho|u|^2 dx + \int_{\mathbb{R}^n} \phi|\Delta u|^2 dx\right\}^{\frac{1}{2}}.$$

This norm is equivalent to

$$\|Au\|_{L^2_\rho} = \left\{\int_{\mathbb{R}^n} \phi|\Delta u|^2 dx\right\}^{\frac{1}{2}}.$$

For the eigenvalue problem,

$$-\phi(x)\Delta u = \mu u, \quad x \in \mathbb{R}^n, \tag{8.13}$$

there exists a complete system of eigensolutions $\{w_n, \mu_n\}$ with the following properties:

$$\begin{cases} -\phi\Delta w_j = \mu w_j, & j = 1, 2 \cdots, \quad w_j \in \mathscr{H}(\mathbb{R}^n), \\ 0 < \mu_1 \leq \mu_2 \leq \cdots, & \mu_j \to \infty, \quad \text{as } j \to \infty. \end{cases} \tag{8.14}$$

It can be shown that every solution of (8.13) is such that

---

1 E. Zeidler. Nonlinear Functional Analysis and its Applications, Vol. II, Monotone Operators. Springer-Verlag, Berlin, 1990.

$$u(x) \longrightarrow 0, \quad \text{as } |x| \longrightarrow \infty \tag{8.15}$$

uniformly with respect to x. We introduce the weighted L^2-spaces

$$L^2_\mu(\mathbb{R}^+; H^2(\mathbb{R}^n)) = \left\{ \xi : \mathbb{R}^+ \to H^2(\mathbb{R}^n) \,\middle|\, \int_0^\infty \mu(s) \|\xi(s)\|_V^2 ds < \infty \right\},$$

where

$$L^2_\mu(\mathbb{R}^+; H^2(\mathbb{R}^n))$$

is the Hilbert space endowed with the inner product

$$(\xi, \xi)_\mu = \int_0^\infty \mu(s)(\xi(s), \xi(s))_V ds,$$

and

$$\|\xi\|_\mu^2 = \int_0^\infty \mu(s)\|\xi\|^2 ds.$$

In order to consider the relative displacement η as a new variable, the following compact embedding is useful.

Lemma 8.1. *Let*

$$\rho \in L^{\frac{n}{2}}(\mathbb{R}^n) \cap L^\infty(\mathbb{R}^n).$$

Then the embedding $\mathscr{H} \subset L^2_\rho$ is compact.

Now, the definition of the weak solutions for (8.6) is given by the following.

Definition 8.1. A weak solution of the problem (8.6) is a function

$$(u_1, u_2, u_3, \eta^1, \eta^2),$$

such that:
(i) $u_i \in L^2[0, T; \mathscr{H}(\mathbb{R}^n)]$, $\frac{\partial u_i}{\partial t} \in L^2[0, T; L^2_\rho(\mathbb{R}^n)]$, $\frac{\partial^2 u_i}{\partial t^2} \in L^2[0, T; D^{-1,2}(\mathbb{R}^n)]$,
(ii) For all

$$\varphi, \phi, \chi \in C_0^\infty([0, T] \times \mathbb{R}^n),$$

then

$$(u_1, u_2, u_3, \eta^1, \eta^2)$$

satisfies the generalized formula

$$\int_0^T \left(\frac{\partial^2 u_1}{\partial t^2}(\tau), \varphi(\tau)\right)_{L^2_\rho} d\tau + \alpha \int_0^T ((u_2 + u_3)(\tau), \varphi(\tau))_{L^2_\rho} d\tau - \mu_1^0 \int_0^T \int_{\mathbb{R}^n} \nabla u_1(\tau) \nabla \varphi(\tau) dx d\tau$$
$$- \int_0^T \int_{\mathbb{R}^n} \int_0^{+\infty} \mu_1(s) \nabla \eta^1(s) \nabla \varphi(\tau) ds dx d\tau = 0, \tag{8.16}$$

$$\int_0^T \left(\frac{\partial^2 u_2}{\partial t^2}(\tau), \phi(\tau)\right)_{L^2_\rho} d\tau + \alpha \int_0^T (u_1 + u_3)(\tau), \phi(\tau))_{L^2_\rho} d\tau - \mu_2 \int_0^T \int_{\mathbb{R}^n} \nabla u_2(\tau) \nabla \phi(\tau) dx d\tau$$
$$- \int_0^T \int_{\mathbb{R}^n} \int_0^{+\infty} \mu_2(s) \nabla \eta^2(s) \nabla \varphi(\tau) ds dx d\tau = 0, \tag{8.17}$$

and

$$\int_0^T \left(\frac{\partial^2 u_3}{\partial t^2}(\tau), \chi(\tau)\right)_{L^2_\rho} d\tau + \alpha \int_0^T (u_1 + u_2)(\tau), \chi(\tau))_{L^2_\rho} d\tau - \int_0^T \int_{\mathbb{R}^n} \nabla u_3(\tau) \nabla \chi(\tau) dx d\tau = 0, \tag{8.18}$$

(iii) $(u_1, u_2, u_3, \eta^1, \eta^2)$ satisfies the initial conditions

$$u_i^0(x) \in \mathscr{H}(\mathbb{R}^n), \quad u_i^1(x) \in L^2_\rho(\mathbb{R}^n), \quad \eta_0^i \in L^2_\mu(\mathbb{R}^+, H^1(\mathbb{R}^n)).$$

To treat the viscoelastic term by considering a new variable η, we can reformulate our initial system in the following:

$$\begin{cases} \partial_t U = \mathscr{A} U, \\ U(x,0) = U^0(x) \end{cases} \tag{8.19}$$

where

$$U = \left(u_1, u_2, u_3, \frac{\partial u_1}{\partial t}, \frac{\partial u_2}{\partial t}, \frac{\partial u_3}{\partial t}, \eta^1, \eta^2\right)^T,$$

and

$$U^0 = (u_1^0, u_2^0, u_3^0, u_1^1, u_2^1, u_3^1, \eta_0^1, \eta_0^2)^T \in K.$$

Here,

$$K = (H^1(\mathbb{R}^n))^3 \times (H)^3 \times (L^2_\mu(\mathbb{R}^+; H^1(\mathbb{R}^n)))^2$$

and

$$L^2_\mu(\mathbb{R}^+;H^1(\mathbb{R}^n)) = \left\{u : \mathbb{R}^+ \longrightarrow H^1(\mathbb{R}^n); \int_0^\infty \mu(s) \int_{\mathbb{R}^n} |\nabla u|^2 dxds < +\infty\right\}$$

and $\mathscr{A}$ is a linear operator given by

$$\mathscr{A}U = \begin{pmatrix} w_4 \\ w_5 \\ w_6 \\ -\mu_1^0\phi\Delta w_1 - aw_2 - aw_3 - \int_0^\infty \mu_1(s)\phi\Delta w_7 ds \\ -aw_1 - \mu_1^0\phi\Delta w_2 - aw_3 - \int_0^\infty \mu_1(s)\phi\Delta w_8 ds \\ -aw_1 - aw_2 - \phi\Delta w_3 \\ w_4 - \partial_s w_7 \\ w_5 - \partial_s w_8 \end{pmatrix}, \tag{8.20}$$

with domain $D(\mathscr{A})$ given by

$$D(\mathscr{A}) = \left\{\begin{matrix}(u_1, u_2, u_3, v_1, v_2, v_3, \eta_1, \eta_2) \in K; \mu_i^0 u_i + \int_0^\infty \mu_i(s)\eta_i^t(s)ds \in H^1 \cap H^2 \\ (\eta^i, \eta_s^i) \in (L^2_{\mu_i}(\mathbb{R}^+;H^1)^2/_{i=1,2}, u_j \in H^1 \cap H^2(\mathbb{R}^n), v_j \in H^1/_{j=\overline{1,3}}, \ \eta^i(0) = 0\end{matrix}\right\}$$

endowed with the inner product

$$(u, v)_{L^2_\mu(\mathbb{R}^+;H^1(\mathbb{R}^n))} = \int_0^\infty \mu(s) \int_{\mathbb{R}^n} \nabla u.\nabla v dxds,$$

and

$$\begin{aligned}\langle U, V\rangle_K = {} & \sum_{i=1}^2 \mu_i^0\langle\nabla w_i, \nabla v_i\rangle + \langle\nabla w_3, \nabla v_3\rangle + a \sum_{i,j,l=1_{i\neq j\neq l}}^3 \langle w_i, v_j + v_l\rangle_{L^2_p} \\ & + \sum_{r=4}^6 \langle w_r, v_r\rangle_{L^2_p} + \sum_{p=7}^8 \langle w_p, v_p\rangle_{L^2_\mu(\mathbb{R}^+;H^1(\mathbb{R}^n))}.\end{aligned} \tag{8.21}$$

Proposition 8.1. *For every $U \in D(\mathscr{A})$, the operator $\mathscr{A}$ is dissipative, and* $\mathrm{Id} - \mathscr{A}$ *is surjective.*

Proof. First, we are going to prove that $\mathscr{A}$ is dissipative:

$$\begin{aligned}\langle\mathscr{A}U, U\rangle = {} & \mu_1^0\langle\phi\nabla w_4, \nabla w_1\rangle + \mu_1^0\langle\phi\nabla w_5, \nabla w_2\rangle \\ & + \langle\nabla w_6, \nabla w_3\rangle + a(\langle w_4, w_2 + w_3\rangle + \langle w_5, w_1 + w_3\rangle + \langle w_6, w_1 + w_2\rangle) \\ & + \langle -\mu_1^0\phi\Delta w_1 - aw_2 - aw_3 - T_1 w_7, w_4\rangle + \langle -aw_1 - aw_2 - \phi\Delta w_3, w_6\rangle \\ & + \langle -aw_1 - \mu_2^0\phi\Delta w_2 - aw_3 - T_2 w_8, w_5\rangle \\ & + \langle w_4 - \partial_s w_7, w_7\rangle_{L^2_\mu(\mathbb{R}^+;H^1(\mathbb{R}^n))} + \langle w_5 - \partial_s w_8, w_8\rangle_{L^2_\mu(\mathbb{R}^+;H^1(\mathbb{R}^n))}\end{aligned}$$

$$= -\langle T_1 w_7, w_4\rangle + \langle w_4, w_7\rangle_{L^2_\mu(\mathbb{R}^+;H^1(\mathbb{R}^n))} - \langle T_2 w_8, w_5\rangle + \langle w_5, w_8\rangle_{L^2_\mu(\mathbb{R}^+;H^1_0(\mathbb{R}^n))}$$
$$- \langle \partial_s w_7, w_7\rangle_{L^2_\mu(\mathbb{R}^+;H^1(\mathbb{R}^n))} - \langle \partial_s w_8, w_8\rangle_{L^2_\mu(\mathbb{R}^+;H^1(\mathbb{R}^n))}, \tag{8.22}$$

where

$$T_i = \int_0^\infty \mu_i(s)\Delta\eta ds.$$

From the definition of T_i, we have

$$-\langle T_1 w_7, w_4\rangle + \langle w_4, w_7\rangle_{L^2_\mu(\mathbb{R}^+;H^1(\mathbb{R}^n))} - \langle T_2 w_8, w_5\rangle + \langle w_5, w_8\rangle_{L^2_\mu(\mathbb{R}^+;H^1(\mathbb{R}^n))} = 0.$$

Then using integrating by part of the last two terms in (8.22) and using the fact that

$$\lim_{s\to\infty} \mu_i(s)\nabla\eta_i(s)ds = 0$$

and

$$\eta_i(0) = 0,$$

we obtain

$$\begin{aligned} -\langle \partial_s w_i, w_i\rangle_{L^2_{\mu_i}(\mathbb{R}^+;H^1(\mathbb{R}^n))} &= \int_0^\infty \mu_i(s)\partial_s w_i.w_i ds \\ &= \int_0^\infty \mu_i'(s)\|w_i\|^2 ds - \int_0^\infty \mu_i(s)\left\langle \frac{\partial w_i}{\partial t}, w_i\right\rangle ds \\ &= \frac{1}{2}\int_0^\infty \mu_i'(s)\|w_i\|^2 ds \le 0. \end{aligned} \tag{8.23}$$

Thus, $\mathscr{A}$ is a dissipative operator.

Second, we should prove that $\mathrm{Id} - \mathscr{A}$ is surjective. For this, let us take

$$F = (f_1, f_2, f_3, f_4, f_5, f_6, f_7, f_8)^T \in K.$$

There exist

$$W(w_1, w_2, w_3, w_4, w_5, w_6, w_7, w_8) \in D(\mathscr{A})$$

satisfying the following equation:

$$(\mathrm{Id} - \mathscr{A})W = F. \tag{8.24}$$

This is equivalent to

$$\begin{cases} w_1 - w_4 = f_1, \\ w_2 - w_5 = f_2, \\ w_3 - w_6 = f_3, \\ w_4 + \mu_1^0 \phi \Delta w_1 - \alpha w_2 + \alpha w_3 + T_1 w_7 = f_4, \\ w_5 + \alpha w_1 + \mu_2^0 \phi \Delta w_2 + \alpha w_3 + T_2 w_8 = f_5, \\ w_6 + \alpha w_1 + \alpha w_2 + \mu_1^0 \phi \Delta w_3 = f_6, \\ w_7 - w_4 + \partial_s w_7 = f_7, \\ w_8 - w_5 + \partial_s w_8 = f_8. \end{cases} \tag{8.25}$$

We note that the first, second, and third equations in (8.25) give

$$w_4 = w_1 - f_1, \quad w_5 = w_2 - f_2 \quad \text{and} \quad w_6 = w_3 - f_3. \tag{8.26}$$

With $w_i(0) = 0$, $i = 7, 8$, the last two equations in (8.25) have a unique solution given by

$$w_i(\tau) = \left(\int_0^\tau (f_i(z) + w_{i-6} - f_{i-6}) e^z \, dz \right) e^{-\tau}, \quad i = 7, 8. \tag{8.27}$$

On the other hand, plugging (8.26) and (8.27) into the fourth, fifth and sixth equations in (8.25), we get

$$\begin{cases} ((1 - \mu_1^1)\Phi\Delta + \mathrm{Id}) w_1 + \alpha w_2 + \alpha w_3 = \tilde{f}_1, \\ ((1 - \mu_2^1)\Phi\Delta + \mathrm{Id}) w_2 + \alpha w_1 + \alpha w_3 = \tilde{f}_2, \\ (\mu_1^0 \Phi\Delta + \mathrm{Id}) w_3 + \alpha w_1 + \alpha w_2 = f_6 + f_3 \end{cases} \tag{8.28}$$

where

$$\tilde{f}_i = f_i + f_{i+3} + \int_0^\infty \mu_i(\tau) e^{-\tau} \int_0^\tau e^z \Phi\Delta(f_i - f_{i+6}(z)) \, dz \, d\tau,$$

and

$$\mu_i^1 = \int_0^\infty \mu_i(\tau) e^{-\tau} \, d\tau.$$

Then we will define the operator

$$\mathscr{L} : (H^1(\mathbb{R}^n) \cap H^2(\mathbb{R}^n))^3 \longrightarrow H \times H \times H,$$

by

$$\mathscr{L} = \begin{pmatrix} (1-\mu_1^1)\Phi\Delta + \mathrm{Id} & \alpha.\mathrm{Id} & \alpha.\mathrm{Id} \\ \alpha.\mathrm{Id} & (1-\mu_2^1)\Phi\Delta + \mathrm{Id} & \alpha.I \\ \alpha.I & \alpha.I & \mu_1^0\Phi\Delta + \mathrm{Id} \end{pmatrix}. \tag{8.29}$$

This operator is linear, continued and coercive (is self-adjoint linear positive definite). Then system (8.28) has a unique solution

$$(w_1, w_2, w_3)^T \in (H^1(\mathbb{R}^n) \cap H^2(\mathbb{R}^n))^3,$$

by applying the Lax–Milgram theorem and other arguments,. Hence, $\mathrm{Id}-\mathscr{A}_1$ is surjective. Finally, $-\mathscr{A}_1$ is a maximal monotone operator. Therefore, using the Lummer–Phillips theorem, we can deduce that $\mathscr{A}_1$ is the infinitesimal generator of a linear contraction C_0-semigroup on H, and then the result of well-posedness is ensured by the semigroups theory. □

8.1.1 Polynomial stability

Theorem 8.1. *Let* (1.33), (1.36) *hold. Assume that*

$$u_i^0 \in K(\mathbb{R}^n),$$
$$u_i^1 \in L_\rho^2(\mathbb{R}^n),$$

and

$$\eta_0^i \in L_\mu^2(\mathbb{R}^+, H^1(\mathbb{R}^n)).$$

Then the solution of (8.6), (8.7) *satisfies*

$$E(t) \leq Ct^{-1}(E(0) + E_2(0) + E_3(0)), \quad \forall t \in \mathbb{R}^+, \tag{8.30}$$

for certain positive constant C.

Now, we carry out the proof of Theorem 8.1. Let us introduce the second- and third-order energy

$$E_2(t) = E\left(\frac{\partial u_1}{\partial t}, \frac{\partial u_2}{\partial t}, \frac{\partial u_3}{\partial t}, \eta_t^1, \eta_t^2\right),$$

and

$$E_3(t) = E\left(\frac{\partial^2 u_1}{\partial t^2}, \frac{\partial^2 u_2}{\partial t^2}, \frac{\partial^2 u_3}{\partial t^2}, \eta_{tt}^1, \eta_{tt}^2\right).$$

Lemma 8.2. *Let*

$$(u_1, u_2, u_3, \eta^1, \eta^2)$$

be the solution of (8.6). *Then the following energy estimate holds for any* $t \geq 0$*:*

$$\frac{d}{dt}E(t) = \frac{1}{2}\sum_{i=1}^{2}\int_0^{+\infty} \mu_i'(s)\|\nabla\eta^i\|_2^2 ds \leq 0, \quad \forall t \geq 0, \tag{8.31}$$

$$\frac{d}{dt}E_2(t) = \frac{1}{2}\sum_{i=1}^{2}\int_0^{+\infty} \mu_i'(s)\|\nabla\eta_t^i\|_2^2 ds \leq 0, \quad \forall t \geq 0, \tag{8.32}$$

$$\frac{d}{dt}E_3(t) = \frac{1}{2}\sum_{i=1}^{2}\int_0^{+\infty} \mu_i'(s)\|\nabla\eta_{tt}^i\|_2^2 ds \leq 0, \quad \forall t \geq 0. \tag{8.33}$$

Proof. Multiplying $(8.6)_1$ by $\rho\frac{\partial u_1}{\partial t}$, and $(8.6)_2$ by $\rho\frac{\partial u_2}{\partial t}$ and $(8.6)_3$ by $\rho\frac{\partial u_3}{\partial t}$, summing the results obtained follows the conclusion of inequality (8.31)–(8.33). □

Lemma 8.3. *Let*

$$(u_1, u_2, u_3, \eta^1, \eta^2)$$

be the solution of (8.1). *Then the functional* F_1 *and* F_2 *defined by*

$$\begin{aligned} F_1(t) &= -\sum_{i=1}^{2}\int_{\mathbb{R}^n} \rho(x)\frac{\partial u_i}{\partial t}\left(\int_0^{+\infty} \mu_i(s)\eta^i(s)ds\right)dx \\ &= F_1^1(t) + F_1^2(t). \end{aligned} \tag{8.34}$$

For any $\varepsilon_1, \varepsilon_2 > 0$, *there exists a positive constant*

$$K_{\varepsilon_1}^i = \frac{\mu_i^0}{2\varepsilon_1} + \mu_i^1 + \frac{\alpha\mu_i^1}{2\varepsilon_1} > 0,$$

such that

$$\begin{aligned} \frac{d}{dt}F_1(t) \leq &-\sum_{i=1}^{2}\left(\mu_i^1 - \frac{\epsilon_1}{2}\right)\left\|\frac{\partial u_i}{\partial t}\right\|_{L_\rho^2}^2 + \sum_{i=1}^{2}\frac{(1-\mu_i^0)\mu_i^0\epsilon_1}{2}\|\nabla u_i(t)\|^2 + c\alpha\epsilon_1\|\nabla u_3(t)\|^2 \\ &+ K_{\varepsilon_1}^1\sum_{i=1}^{2}\int_0^{+\infty} \mu_i(s)\|\nabla\eta^i(t,s)\|^2 ds, \end{aligned} \tag{8.35}$$

where

$$\mu_i^1 = \int_0^\infty \mu_i(s)ds.$$

Proof. Multiplying $(8.1)_1$ by

$$\int_0^{+\infty} \rho(x)\mu_1(s)\eta^1(t,s)ds,$$

to get

$$\begin{aligned}
&\int_{\mathbb{R}^n} \frac{\partial^2 u_1}{\partial t^2}\left(\int_0^\infty \rho(x)\mu_1(s)\eta^1(t,s)ds\right)dx - \int_{\mathbb{R}^n}(1+\mu_1)\nabla u_1\left(\int_0^\infty \mu_1(s)\nabla\eta^1(t,s)ds\right)dx \\
&+\int_{\mathbb{R}^n} \mu_1\nabla\eta^1(t,s)ds\left(\int_0^\infty \mu_1(s)\nabla\eta^1(t,s)ds\right)dx \qquad (8.36)\\
&+\alpha\int_{\mathbb{R}^n}(u_2+u_3)\left(\int_0^\infty \rho(x)\mu_1(s)\eta^1(t,s)ds\right)dx = 0.
\end{aligned}$$

On the other hand, by using the fact that

$$\eta_t^1(t,s) = \frac{\partial u_1}{\partial t}(t) - \eta_s^1(t,s),$$

we find

$$\begin{aligned}
&\int_{\mathbb{R}^n} \frac{\partial^2 u_1}{\partial t^2}(t)\left(\int_0^\infty \rho(x)\mu_1(s)\eta^1(t,s)ds\right)dx \\
&= \partial_t\int_{\mathbb{R}^n} \frac{\partial u_1}{\partial t}(t)\left(\int_0^\infty \rho(x)\mu_1(s)\eta_1(t,s)ds\right)dx \\
&- \int_{\mathbb{R}^n} \frac{\partial u_1}{\partial t}(t)\left(\int_0^\infty \rho(x)\mu_1(s)\eta_t^1(t,s)ds\right)dx \qquad (8.37)\\
&= -\frac{d}{dt}F_1^1(t) - \mu_1\left\|\frac{\partial u_1}{\partial t}(t)\right\|_{L^2_\rho} \\
&+ \int_{\mathbb{R}^n} \frac{\partial u_1}{\partial t}(t)\left(\int_0^\infty \rho(x)\mu_1(s)\eta_s^1(t,s)ds\right)dx.
\end{aligned}$$

Integrating by parts with respect to s in the infinite memory integral, and using the fact that

$$\lim_{s\longrightarrow\infty}\mu_1(s)\eta^1(t,s)=0,$$

and

$$\eta^1(t,0)=0,$$

we get

$$\begin{aligned}\int_{\mathbb{R}^n}\frac{\partial^2u_1}{\partial t^2}(t)\left(\int_0^{\infty}\rho(x)\mu_1(s)\eta^1(t,s)ds\right)dx &= -\frac{d}{dt}F_1^1(t)-\mu_1\left\|\frac{\partial u_1}{\partial t}(t)\right\|_{L^2_\rho}\\ &\quad-\int_{\mathbb{R}^n}\frac{\partial u_1}{\partial t}\left(\int_0^{\infty}\rho(x)\mu_1'(s).\eta^1(t,s)ds\right)dx.\end{aligned}\tag{8.38}$$

Exploiting (8.36) and (8.38), we deduce

$$\begin{aligned}\frac{d}{dt}F_1^1(t) &= -\mu_1\left\|\frac{\partial u_1}{\partial t}\right\|_{L^2_\rho}-\int_{\mathbb{R}^n}\frac{\partial u_1}{\partial t}\left(\int_0^{\infty}\rho(x)\mu_1'(s).\eta^1(t,s)ds\right)dx\\ &\quad+\int_{\mathbb{R}^n}(1-\mu_1)\nabla u_1\left(\int_0^{\infty}\mu_1(s)\nabla\eta^1(t,s)ds\right)dx\\ &\quad+\int_{\mathbb{R}^n}\int_0^{\infty}\mu_1(s)\nabla\eta^1(t,s)ds\left(\int_0^{\infty}\mu_1(s)\nabla\eta^1(t,s)ds\right)dx\\ &\quad+\alpha\int_{\mathbb{R}^n}(u_2+u_3)\left(\int_0^{\infty}\rho(x)\mu_1(s)\eta^1(t,s)ds\right)dx.\end{aligned}$$

Using Cauchy–Schwarz and Young's inequalities for the last foure terms of this equality, we get

$$\begin{aligned}\frac{d}{dt}F_1^1(t) &\leq -\left(\mu_1-\frac{\epsilon_1}{2}\right)\left\|\frac{\partial u_1}{\partial t}\right\|^2_{L^2_\rho}+\frac{\alpha\epsilon_1}{2}\|u_2\|^2_{L^2_\rho}+\frac{(1-\mu_1)\mu_1\epsilon_1}{2}\|\nabla u_1\|^2+\frac{\alpha\epsilon_1}{2}\|\nabla u_3\|^2\\ &\quad+K_{\epsilon_1}\int_0^{+\infty}\mu_1(s)\|\nabla\eta^1(t,s)\|^2_{L^2_\rho}ds.\end{aligned}$$

Similarly, $(8.1)_2$ by

$$\int_0^{+\infty} \rho(x)\mu_2(s)\eta^2(t,s)ds,$$

we have

$$\frac{d}{dt}F_1^2(t) \leq -\left(\mu_2 - \frac{\epsilon_1}{2}\right)\left\|\frac{\partial u_2}{\partial t}\right\|_{L_\rho^2}^2 + \frac{\alpha\epsilon_1}{2}\|u_1\|_{L_\rho^2}^2 + \frac{\mu_2\epsilon_1}{2}\|\nabla u_2(t)\|^2 + \frac{\alpha\epsilon_1}{2}\|u_3\|_{L_\rho^2}^2$$
$$+ K_{\epsilon_1}^2 \int_0^{+\infty} \mu_2(s)\|\nabla\eta^2(t,s)\|_{L_\rho^2}^2 ds.$$

Summing the two above identities, we have a conclusion (8.35). □

Let us denote by $F_2(t)$ the following functional:

$$F_2(t) = F_1\left(\frac{\partial u_1}{\partial t}, \frac{\partial u_2}{\partial t}, \frac{\partial u_3}{\partial t}, \eta_t^1, \eta_t^2\right). \tag{8.39}$$

Lemma 8.4. *Define the functional*

$$Z_1(t) = Z_1^1(t) + Z_1^2(t) = \sum_{i=1}^{2} \int_{\mathbb{R}^n} \rho(x)\frac{\partial u_i}{\partial t}.u_i dx. \tag{8.40}$$

Then, for any $\epsilon_3, \epsilon_4 > 0$*, there exist* $K_{\epsilon_3}, K_{\epsilon_4} > 0$ *such that*

$$\begin{aligned}\frac{d}{dt}Z_1(t) \leq &\sum_{i=1}^{2}\left\|\frac{\partial u_i}{\partial t}\right\|_{L_\rho^2}^2 - \sum_{i=1}^{2}\left(\mu_i^0 - \frac{\epsilon_3}{2}\right)\|\nabla u_i\|^2 + c\alpha\|\nabla u_3\|^2\\ &+ \int_0^{\infty} K_{\epsilon_3}\sum_{i=1}^{2}\mu_i(s)\|\nabla\eta^i(t,s)\|^2 ds,\end{aligned} \tag{8.41}$$

where

$$K_{\epsilon_3} = \frac{\mu_i^1}{2\epsilon_3}.$$

Proof. Multiplying the first equation of (8.6) by ρu_1, we find

$$\begin{aligned}&\frac{d}{dt}\left(\int_{\mathbb{R}^n} \frac{\partial u_1}{\partial t}.\rho u_1 dx\right) - \left\|\frac{\partial u_1}{\partial t}\right\|_{L_\rho^2}^2 + \mu_1^0\|\nabla u_1\|^2 + \int_{\mathbb{R}^n}\left(\int_0^{\infty}\mu_1(s)\nabla\eta^1(t,s)ds\right).\nabla u_1 dx\\ &+ \alpha\int_{\mathbb{R}^n}(u_2 + u_3).\rho u_1 dx = 0.\end{aligned} \tag{8.42}$$

By using Cauchy–Schwarz, Young's inequalities and Poincaré's inequality, we find

$$\frac{d}{dt}Z_1^1(t) = \left\|\frac{\partial u_1}{\partial t}\right\|_{L_\rho^2}^2 - \mu_1^0\|\nabla u_1\|^2 - \int_{\mathbb{R}^n}\left(\int_0^\infty \mu_1(s)\nabla\eta^1(s)ds\right).\nabla u_1 dx - \alpha\int_{\mathbb{R}^n}(u_2+u_3).\rho u_1 dx$$
$$\leq \left\|\frac{\partial u_1}{\partial t}\right\|_{L_\rho^2}^2 - \left(\mu_1^0 - \frac{\epsilon_3}{2} - \frac{c\alpha}{2}\right)\|\nabla u_1\|^2 + \frac{c\alpha}{2}(\|\nabla u_2\|^2 + \|\nabla u_3\|^2)$$
$$+ \frac{\mu_1}{2\epsilon_3}\int_0^\infty \mu_1(s)\|\nabla\eta^1(t,s)\|^2 ds. \tag{8.43}$$

Similarly, multiplying the second equation of (8.6) by ρu_2, we obtain

$$\frac{d}{dt}Z_1^2(t) \leq \left\|\frac{\partial u_2}{\partial t}\right\|_{L_\rho^2}^2 - \left(\mu_2^0 - \frac{\epsilon_3}{2} - \frac{c\alpha}{2}\right)\|\nabla u_2\|^2 + \frac{c\alpha}{2}(\|\nabla u_1\|^2 + \|\nabla u_3\|^2)$$
$$+ \frac{\mu_2}{2\epsilon_3}\int_0^\infty \mu_2(s)\|\nabla\eta^2(t,s)\|^2 ds.$$

Let us denote by $Z_2(t)$ the following functional:

$$Z_2(t) = Z_1\left(\frac{\partial u_1}{\partial t}, \frac{\partial u_2}{\partial t}, \frac{\partial u_3}{\partial t}, \eta_t^1, \eta_t^2\right). \tag{8.44}$$

□

Lemma 8.5 shows the dissipative property of u_3. Before this, let us consider the following functional:

$$R_2(t) = \sum_{i=1}^{2}\left(\int_{\mathbb{R}^n}\rho(x)\frac{\partial^2 u_i}{\partial t^2}.\frac{\partial u_3}{\partial t}dx + \mu_i^0\int_{\mathbb{R}^n}\nabla\frac{\partial u_i}{\partial t}.\nabla u_3 dx + \int_{\mathbb{R}^n}\left(\int_0^\infty \mu_i(s)\nabla\eta_i'(t,s).\nabla u_3(t)ds\right)dx\right)$$
$$= R_2^1(t) + R_2^2(t) \tag{8.45}$$

Lemma 8.5. *For any $\varepsilon_2 > 0$, we have*

$$\frac{d}{dt}R_2(t) \leq -\alpha\left\|\frac{\partial u_3}{\partial t}\right\|_{L_\rho^2}^2 + \sum_{i=1}^{2}\left(\frac{1}{2\varepsilon_2} + \frac{\mu_i^0}{2\varepsilon_2}\right)\int_{\mathbb{R}^n}\left(\int_0^\infty \mu_i(s)|\nabla\eta_{tt}^i|^2 ds\right)dx$$
$$+ \sum_{i=1}^{2}\left(\frac{k_2}{2\varepsilon_2} + \frac{\mu_i^1}{2\varepsilon_2} + \frac{k_2\mu_i^0}{2\varepsilon_2}\right)\int_{\mathbb{R}^n}\left(\int_0^\infty \mu_i(s)|\nabla\eta^i|^2 ds\right)dx + \sum_{i=1}^{2}\frac{\mu_i(0)}{2\varepsilon_2}\left\|\frac{\partial u_i}{\partial t}\right\|^2 dx$$
$$+ \sum_{i=1}^{2}\left(\frac{1}{\mu_i^1} + \frac{\mu_i(0)+1}{2} + \frac{\mu_i^1}{\mu_i^0}\right)\varepsilon_2\|\nabla u_3\|_2^2 + \frac{\alpha}{2}\sum_{i=1}^{2}\left\|\frac{\partial^2 u_i}{\partial t^2}\right\|_{L_\rho^2}^2 dx + \frac{\alpha C_0}{2}\sum_{i=1}^{2}\|\nabla u_i\|_2^2.$$

Proof. Differentiating $R_2^1(t)$ with respect to t and using of $(8.1)_1$, we obtain

$$\begin{aligned}\frac{d}{dt}R_2^1(t) &= -\alpha\int_{\mathbb{R}^n}\rho(x)\left|\frac{\partial u_3}{\partial t}\right|^2 dx - \int_{\mathbb{R}^n}\nabla\frac{\partial^2 u_1}{\partial t^2}.\nabla u_3 dx\\ &\quad -\alpha\int_{\mathbb{R}^n}\rho(x)\left(\frac{\partial u_2}{\partial t}.\frac{\partial u_3}{\partial t} + u_1.\frac{\partial^2 u_1}{\partial t^2} + u_2.\frac{\partial^2 u_1}{\partial t^2}\right)dx\\ &\quad + \int_{\mathbb{R}^n}\left(\int_0^\infty \mu_1(s)\nabla\eta_t^1 ds\right).\nabla u_3 dx + \mu_1^0\int_{\mathbb{R}^n}\nabla\frac{\partial^2 u_1}{\partial t^2}.\nabla u_3 dx.\end{aligned}\tag{8.46}$$

On the other hand, from (8.3) we deduce

$$\begin{aligned}&\eta_t^t(0) = 0, \quad \eta_s^t(0) = \frac{\partial u}{\partial t}(t),\\ &\frac{\partial^2 u}{\partial t^2} = \eta_{tt}^t - \eta_{ss}^t.\end{aligned}\tag{8.47}$$

Recalling that

$$\mu_1^1 = \int_0^\infty \mu_1(s)ds,$$

and using (8.47), we obtain

$$\begin{aligned}\mu_1^1\int_{\mathbb{R}^n}\nabla\frac{\partial^2 u_1}{\partial t^2}.\nabla u_3 dx &= \int_{\mathbb{R}^n}\left(\int_0^\infty \mu_1(s)\nabla\frac{\partial^2 u_1}{\partial t^2} ds\right).\nabla u_3 dx\\ &= \int_{\mathbb{R}^n}\left(\int_0^\infty \mu_1(s)\nabla\eta_{tt}^1\right).\nabla u_3 dx + \mu_1(0)\int_{\mathbb{R}^n}\nabla\frac{\partial u}{\partial t}\nabla u_3 dx\\ &\quad - \int_{\mathbb{R}^n}\left(\int_0^\infty \mu_1'(s)\nabla\eta^1\right).\nabla u_3 dx\\ &\le \int_{\mathbb{R}^n}\left(\int_0^\infty \mu_1(s)\nabla\eta_{tt}^1\right).\nabla u_3 dx + \mu_1(0)\int_{\mathbb{R}^n}\nabla\frac{\partial u_1}{\partial t}\nabla u_3 dx\\ &\quad - k_2\int_{\mathbb{R}^n}\left(\int_0^\infty \mu_1(s)\nabla\eta^1 ds\right).\nabla u_3 dx.\end{aligned}\tag{8.48}$$

Substituting (8.48) into (8.46) and using Poincaré and Young inequalities, we arrive at

$$\frac{d}{dt}R_2^1(t) \le -\alpha\left\|\frac{\partial u_3}{\partial t}\right\|_{L_\rho^2}^2 + \frac{1}{2\varepsilon_2\mu_1^1}\int_{\mathbb{R}^n}\left(\int_0^\infty \mu_1(s)\nabla\eta_{tt}^1 ds\right)^2 dx$$

$$
\begin{aligned}
&+\frac{\varepsilon_2}{2\mu_1^1}\int_{\mathbb{R}^n}|\nabla u_3|^2dx+\frac{\alpha}{2}\int_{\mathbb{R}^n}\rho(x)\left|\frac{\partial^2 u_1}{\partial t^2}\right|^2dx+\frac{\alpha C_0}{2}\int_{\mathbb{R}^n}|\nabla u_1|^2dx\\
&+\frac{k_2}{2\varepsilon_2\mu_1^1}\int_{\mathbb{R}^n}\left(\int_0^\infty\mu_1(s)|\nabla\eta^1|ds\right)^2dx+\frac{\varepsilon_2}{2\mu_1^1}\int_{\mathbb{R}^n}|\nabla u_3|^2dx\\
&+\frac{k_2}{2\varepsilon_2}\int_{\mathbb{R}^n}\left(\int_0^\infty\mu_1(s)|\nabla\eta^1|ds\right)^2dx+\frac{\mu_1(0)+1}{2}\varepsilon_2\int_{\mathbb{R}^n}|\nabla u_3|^2dx+\frac{\mu_1(0)}{2\varepsilon_2}\int_{\mathbb{R}^n}\left|\frac{\partial u_1}{\partial t}\right|^2dx\\
&+\frac{\mu_1^0}{2\varepsilon_2\mu_1^1}\int_{\mathbb{R}^n}\left(\int_0^\infty\mu_1(s)\nabla\eta_{tt}^1ds\right)^2dx+\frac{\varepsilon_2\mu_1^0}{2\mu_1^1}\int_{\mathbb{R}^n}|\nabla u_3|^2dx\\
&+\frac{k_2\mu_1^0}{2\varepsilon_2\mu_1^1}\int_{\mathbb{R}^n}\left(\int_0^\infty\mu_1(s)|\nabla\eta^1|ds\right)^2dx+\frac{\varepsilon_2\mu_1^0}{2\mu_1^1}\int_{\mathbb{R}^n}|\nabla u_3|^2dx. \qquad (8.49)
\end{aligned}
$$

Noting that

$$\int_{\mathbb{R}^n}\left(\int_0^\infty\mu_1(s)\nabla\eta_{tt}^1\right)^2dx\le\mu_1^1\int_{\mathbb{R}^n}\left(\int_0^\infty\mu_1(s)|\nabla\eta_{tt}^1|^2ds\right)dx \qquad (8.50)$$

and

$$\int_{\mathbb{R}^n}\left(\int_0^\infty\mu_1(s)\nabla\eta_{t}^1\right)^2dx\le\mu_1^1\int_{\mathbb{R}^n}\left(\int_0^\infty\mu_1(s)|\nabla\eta_{t}^1|^2ds\right)dx, \qquad (8.51)$$

our conclusion follows. □

Finally, we define the functional

$$Z_3(t)=\int_{\mathbb{R}^n}\rho(x)\frac{\partial u}{\partial t}{}_3.u_3dx. \qquad (8.52)$$

From the third equation, we get

$$\frac{d}{dt}Z_3(t)\le\left\|\frac{\partial u_3}{\partial t}\right\|_{L_\rho^2}^2+\frac{c.\alpha}{2}\sum_{i=1}^2\|\nabla u_i\|^2-(1-c\alpha)\|\nabla u_3\|^2. \qquad (8.53)$$

Proof of Theorem 10.1. We define $\mathcal{L}$ as

$$\mathcal{L}(t)=N_1(E(t)+E_2(t)+E_3(t))+N_2\sum_{i=1}^2F_i(t)+N_3\sum_{i=1}^2Z_i(t)+N_4R_2(t)+Z_3(t), \qquad (8.54)$$

where N_1, N_2, N_3 and N_4 are positive constants. Then from Lemma 8.2–Lemma 8.5, we get

$$
\begin{aligned}
\frac{d}{dt}\mathscr{L}(t) \leq & -\Lambda_1 \sum_{i=1}^{2} \int_{\mathbb{R}^n} \left(\int_0^{\infty} \mu_i(s)|\nabla\eta^i|^2 ds \right) dx - \Lambda_1 \sum_{i=1}^{2} \int_{\mathbb{R}^n} \left(\int_0^{\infty} \mu_i(s)|\nabla\eta_t^i|^2 ds \right) dx \\
& - \Lambda_1 \sum_{i=1}^{2} \int_{\mathbb{R}^n} \left(\int_0^{\infty} \mu_i(s)|\nabla\eta_{tt}^i|^2 ds \right) dx - \Lambda_2 \int_{\mathbb{R}^n} |\nabla u_3|^2 dx \\
& - \left(\frac{\mu_1}{2} N_2 - N_3 - \frac{\alpha}{2} N_4 \right) \sum_{i=1}^{2} \int_{\mathbb{R}^n} \frac{\partial u_i}{\partial t}^2 dx - \left(\frac{\mu_1}{2} N_2 - N_3 - \frac{\alpha}{2} N_4 \right) \int_{\mathbb{R}^n} \frac{\partial^2 u}{\partial t^2}^2 dx \\
& - \left(\alpha N_4 - \frac{\alpha\varepsilon_1}{2} N_2 - 1 \right) \int_{\mathbb{R}^n} \frac{\partial u_3}{\partial t}^2 dx - \left(\frac{\mu_0}{2} N_3 - \frac{\mu_0\varepsilon_1}{2} N_2 - \frac{\alpha}{2} N_4 \right) \sum_{i=1}^{2} \int_{\mathbb{R}^n} |\nabla u_i|^2 dx \\
& - \left(\frac{\mu_0}{2} N_3 - \frac{\mu_0\varepsilon_1}{2} N_2 - \frac{\mu(0)}{2\varepsilon_2} N_4 \right) \sum_{i=1}^{2} \int_{\mathbb{R}^n} \left| \nabla \frac{\partial u_i}{\partial t} \right|^2 dx,
\end{aligned}
\tag{8.55}
$$

where

$$
\Lambda_1 = -\left(N_1 \frac{k_1}{2} - \left[\left(\frac{\mu_0\mu_1}{2\varepsilon_1} + \mu_1 + \frac{\alpha\mu_1}{2\varepsilon_1} + \frac{k_0 C}{2} \right) N_2 + \frac{\mu_1}{2\mu_0} N_3 + \left(\frac{k_2}{2\varepsilon_2} + \frac{\mu_1}{2\varepsilon_2} + \frac{k_2\mu_0}{\varepsilon_2} \right) N_4 \right] \right)
$$

and

$$
\Lambda_2 = \left(1 - \frac{\alpha C\varepsilon_1}{2} N_2 - \left(\frac{1}{\mu_1} + \frac{\mu(0)+1}{2} + \frac{\mu_0}{\mu_1} \right) \varepsilon_2 N_4 \right).
$$

Now, choosing ε_1, ε_2 small enough such that

$$
0 < \varepsilon_1 < \varepsilon_2 < \frac{\mu_1}{2},
$$

and $N_4 > 0$ large enough we get

$$
\alpha > \frac{\mu_0\mu_1(\mu_1 - 2\varepsilon_1)}{N_4(\mu_0\mu_1(\mu_1 - 2\varepsilon_1) - 2\varepsilon_1(\varepsilon_1(\mu_0 + \mu_1 C)))} > 0
$$

is small enough. Then taking

$$
\begin{aligned}
N_1 > & \frac{2}{k_1} \left[\left(\frac{\mu_0\mu_1 + \alpha\mu_1}{2\varepsilon_1} + \mu_1 + \frac{k_0 C}{2} \right) + \left(2\frac{(\varepsilon_1 + \alpha)\mu_0\varepsilon_2 + \mu(0)\mu_1}{\mu_0\mu_1\varepsilon_2(\mu_1 - 2\varepsilon_1)} + \alpha \right) \right. \\
& \left. + \frac{\mu_1^2(\varepsilon_1\varepsilon_2\alpha + \mu(0)\mu_1)}{2\mu_0(\mu_1 - 2\varepsilon_1)} + \frac{k_2 + \mu_1 + 2k_2\mu_0}{2\varepsilon_2} \right] N_4, \\
N_2 > & \max\left\{ \left(\frac{2(\varepsilon_1\mu_0 + \mu_1 C\varepsilon_1)}{\mu_0\mu_1(\mu_1 - 2\varepsilon_1)} + \alpha \right) N_4, \left(\frac{(\varepsilon_1 + \alpha)\mu_0\varepsilon_2 + \mu(0)\mu_1}{\mu_0\mu_1(\mu_1 - 2\varepsilon_1)} + \alpha \right) N_4 \right\}
\end{aligned}
$$

and

$$N_3 > \max\left\{\left(\frac{(\varepsilon_1\mu_0 + \mu_1 C\varepsilon_1)}{\mu_0\mu_1(\mu_1 - 2\varepsilon_1)}\right)N_4, \left(\frac{\varepsilon_1\varepsilon_2\alpha\mu_0 + \mu(0)\mu_1}{\mu_0\mu_1(\mu_1 - 2\varepsilon_1)}\right)N_4\right\},$$

we get that there is a positive constant $\kappa > 0$ such that

$$\mathcal{L}'(t) \leq -\kappa E(t). \tag{8.56}$$

Therefore,

$$\kappa \int_0^t E(s)ds \leq \mathcal{L}(0) - \mathcal{L}(t), \quad \forall t > 0. \tag{8.57}$$

On the other hand, it is not hard to prove that for $\zeta > 0$, we have

$$\mathcal{L}(0) - \mathcal{L}(t) \leq \frac{\zeta}{\kappa}(E(0) + E_2(0) + E_3(0)), \quad \forall t > 0. \tag{8.58}$$

From (8.57) and (8.58), we obtain

$$\int_{\mathbb{R}^n} E(s)ds \leq \frac{\zeta}{\kappa}(E(0) + E_2(0) + E_3(0)). \tag{8.59}$$

Finally, since

$$(tE(t))' = E(t) + tE'(t) \leq E(t), \tag{8.60}$$

from (8.51) we get

$$E(t) \leq Ct^{-1}(E(0) + E_2(0) + E_3(0)), \tag{8.61}$$

where $C = \frac{\zeta}{w}$. Finally, we obtain (8.30), which completes the proof. □

8.2 Three nonlinear wave equations depending on the relaxation functions

We consider the following system:

$$\begin{cases} \rho(x)(\frac{\partial^2 u}{\partial t^2} + \alpha\frac{\partial u}{\partial t}) - \beta\Delta\frac{\partial u}{\partial t} = \Delta u - \int_0^t \mu_1(t-s)\Delta u(s)\,ds + \rho(x)h_1(u,v,w), \\ \rho(x)(\frac{\partial^2 v}{\partial t^2} + \alpha\frac{\partial v}{\partial t}) - \beta\Delta\frac{\partial v}{\partial t} = \Delta v - \int_0^t \mu_2(t-s)\Delta v(s)\,ds + \rho(x)h_2(u,v,w), \\ \rho(x)(\frac{\partial^2 w}{\partial t^2} + \alpha\frac{\partial w}{\partial t}) - \beta\Delta\frac{\partial w}{\partial t} = \Delta w - \int_0^t \mu_3(t-s)\Delta w(s)\,ds + \rho(x)h_3(u,v,w), \\ u(x,0) = u_0(x), v(x,0) = v_0(x), w(x,0) = w_0(x), \\ \frac{\partial u}{\partial t}(x,0) = u_1(x), \frac{\partial v}{\partial t}(x,0) = v_1(x), \frac{\partial w}{\partial t}(x,0) = w_1(x), \end{cases} \tag{8.62}$$

where $\alpha \in \mathbb{R}, \beta > 0, n \geq 3$, the functions $h_i(\cdot,\cdot,\cdot) \in (\mathbb{R}^3, \mathbb{R}), i = 1,2,3$ are given by

$$h_1(\xi_1,\xi_2,\xi_3) = (p+1)[d|\xi_1+\xi_2+\xi_3|^{(p-1)}(\xi_1+\xi_2+\xi_3) + e|\xi_1|^{(p-3)/2}\xi_1|\xi_2|^{(p+1)/2}],$$
$$h_2(\xi_1,\xi_2,\xi_3) = (p+1)[d|\xi_1+\xi_2+\xi_3|^{(p-1)}(\xi_1+\xi_2+\xi_3) + e|\xi_2|^{(p-3)/2}\xi_2|\xi_3|^{(p+1)/2}],$$
$$h_3(\xi_1,\xi_2,\xi_3) = (p+1)[d|\xi_1+\xi_2+\xi_3|^{(p-1)}(\xi_1+\xi_2+\xi_3) + e|\xi_3|^{(p-3)/2}\xi_3|\xi_1|^{(p+1)/2}],$$

with $d, e > 0, p > 3$. The function $\rho(x) > 0$ is a density given as (1.58). It is hard to see that there exists a function $G \in C^1(\mathbb{R}^3, \mathbb{R})$ such that

$$uh_1(u,v,w) + vh_2(u,v,w) + wh_3(u,v,w) = (p+1)G(u,v,w), \quad \forall (u,v,w) \in \mathbb{R}^3 \tag{8.63}$$

satisfies

$$(p+1)G(u,v,w) = |u+v+w|^{p+1} + 2|uv|^{(p+1)/2} + 2|vw|^{(p+1)/2} + 2|wu|^{(p+1)/2}. \tag{8.64}$$

We assume that the kernel functions $\mu_1, \mu_2, \mu_3 \in C^1(\mathbb{R}^+, \mathbb{R}^+)$ satisfying

$$1 - \overline{\mu_1} = \mu_0 > 0, \quad \mu_1'(t) \leq 0, \tag{8.65}$$
$$1 - \overline{\mu_2} = m > 0, \quad \mu_2'(t) \leq 0, \tag{8.66}$$
$$1 - \overline{\mu_3} = \nu > 0, \quad \mu_3'(t) \leq 0. \tag{8.67}$$

Noting by

$$\mu(t) = \max_{t\geq 0}\{\mu_1(t), \mu_2(t), \mu_3(t)\}, \tag{8.68}$$

and

$$\mu_0(t) = \min_{t\geq 0}\left\{\int_0^t \mu_1(s)ds, \int_0^t \mu_2(s)ds, \int_0^t \mu_3(s)ds\right\}. \tag{8.69}$$

We assume that there is a function $\chi \in C^1(\mathbb{R}^+, \mathbb{R}^+)$ such that

$$\mu_i'(t) + \chi(\mu_i(t)) \leq 0, \quad \chi(0) = 0, \quad \chi'(0) > 0, \quad i = 1,2,3, \tag{8.70}$$

for any $\xi \geq 0$.

Hölder and Young's inequalities give

$$\begin{aligned}\|uv\|^{(p+1)/2}_{L^{(p+1)/2}_\rho} &\leq \left(\|u\|^2_{L^{(p+1)}_\rho} + \|v\|^2_{L^{(p+1)}_\rho}\right)^{(p+1)/2} \\ &\leq \left(\mu_0\|u\|^2_{\mathcal{H}} + m\|v\|^2_{\mathcal{H}}\right)^{(p+1)/2}\end{aligned} \tag{8.71}$$

and

$$\|vw\|_{L_\rho^{(p+1)/2}}^{(p+1)/2} \le (m\|v\|_{\mathcal{H}}^2 + \nu\|w\|_{\mathcal{H}}^2)^{(p+1)/2}, \tag{8.72}$$

and

$$\|wu\|_{L_\rho^{(p+1)/2}}^{(p+1)/2} \le (\nu\|w\|_{\mathcal{H}}^2 + \mu_0\|u\|_{\mathcal{H}}^2)^{(p+1)/2}. \tag{8.73}$$

Thanks to Minkowski's inequality to give

$$\begin{aligned}\|u+v+w\|_{L_\rho^{(p+1)}}^{(p+1)} &\le c(\|u\|_{L_\rho^{(p+1)}}^2 + \|v\|_{L_\rho^{(p+1)}}^2 + \|w\|_{L_\rho^{(p+1)}}^2)^{(p+1)/2} \\ &\le c(\|u\|_{\mathcal{H}}^2 + \|v\|_{\mathcal{H}}^2 + \|w\|_{\mathcal{H}}^2)^{(p+1)/2}.\end{aligned}$$

Then there exist $\eta > 0$ such that

$$\begin{aligned}&\|u+v+w\|_{L_\rho^{(p+1)}}^{(p+1)} + 2\|uv\|_{L_\rho^{(p+1)/2}}^{(p+1)/2} + 2\|vw\|_{L_\rho^{(p+1)/2}}^{(p+1)/2} + 2\|wu\|_{L_\rho^{(p+1)/2}}^{(p+1)/2} \\ &\le \eta(\mu_0\|u\|_{\mathcal{H}}^2 + m\|v\|_{\mathcal{H}}^2 + \nu\|w\|_{\mathcal{H}}^2)^{(p+1)/2}.\end{aligned} \tag{8.74}$$

We need to define positive constants λ_0 and $\mathcal{E}_0$ by

$$\lambda_0 \equiv \eta^{-1/(p-1)} \quad \text{and} \quad \mathcal{E}_0 = \left(\frac{1}{2} - \frac{1}{p+1}\right)\eta^{-2/(p-1)}. \tag{8.75}$$

The aim is to obtain a novel decay rate of solution from the convexity property of the function χ given in Theorem 8.7.

We denote an eigenpair

$$\{(\lambda_i, e_i)\}_{i\in\mathbb{N}} \subset \mathbb{R} \times \mathcal{H},$$

of

$$-\Theta(x)\Delta e_i = \lambda_i e_i \quad x \in \mathbb{R}^n,$$

for any $i \in \mathbb{N}$, $(\Theta(x))^{-1} \equiv \rho(x)$. Then

$$0 < \lambda_1 \le \lambda_2 \le \cdots \le \lambda_i \le \cdots \uparrow +\infty,$$

holds and $\{e_i\}$ is a complete orthonormal system in $\mathcal{H}$.

Definition 8.2. The triplet functions (u, v, w) is said to be a weak solution to (8.62) on $[0, T]$ if it satisfies

$$\begin{cases} \int_{\mathbb{R}^n} \rho(x)(\frac{\partial^2 u}{\partial t^2} + \alpha \frac{\partial u}{\partial t})\varphi dx + \int_{\mathbb{R}^n} (\nabla u + \beta \nabla \frac{\partial u}{\partial t})\nabla \varphi dx = \int_0^t \mu_1(t-s)\nabla u(s)\, ds \nabla \varphi dx \\ \quad + \int_{\mathbb{R}^n} \rho(x) h_1(u, v, w)\varphi dx, \\ \int_{\mathbb{R}^n} \rho(x)(\frac{\partial^2 v}{\partial t^2} + \alpha \frac{\partial v}{\partial t})\psi dx + \int_{\mathbb{R}^n} (\nabla v + \beta \nabla \frac{\partial v}{\partial t})\nabla \psi dx = \int_0^t \mu_2(t-s)\nabla v(s)\, ds \nabla \psi dx \\ \quad + \int_{\mathbb{R}^n} \rho(x) h_2(u, v, w)\psi dx, \\ \int_{\mathbb{R}^n} \rho(x)(\frac{\partial^2 w}{\partial t^2} + \alpha \frac{\partial w}{\partial t})\Psi dx + \int_{\mathbb{R}^n} (\nabla v + \beta \nabla \frac{\partial v}{\partial t})\nabla \Psi dx = \int_0^t \mu_3(t-s)\nabla w(s)\, ds \nabla \Psi dx \\ \quad + \int_{\mathbb{R}^n} \rho(x) h_3(u, v, w)\Psi dx, \end{cases} \tag{8.76}$$

for all test functions $\varphi, \psi, \Psi \in \mathscr{H}$ for almost all $t \in [0, T]$.

8.2.1 Local and global existence

The next theorem is concerned on the local solution (in time $[0, T]$).

Theorem 8.2 (Local existence). *Assume that*

$$1 < p \le \frac{n+2}{n-2} \quad \textit{and that} \quad n \ge 3. \tag{8.77}$$

Let

$$(u_0, v_0, w_0) \in \mathscr{H}^3,$$

and

$$(u_1, v_1, w_3) \in L^2_\rho(\mathbb{R}^n) \times L^2_\rho(\mathbb{R}^n) \times L^2_\rho(\mathbb{R}^n).$$

Under the assumptions (1.58) *and* (8.65)–(8.70), *suppose that*

$$\alpha + \lambda_1 \beta > 0. \tag{8.78}$$

Then (8.62) *admits a unique local solution* (u, v, w) *such that*

$$(u, v, w) \in X_T^3,$$

where

$$X_T \equiv C([0, T]; \mathscr{H}) \cap C^1([0, T]; L^2_\rho(\mathbb{R}^n)),$$

for sufficiently small $T > 0$.

Remark 8.1. The constant λ_1 introduced in (8.78) is the first eigenvalue of the operator $-\Delta$.

We prove the existence of the global solution in time. Let us introduce the potential energy

$$J : \mathscr{H}^3 \to \mathbb{R},$$

defined by

$$\begin{aligned} J(u, v, w) = & \left(1 - \int_0^t \mu_1(s)\, ds\right)\|u\|^2_{\mathscr{H}} + (\mu_1 \circ u) \\ & + \left(1 - \int_0^t \mu_2(s)\, ds\right)\|v\|^2_{\mathscr{H}} + (\mu_2 \circ v) \\ & + \left(1 - \int_0^t \mu_3(s)\, ds\right)\|w\|^2_{\mathscr{H}} + (\mu_3 \circ w). \end{aligned} \tag{8.79}$$

The modified energy is defined by

$$\begin{aligned} \mathscr{E}(t) = & \frac{1}{2}\left(\left\|\frac{\partial u}{\partial t}\right\|^2_{L^2_\rho} + \left\|\frac{\partial v}{\partial t}\right\|^2_{L^2_\rho} + \left\|\frac{\partial w}{\partial t}\right\|^2_{L^2_\rho}\right) \\ & + \frac{1}{2}J(u, v, w) - \int_{\mathbb{R}^n} \rho(x) G(u, v, w) dx. \end{aligned} \tag{8.80}$$

The next lemma will play an important role in the sequel.

Lemma 8.6. *For* $(u, v, w) \in X^3_T$, *the functional* $\mathscr{E}(t)$ *associated with problem* (8.62) *is a decreasing energy.*

Proof. For $0 \le t_1 < t_2 \le T$, we have

$$\begin{aligned} & \mathscr{E}(t_2) - \mathscr{E}(t_1) \\ & = \int_{t_1}^{t_2} \frac{d}{dt} E(t)\, dt \\ & = -\int_{t_1}^{t_2} \left(a\left\|\frac{\partial u}{\partial t}\right\|^2_{L^2_\rho} + \beta\left\|\frac{\partial u}{\partial t}\right\|^2_{\mathscr{H}} + \frac{1}{2}\mu_1(t)\|u\|^2_{\mathscr{H}} - \frac{1}{2}(\mu_1' \circ u)\right) dt \\ & \quad - \int_{t_1}^{t_2} \left(a\left\|\frac{\partial v}{\partial t}\right\|^2_{L^2_\rho} + \beta\left\|\frac{\partial v}{\partial t}\right\|^2_{\mathscr{H}} + \frac{1}{2}\mu_2(t)\|v\|^2_{\mathscr{H}} - \frac{1}{2}(\mu_2' \circ v)\right) dt \end{aligned}$$

$$-\int_{t_1}^{t_2}\left(a\left\|\frac{\partial w}{\partial t}\right\|_{L^2_\rho}^2+\beta\left\|\frac{\partial w}{\partial t}\right\|_{\mathscr{H}}^2+\frac{1}{2}\mu_3(t)\|w\|_{\mathscr{H}}^2-\frac{1}{2}(\mu_3'\circ w)\right)dt$$
$$\leq 0,$$

owing to (8.65)–(8.70). □

The inner product is given as

$$(v,w)_* = \beta\int_{\mathbb{R}^n}\nabla v\cdot\nabla w\,dx + a\int_{\mathbb{R}^n}\rho v w\,dx,$$

and the associated norm is given by

$$\|v\|_* = \sqrt{(v,v)_*},$$

$\forall v,w\in\mathscr{H}$. By (8.78), we get

$$(v,v)_* = \beta\int_{\mathbb{R}^n}|\nabla v|^2\,dx + a\int_{\mathbb{R}^n}\rho v^2\,dx \geq (\beta\lambda_1+a)\int_{\mathbb{R}^n}\rho v^2\,dx \geq 0.$$

The following lemma yields.

Lemma 8.7. *Let* ρ *satisfy* (1.58). *Under condition* (8.78), *we get*

$$\sqrt{\beta}\|v\|_{\mathscr{H}} \leq \|v\|_* \leq \sqrt{\beta + C_P^2}\|v\|_{\mathscr{H}},$$

for $v\in\mathscr{H}$.

Here, we sketch the outline of the proof for the local solution by a standard procedure.

Proof of Theorem 8.2. Let

$$(u_0,u_1),(v_0,v_1),(w_0,w_1)\in\mathscr{H}\times L^2_\rho(\mathbb{R}^n).$$

For any

$$(u,v,w)\in X_T^3,$$

we can obtain a weak solution of the related system

$$\begin{cases} \rho(x)\frac{\partial^2 z}{\partial t^2} + a\rho(x)\frac{\partial z}{\partial t} - \Delta(z + \beta\frac{\partial z}{\partial t}) = -\int_0^t \mu_1(t-s)\Delta u(s)\,ds + \rho(x)h_1(u,v,w), \\ \rho(x)\frac{\partial^2 y}{\partial t^2} + a\rho(x)\frac{\partial y}{\partial t} - \Delta(y + \beta\frac{\partial y}{\partial t}) = -\int_0^t \mu_2(t-s)\Delta v(s)\,ds + \rho(x)h_2(u,v,w), \\ \rho(x)\frac{\partial^2 \zeta}{\partial t^2} + a\rho(x)\frac{\partial \zeta}{\partial t} - \Delta(\zeta + \beta\frac{\partial \zeta}{\partial t}) = -\int_0^t \mu_3(t-s)\Delta w(s)\,ds + \rho(x)h_3(u,v,w), \\ z(x,0) = u_0(x), y(x,0) = v_0(x), \zeta(x,0) = w_0(x), \\ \frac{\partial^2 z}{\partial t^2}(x,0) = u_1(x), \frac{\partial y}{\partial t}(x,0) = v_1(x), \zeta_t(x,0) = w_1(x). \end{cases} \tag{8.81}$$

We reduce problem (8.81) to a Cauchy problem for the system of ODE by using the Faedo–Galerkin approximation. We then find a solution map

$$\mathsf{T} : (u,v,w) \mapsto (z,y,\zeta),$$

from X_T^3 to X_T^3. We are now ready to show that T is a contraction mapping in an appropriate subset of X_T^3 for a small $T > 0$. Hence, T has a fixed point

$$\mathsf{T}(u,v,w) = (u,v,w),$$

which gives a unique solution in X_T^3. □

Theorem 8.3 (Global existence). *Let* (1.58) *and* (8.65)–(8.70) *hold. Under* (8.77), (8.78) *and for sufficiently small*

$$(u_0,u_1), (v_0,v_1), (w_0,w_1) \in \mathcal{H} \times L^2_\rho(\mathbb{R}^n).$$

Problem (8.62) *admits a unique global solution* (u,v,w) *such that*

$$(u,v,w) \in X^3, \tag{8.82}$$

where

$$X \equiv C([0,+\infty);\mathcal{H}) \cap C^1([0,+\infty);L^2_\rho(\mathbb{R}^n)).$$

We will show the global solution. By using conditions on functions μ_1, μ_2, μ_3, we have

$$\begin{aligned} \mathcal{E}(t) &\geq \frac{1}{2}J(u,v,w) - \int_{\mathbb{R}^n} \rho(x)G(u,v,w)dx \\ &\geq \frac{1}{2}J(u,v,w) - \frac{1}{p+1}\|u+v+w\|^{(p+1)}_{L^{(p+1)}_\rho} \\ &\quad - \frac{2}{p+1}\left(\|uv\|^{(p+1)/2}_{L^{(p+1)/2}_\rho} + \|vw\|^{(p+1)/2}_{L^{(p+1)/2}_\rho} + \|wu\|^{(p+1)/2}_{L^{(p+1)/2}_\rho}\right) \\ &\geq \frac{1}{2}J(u,v,w) - \frac{\eta}{p+1}\left[\mu_0\|u\|^2_{\mathcal{H}} + m\|v\|^2_{\mathcal{H}} + v\|w\|^2_{\mathcal{H}}\right]^{(p+1)/2} \\ &\geq \frac{1}{2}J(u,v,w) - \frac{\eta}{p+1}(J(u,v,w))^{(p+1)/2} \\ &= G(\varsigma). \end{aligned} \tag{8.83}$$

Here,

$$\varsigma^2 = J(u, v, w),$$

for $t \in [0, T)$, where

$$G(\xi) = \frac{1}{2}\xi^2 - \frac{\eta}{p+1}\xi^{(p+1)}.$$

Noting that

$$\mathscr{E}_0 = G(\lambda_0),$$

given in (8.75). Then

$$\begin{cases} G(\xi) \geq 0 & \text{in } \xi \in [0, \lambda_0], \\ G(\xi) < 0 & \text{in } \xi > \lambda_0. \end{cases} \tag{8.84}$$

Moreover,

$$\lim_{\xi \to +\infty} G(\xi) \to -\infty.$$

Then we have the following lemma.

Lemma 8.8. *Let*

$$0 \leq \mathscr{E}(0) < \mathscr{E}_0.$$

(i) *If*

$$\|u_0\|^2_{\mathscr{H}} + \|v_0\|^2_{\mathscr{H}} + \|w_0\|^2_{\mathscr{H}} < \lambda_0^2,$$

then the local solution of (8.62) *satisfies*

$$J(u, v, w) < \lambda_0^2, \quad \forall t \in [0, T).$$

(ii) *If*

$$\|u_0\|^2_{\mathscr{H}} + \|v_0\|^2_{\mathscr{H}} + \|w_0\|^2_{\mathscr{H}} > \lambda_0^2,$$

then the local solution of (8.62) *satisfies*

$$\|u\|^2_{\mathscr{H}} + \|v\|^2_{\mathscr{H}} + \|w\|^2_{\mathscr{H}} > \lambda_1^2, \quad \forall t \in [0, T), \lambda_1 > \lambda_0.$$

Proof. Since

$$0 \le \mathscr{E}(0) < \mathscr{E}_0 = G(\lambda_0),$$

there exist ξ_1 and ξ_2 such that

$$G(\xi_1) = G(\xi_2) = \mathscr{E}(0),$$

with

$$0 < \xi_1 < \lambda_0 < \xi_2.$$

The case (i). By (8.83), we have

$$G(J(u_0, v_0, w_0)) \le \mathscr{E}(0) = G(\xi_1),$$

which implies that

$$J(u_0, v_0, w_0) \le \xi_1^2.$$

Then we claim that

$$J(u, v, w) \le \xi_1^2, \quad \forall t \in [0, T).$$

Moreover, there exists $t_0 \in (0, T)$ such that

$$\xi_1^2 < J(u(t_0), v(t_0), w(t_0)) < \xi_2^2.$$

Then

$$G(J(u(t_0), v(t_0), w(t_0))) > \mathscr{E}(0) \ge \mathscr{E}(t_0),$$

by Lemma 8.6, which contradicts (8.83). Hence, we have

$$J(u, v, w) \le \xi_1^2 < \lambda_0^2, \quad \forall t \in [0, T).$$

The case (ii). We can now show that

$$\|u_0\|_{\mathscr{H}}^2 + \|v_0\|_{\mathscr{H}}^2 + \|w_0\|_{\mathscr{H}}^2 \ge \xi_2^2,$$

and that

$$\|u\|_{\mathscr{H}}^2 + \|v\|_{\mathscr{H}}^2 + \|w\|_{\mathscr{H}}^2 \ge \xi_2^2 > \lambda_0^2,$$

in the same way as (i). □

Proof of Theorem 8.3. Let

$$(u_0,u_1),(v_0,v_1),(w_0,w_1)\in\mathscr{H}\times L^2_\rho(\mathbb{R}^n),$$

satisfy both

$$0\le\mathscr{E}(0)<\mathscr{E}_0,$$

and

$$\|u_0\|^2_{\mathscr{H}}+\|v_0\|^2_{\mathscr{H}}+\|w_0\|^2_{\mathscr{H}}<\lambda_0^2.$$

By Lemma 8.6 and Lemma 8.8, we have

$$\begin{aligned}
&\frac{1}{2}\left(\left\|\frac{\partial u}{\partial t}\right\|^2_{L^2_\rho}+\left\|\frac{\partial v}{\partial t}\right\|^2_{L^2_\rho}+\left\|\frac{\partial w}{\partial t}\right\|^2_{L^2_\rho}\right)+l\|u\|^2_{\mathscr{H}}+m\|v\|^2_{\mathscr{H}}+v\|w\|^2_{\mathscr{H}}\\
&\le\frac{1}{2}\left(\left\|\frac{\partial u}{\partial t}\right\|^2_{L^2_\rho}+\left\|\frac{\partial v}{\partial t}\right\|^2_{L^2_\rho}+\left\|\frac{\partial w}{\partial t}\right\|^2_{L^2_\rho}\right)+\left(1-\int_0^t\mu_1(s)\,ds\right)\|u\|^2_{\mathscr{H}}+(\mu_1\circ u)\\
&\quad+\left(1-\int_0^t\mu_2(s)\,ds\right)\|u\|^2_{\mathscr{H}}+(\mu_2\circ v)+\left(1-\int_0^t\mu_3(s)\,ds\right)\|w\|^2_{\mathscr{H}}+(\mu_3\circ w)\\
&\le 2\mathscr{E}(t)+\frac{2\eta}{p+1}\left[\mu_0\|u\|^2_{\mathscr{H}}+m\|u\|^2_{\mathscr{H}}+v\|w\|^2_{\mathscr{H}}\right]^{(p+1)/2}\\
&\le 2\mathscr{E}(0)+\frac{2\eta}{p+1}(J(u,v,w))^{(p+1)/2}\\
&<2\mathscr{E}_0+\frac{2\eta}{p+1}\lambda_0^{p+1}\\
&=\eta^{-2/(p-1)}.
\end{aligned}\tag{8.85}$$

This completes the proof. □

Let

$$\begin{aligned}
\Lambda(u,v,w)&=\frac{1}{2}\left(1-\int_0^t\mu_1(s)\,ds\right)\|u\|^2_{\mathscr{H}}+\frac{1}{2}(\mu_1\circ u)\\
&\quad+\frac{1}{2}\left(1-\int_0^t\mu_2(s)\,ds\right)\|v\|^2_{\mathscr{H}}+\frac{1}{2}(\mu_2\circ v)\\
&\quad+\frac{1}{2}\left(1-\int_0^t\mu_3(s)\,ds\right)\|w\|^2_{\mathscr{H}}+\frac{1}{2}(\mu_3\circ w)-\int_{\mathbb{R}^n}\rho(x)G(u,v,w)dx,
\end{aligned}\tag{8.86}$$

$$\Pi(u,v,w) = \left(1 - \int_0^t \mu_1(s)\,ds\right)\|u\|_{\mathcal{H}}^2 + (\mu_1 \circ u) \tag{8.87}$$
$$+ \left(1 - \int_0^t \mu_2(s)\,ds\right)\|v\|_{\mathcal{H}}^2 + (\mu_2 \circ v)$$
$$+ \left(1 - \int_0^t \mu_3(s)\,ds\right)\|w\|_{\mathcal{H}}^2 + (\mu_3 \circ w) - (p+1)\int_{\mathbb{R}^n} \rho(x)G(u,v,w)dx.$$

Lemma 8.9. *Let* (u,v,w) *be the solution of problem* (8.62). *If*

$$\|u_0\|_{\mathcal{H}}^2 + \|v_0\|_{\mathcal{H}}^2 + \|w_0\|_{\mathcal{H}}^2 - (p+1)\int_{\mathbb{R}^n} \rho(x)G(u_0,v_0,w_0)dx > 0. \tag{8.88}$$

Then the functional

$$\Pi(u,v,w) > 0, \quad \forall t > 0.$$

Proof. By (8.88) and continuity, there exists a time $t_1 > 0$ such that

$$\Pi(u,v,w) \geq 0, \quad \forall t < t_1.$$

Let

$$Y = \{(u,v,w) \mid \Pi(u(t_0),v(t_0),w(t_0)) = 0,\ \Pi(u,v,w) > 0, \forall t \in [0,t_0)\}. \tag{8.89}$$

Then, by (8.86), (8.87), we have for all $(u,v,w) \in Y$,

$$\begin{aligned}
&\Lambda(u,v,w)\\
&\quad - \frac{p-1}{2(p+1)}\Bigg[\left(1 - \int_0^t \mu_1(s)\,ds\right)\|u\|_{\mathcal{H}}^2\\
&\quad + \left(1 - \int_0^t \mu_2(s)\,ds\right)\|v\|_{\mathcal{H}}^2 + \left(1 - \int_0^t \mu_3(s)\,ds\right)\|w\|_{\mathcal{H}}^2\Bigg]\\
&\quad + \frac{p-1}{2(p+1)}[(\mu_1 \circ u) + (\mu_2 \circ v) + (\mu_3 \circ w)] + \frac{1}{p+1}\Pi(u,v,w)\\
&\geq \frac{p-1}{2(p+1)}[\mu_0\|u\|_{\mathcal{H}}^2 + m\|v\|_{\mathcal{H}}^2 + \nu\|w\|_{\mathcal{H}}^2 + (\mu_1 \circ u) + (\mu_2 \circ v) + (\mu_3 \circ w)].
\end{aligned}$$

Owing to (8.80), it follows for $(u,v,w) \in Y$,

$$\mu_0\|u\|_{\mathcal{H}}^2 + m\|v\|_{\mathcal{H}}^2 + \nu\|w\|_{\mathcal{H}}^2 \leq \frac{2(p+1)}{p-1}\Lambda(u,v,w)$$

$$\leq \frac{2(p+1)}{p-1}\mathscr{E}(t)$$
$$\leq \frac{2(p+1)}{p-1}\mathscr{E}(0). \tag{8.90}$$

By (8.92), we have

$$\begin{aligned}
&(p+1)\int_{\mathbb{R}^n} G(u(t_0), v(t_0), w(t_0)) \\
&\leq \eta(\mu_0\|u(t_0)\|^2_{\mathscr{H}} + m\|v(t_0)\|^2_{\mathscr{H}} + \nu\|w(t_0)\|^2_{\mathscr{H}})^{(p+1)/2} \\
&\leq \eta\left(\frac{2(p+1)}{p-1}E(0)\right)^{(p-1)/2}(\mu_0\|u(t_0)\|^2_{\mathscr{H}} + m\|v(t_0)\|^2_{\mathscr{H}} + \nu\|w(t_0)\|^2_{\mathscr{H}}) \\
&\leq \gamma(\mu_0\|u(t_0)\|^2_{\mathscr{H}} + m\|v(t_0)\|^2_{\mathscr{H}} + \nu\|w(t_0)\|^2_{\mathscr{H}}) \\
&< \left(1 - \int_0^{t_0}\mu_1(s)ds\right)\|u(t_0)\|^2_{\mathscr{H}} + \left(1 - \int_0^{t_0}\mu_2(s)ds\right)\|v(t_0)\|^2_{\mathscr{H}} \\
&\quad + \left(1 - \int_0^{t_0}\mu_3(s)ds\right)\|w(t_0)\|^2_{\mathscr{H}} \\
&< \left(1 - \int_0^{t_0}\mu_1(s)ds\right)\|u(t_0)\|^2_{\mathscr{H}} + \left(1 - \int_0^{t_0}\mu_2(s)ds\right)\|v(t_0)\|^2_{\mathscr{H}} \\
&\quad + \left(1 - \int_0^{t_0}\mu_3(s)ds\right)\|w(t_0)\|^2_{\mathscr{H}} \\
&\quad + (\mu_1 \circ u) + (\mu_2 \circ v) + (\mu_3 \circ w).
\end{aligned} \tag{8.91}$$

Hence,

$$\Pi(u(t_0), v(t_0), w(t_0)) > 0$$

on Y, which contradicts the definition of Y since

$$\Pi(u(t_0), v(t_0), w(t_0)) = 0.$$

Thus,

$$\Pi(u, v, w) > 0, \quad \forall t > 0.$$

This completes the proof. □

8.2.2 Decay of solution

Theorem 8.4 (Decay of solution). *Let* (1.58) *and* (8.65)–(8.70) *hold. Under conditions* (8.77), (8.78) *and*

$$\gamma = \eta\left(\frac{2(p+1)}{p-1}\mathscr{E}(0)\right)^{(p-1)/2} < 1, \tag{8.92}$$

there exists $t_0 > 0$ *depending only on*

$$\mu_1, \mu_2, \mu_3, a, \beta, \lambda_1, \chi'(0),$$

such that

$$0 \le \mathscr{E}(t) < \mathscr{E}(t_0)\exp\left(-\int_{t_0}^{t} \frac{\mu(s)}{1-\mu_0(t)}\right), \tag{8.93}$$

holds for all $t \ge t_0$.

In particular, by the positivity of μ in (8.68), we have

$$0 \le \mathscr{E}(t) < \mathscr{E}(t_0)\exp\left(-\int_{t_0}^{t} \mu(s)\,ds\right),$$

for a single wave equation. Condition (8.102) is imposed to lead

$$(\mu' + \nu\mu) \circ u;$$

here, $\nu \in \mathbb{R}$.

Proof of Theorem 8.4. By (8.74) and (8.90), we have for $t \ge 0$,

$$0 < \mu_0\|u\|_{\mathscr{H}}^2 + m\|v\|_{\mathscr{H}}^2 + \nu\|w\|_{\mathscr{H}}^2 \le \frac{2(p+1)}{p-1}\mathscr{E}(t). \tag{8.94}$$

Let

$$I(t) = \frac{\mu(t)}{1-\mu_0(t)},$$

where μ and μ_0 are defined in (8.68) and (8.69).

Noting that

$$\lim_{t\to+\infty} \mu(t) = 0,$$

by (8.65)–(8.69), we have

$$\lim_{t\to+\infty} I(t) = 0, \quad I(t) > 0, \ \forall t \geq 0.$$

Then we take $t_0 > 0$ such that

$$0 < \frac{1}{2} I(t) < \min\{2(\beta\lambda_1 + \alpha), \chi'(0)\},$$

with (8.70) for all $t > t_0$. Due to (8.80), we have

$$\begin{aligned}
\mathcal{E}(t) &\leq \frac{1}{2}\left(\left\|\frac{\partial u}{\partial t}\right\|_{L^2_\rho}^2 + \left\|\frac{\partial v}{\partial t}\right\|_{L^2_\rho}^2 + \left\|\frac{\partial w}{\partial t}\right\|_{L^2_\rho}^2\right) + \frac{1}{2}[(\mu_1 \circ u) + (\mu_2 \circ v) + (\mu_3 \circ w)] \\
&\quad + \frac{1}{2}\left(1 - \int_0^t \mu_1(s)\,ds\right)\|u\|_{\mathcal{H}}^2 + \frac{1}{2}\left(1 - \int_0^t \mu_2(s)\,ds\right)\|v\|_{\mathcal{H}}^2 + \frac{1}{2}\left(1 - \int_0^t \mu_3(s)\,ds\right)\|w\|_{\mathcal{H}}^2 \\
&\leq \frac{1}{2}\left(\left\|\frac{\partial u}{\partial t}\right\|_{L^2_\rho}^2 + \left\|\frac{\partial v}{\partial t}\right\|_{L^2_\rho}^2 + \left\|\frac{\partial w}{\partial t}\right\|_{L^2_\rho}^2\right) + \frac{1}{2}[(\mu_1 \circ u) + (\mu_2 \circ v) + (\mu_3 \circ w)] \\
&\quad + \frac{1}{2}(1 - \mu_0(t))[\|u\|_{\mathcal{H}}^2 + \|v\|_{\mathcal{H}}^2 + \|w\|_{\mathcal{H}}^2].
\end{aligned}$$

Then by definition of $I(t)$, we have

$$\begin{aligned}
I(t)\mathcal{E}(t) &\leq \frac{1}{2} I(t)\left(\left\|\frac{\partial u}{\partial t}\right\|_{L^2_\rho}^2 + \left\|\frac{\partial v}{\partial t}\right\|_{L^2_\rho}^2 + \left\|\frac{\partial w}{\partial t}\right\|_{L^2_\rho}^2\right) + \frac{1}{2}\mu(t)[\|u\|_{\mathcal{H}}^2 + \|v\|_{\mathcal{H}}^2 + \|w\|_{\mathcal{H}}^2] \\
&\quad + \frac{1}{2} I(t)[(\mu_1 \circ u) + (\mu_2 \circ v) + (\mu_3 \circ w)],
\end{aligned} \tag{8.95}$$

and Lemma 8.6. We have for all $t_1, t_2 \geq 0$,

$$\begin{aligned}
&\mathcal{E}(t_2) - \mathcal{E}(t_1) \\
&\quad \leq -\int_{t_1}^{t_2}\left(\alpha\left\|\frac{\partial w}{\partial t}\right\|_{L^2_\rho}^2 + \alpha\left\|\frac{\partial u}{\partial t}\right\|_{L^2_\rho}^2 + \beta\left\|\frac{\partial u}{\partial t}\right\|_{\mathcal{H}}^2 + \frac{1}{2}\mu(t)[\|u\|_{\mathcal{H}}^2 + \|v\|_{\mathcal{H}}^2 + \|w\|_{\mathcal{H}}^2]\right)dt \\
&\quad\quad - \int_{t_1}^{t_2}\left(\alpha\left\|\frac{\partial v}{\partial t}\right\|_{L^2_\rho}^2 + \beta\left\|\frac{\partial v}{\partial t}\right\|_{\mathcal{H}}^2 + \beta\left\|\frac{\partial w}{\partial t}\right\|_{\mathcal{H}}^2 - \frac{1}{2}(\mu_1' \circ u) - \frac{1}{2}(\mu_2' \circ v) - \frac{1}{2}(\mu_3' \circ w)\right)dt.
\end{aligned}$$

Then, by generalized Poincaré's inequalities, we get

$$\begin{aligned}
\mathcal{E}'(t) &\leq -(\beta\lambda_1 + \alpha)\left[\left\|\frac{\partial u}{\partial t}\right\|_{L^2_\rho}^2 + \left\|\frac{\partial v}{\partial t}\right\|_{L^2_\rho}^2 + \left\|\frac{\partial w}{\partial t}\right\|_{L^2_\rho}^2\right] \\
&\quad - \frac{1}{2}\mu(t)[\|u\|_{\mathcal{H}}^2 + \|v\|_{\mathcal{H}}^2 + \|w\|_{\mathcal{H}}^2] \\
&\quad + \frac{1}{2}[(\mu_1' \circ u) + (\mu_2' \circ v) + (\mu_3' \circ w)],
\end{aligned}$$

Finally, $\forall t \geq t_0$, we have

$$\begin{aligned}
&\mathscr{E}'(t) + I(t)\mathscr{E}(t) \\
&\leq \left\{\frac{1}{2}I(t) - (\beta\lambda_1 + \alpha)\right\}\left(\left\|\frac{\partial u}{\partial t}\right\|_{L^2_\rho}^2 + \left\|\frac{\partial v}{\partial t}\right\|_{L^2_\rho}^2 + \left\|\frac{\partial w}{\partial t}\right\|_{L^2_\rho}^2\right) \\
&\quad + \frac{1}{2}[(\mu_1' \circ u) + (\mu_2' \circ v) + (\mu_3' \circ w)] + \frac{1}{2}I(t)((\mu_1 \circ u) + (\mu_2 \circ v) + (\mu_3 \circ w)) \\
&\leq \frac{1}{2}\int_0^t \{\mu_1'(t-\tau) + I(t)\mu_2(t-\tau)\}\|u(t) - u(\tau)\|_{\mathscr{H}}^2\, d\tau \\
&\quad + \frac{1}{2}\int_0^t \{\mu_2'(t-\tau) + I(t)\mu_2(t-\tau)\}\|v(t) - v(\tau)\|_{\mathscr{H}}^2\, d\tau \\
&\quad + \frac{1}{2}\int_0^t \{\mu_3'(t-\tau) + I(t)\mu_3(t-\tau)\}\|w(t) - w(\tau)\|_{\mathscr{H}}^2\, d\tau \\
&\leq \frac{1}{2}\int_0^t \{\mu_1'(\tau) + I(t)\mu_1(\tau)\}\|u(t) - u(t-\tau)\|_{\mathscr{H}}^2\, d\tau \\
&\quad + \frac{1}{2}\int_0^t \{\mu_2'(\tau) + I(t)\mu_2(\tau)\}\|v(t) - v(t-\tau)\|_{\mathscr{H}}^2\, d\tau \\
&\quad + \frac{1}{2}\int_0^t \{\mu_3'(\tau) + I(t)\mu_3(\tau)\}\|w(t) - w(t-\tau)\|_{\mathscr{H}}^2\, d\tau \\
&\leq \frac{1}{2}\int_0^t \{-\chi(\mu_1(\tau)) + \chi'(0)\mu_1(\tau)\}\|u(t) - u(t-\tau)\|_{\mathscr{H}}^2\, d\tau \\
&\quad + \frac{1}{2}\int_0^t \{-\chi(\mu_2(\tau)) + \chi'(0)\mu_2(\tau)\}\|v(t) - v(t-\tau)\|_{\mathscr{H}}^2\, d\tau \\
&\quad + \frac{1}{2}\int_0^t \{-\chi(\mu_3(\tau)) + \chi'(0)\mu_3(\tau)\}\|w(t) - w(t-\tau)\|_{\mathscr{H}}^2\, d\tau. \\
&\leq 0,
\end{aligned}$$

By the convexity of χ and (8.70), we have

$$\chi(\xi) \geq \chi(0) + \chi'(0)\xi = \chi'(0)\xi.$$

Then

$$\mathscr{E}(t) \le \mathscr{E}(t_0) \exp\left(-\int_{t_0}^{t} I(s)ds \right),$$

which completes the proof. □

8.3 Systems of longitudinal m-nonlinear wave equations

Let $j = 1, 2, \dots, m$. We consider the following system of m damped wave equations:

$$\begin{cases} (|\partial_t u_j|^{\kappa-2}\partial_t u_j)_t + a\partial_t u_j - \Theta(x)\Delta_x(u_j + \omega\partial_t u_j - \int_0^t \mu_j(t-s)u_j(s)\,ds) \\ \quad = f_j(u_1, u_2, \dots, u_m), \\ u_j(x,0) = u_{j0}(x), \\ \partial_t u_j(x,0) = u_{j1}(x), \end{cases} \tag{8.96}$$

where $a \in \mathbb{R}$, $\omega > 0$, $n \ge 3$, $\kappa \ge 2$. The functions

$$f_j(u_1, u_2, \dots, u_m) \in (\mathbb{R}^m, \mathbb{R})$$

are given by

$$f_j(u_1, u_2, \dots, u_m) = (p+1)\left[d\left|\sum_{i=1}^{m} u_i\right|^{(p-1)} \sum_{i=1}^{m} u_i + e|u_j|^{(p-3)/2}u_j|u_{j+1}|^{(p+1)/2} \right],$$

and

$$f_m(u_1, u_2, \dots, u_m) - (p+1)\left[d\left|\sum_{i=1}^{m} u_i\right|^{(p-1)} \sum_{i=1}^{m} u_i + e|u_m|^{(p-3)/2}u_m|u_1|^{(p+1)/2} \right],$$

with $d, e > 0$, and $p > 3$.

One can find a function

$$F \in C^1(\mathbb{R}^3, \mathbb{R}),$$

where

$$\forall (u_1, u_2, \dots, u_m) \in \mathbb{R}^m.$$

We have

$$\sum_{j=1}^{m} u_j f_j(u_1, u_2, \dots, u_m) = (p+1)F(u_1, u_2, \dots, u_m), \tag{8.97}$$

which satisfies

$$F(u_1, u_2, \dots, u_m) = \frac{1}{(p+1)}\left|\sum_{j=1}^{m} u_j\right|^{p+1} + \frac{2}{(p+1)}\left|\sum_{j=1}^{m-1} u_j u_{j+1}\right|^{(p+1)/2} + \frac{2}{(p+1)}|u_m u_1|^{(p+1)/2}. \tag{8.98}$$

The function $\Theta(x) > 0$ is a density and $(\Theta)^{-1} = 1/\Theta(x) \equiv \rho(x)$ as in (1.58). The kernel functions $\mu_j \in C^1(\mathbb{R}^+, \mathbb{R}^+)$ are assumed to satisfy

$$1 - \overline{\mu_j} = \rho_j > 0, \quad \mu_j'(t) \le 0. \tag{8.99}$$

Let us note by

$$\mu(t) = \max_{t \ge 0}\{\mu_1(t), \mu_2(t), \dots, \mu_m(t)\} \tag{8.100}$$

and

$$\mu_0(t) = \min_{t \ge 0}\left\{\int_0^t \mu_1(s)ds, \int_0^t \mu_2(s)ds, \dots, \int_0^t \mu_m(s)ds\right\}. \tag{8.101}$$

There is a function

$$\chi \in C^1(\mathbb{R}^+, \mathbb{R}^+),$$

that have the properties

$$\mu_j'(t) + \chi(\mu_j(t)) \le 0, \quad \chi(0) = 0, \quad \chi'(0) > 0 \quad \text{and} \quad \chi''(\xi) \ge 0, \quad i = 1, 2, \dots, m, \tag{8.102}$$

satisfied for any $\xi \ge 0$.

Hölder and Young's inequalities give

$$\begin{aligned}\|u_i u_j\|_{L_\rho^{(p+1)/2}}^{(p+1)/2} &\le \left(\|u_i\|_{L_\rho^{(p+1)}}^2 + \|u_j\|_{L_\rho^{(p+1)}}^2\right)^{(p+1)/2} \\ &\le \left(\rho_i\|u_i\|_{\mathcal{H}}^2 + \rho_j\|u_j\|_{\mathcal{H}}^2\right)^{(p+1)/2}.\end{aligned} \tag{8.103}$$

Thanks to Minkowski's inequality to give

$$\begin{aligned}\left\|\sum_{j=1}^{m} u_j\right\|_{L_\rho^{(p+1)}}^{(p+1)} &\le c\left(\sum_{j=1}^{m}\|u_j\|_{L_\rho^{(p+1)}}^2\right)^{(p+1)/2} \\ &\le c\left(\sum_{j=1}^{m}\|u_j\|_{\mathcal{H}}^2\right)^{(p+1)/2}.\end{aligned}$$

Thus, there exists a positive constant $\eta > 0$ such that

$$\left\|\sum_{j=1}^{m} u_j\right\|_{L_\rho^{(p+1)}}^{(p+1)} + 2\left\|\sum_{j=1}^{m-1} u_j u_{j+1}\right\|_{L_\rho^{(p+1)/2}}^{(p+1)/2} + 2\|u_m u_1\|_{L_\rho^{(p+1)/2}}^{(p+1)/2}$$
$$\leq \eta\left(\sum_{j=1}^{m} \rho_j \|u_j\|_{\mathscr{H}}^2\right)^{(p+1)/2}. \tag{8.104}$$

We need to define positive constants λ_0 and $\mathscr{E}_0$ by

$$\lambda_0 \equiv \eta^{-1/(p-1)} \quad \text{and} \quad \mathscr{E}_0 = \left(\frac{1}{2} - \frac{1}{p+1}\right)\eta^{-2/(p-1)}. \tag{8.105}$$

Definition 8.3. The vectors $(u_1, u_2, \ldots, u_m)$ is said to be a weak solution to (8.96) on $[0, T]$ if it satisfies for $x \in \mathbb{R}^n$,

$$\int_{\mathbb{R}^n} (|\partial_t u_j|^{\kappa-2}\partial_t u_j)_t \varphi_j dx + a\int_{\mathbb{R}^n} \partial_t u_j \varphi_j dx$$
$$- \int_{\mathbb{R}^n} \Theta(x)\Delta_x\left(u_j + \omega\partial_t u_j - \int_0^t \mu_j(t-s)u_j(s)\,ds\right)\varphi_j dx$$
$$= \int_{\mathbb{R}^n} f_j(u_1, u_2, \ldots, u_m)\varphi_j dx, \tag{8.106}$$

for all test functions $\varphi_j \in \mathscr{H}, j = 1, 2, \ldots, m$ for almost all $t \in [0, T]$.

8.3.1 Existence of solutions

Theorem 8.5 (Local existence result). *Let*

$$1 < p \leq \frac{n+2}{n-2} \quad \textit{and} \quad n \geq 3. \tag{8.107}$$

Let

$$(u_{10}, u_{20}, \ldots u_{m0}) \in \mathscr{H}^m,$$

and

$$(u_1, u_1, \ldots, u_m) \in [L_\rho^\kappa(\mathbb{R}^n)]^m.$$

Under the assumptions (1.58), (8.97)–(8.102), *suppose that*

$$a + \lambda_1\omega > 0. \tag{8.108}$$

Then (8.96) *has a unique local solution* $(u_1, u_2, \ldots, u_m)$ *such that*

$$(u_1, u_2, \ldots, u_m) \in X_T^m,$$

where

$$X_T \equiv C([0,T]; \mathscr{H}) \cap C^1([0,T]; L_\rho^\kappa(\mathbb{R}^n)),$$

for sufficiently small $T > 0$.

We show now that the solution established in Theorem 8.5 is global in time without any additional conditions.

Proof of Theorem 8.5. Let

$$(u_{10}, u_{11}), (u_{20}, u_{21}), \ldots, (u_{m0}, u_{m1}) \in \mathscr{H} \times L_\rho^\kappa(\mathbb{R}^n).$$

The presence of the nonlinear terms in the right-hand side of our problem (8.96) gives us negative values of the energy. For this purpose, for any fixed

$$(u_1, u_2, \ldots, u_m) \in X_T^m,$$

we can obtain first a weak solution of the related system

$$\begin{aligned}
&(|\partial_t z_j|^{\kappa-2}\partial_t z_j)_t + a\partial_t z_j - \Theta(x)\Delta_x(z_j + \omega\partial_t z_j) + \Theta(x)\Delta_x \int_0^t \mu_j(t-s) z_j(s)\, ds \\
&\quad = f_j(u_1, u_2, \ldots, u_m), \\
&z_j(x,0) = u_{j0}(x), \\
&\partial_t z_j(x,0) = u_{j1}(x).
\end{aligned} \tag{8.109}$$

The Faedo–Galerkin's method consists to construct approximations of solutions $(z_{1n}, z_{2n}, \ldots, z_{mn})$ for (8.109). Then we obtain prior estimates necessary to guarantee the convergence of approximations.

Let

$$W_{jn} = \operatorname{span}\{e_{j1}, e_{j2}, \ldots, e_{jn}\}, \quad j = 1, \ldots, m.$$

Given initial data

$$u_{j0} \in \mathscr{H}, \quad u_{j1} \in L_\rho^\kappa(\mathbb{R}^n),$$

we define

$$z_{jn} = \sum_{i=1}^{n} g_{jin}(t)e_{ji}(x), \tag{8.110}$$

which satisfy the following approximate problem:

$$\begin{aligned}
&((|\partial_t z_{jn}|^{\kappa-2}\partial_t z_{jn})_t, e_{ji}) + (a\partial_t z_{jn}, e_{ji}) - (\Theta(x)\Delta_x(z_{jn} + \omega\partial_t z_{jn}), e_{ji}) \\
&= -\left(\Theta(x)\Delta_x \int_0^t \mu_j(t-s)z_{jn}(s)\,ds, e_{ji}\right) + (f_j(u_1, u_2, \dots, u_m), e_{ji}),
\end{aligned} \tag{8.111}$$

with initial conditions

$$z_{jn}(x,0) = u_{j0}^n(x), \quad \partial_t z_{jn}(x,0) = u_{j1}^n(x), \tag{8.112}$$

which satisfies

$$\begin{aligned}
u_{j0}^n &\to u_{j0}, \quad \text{strongly in } \mathscr{H}, \\
u_1^{jn} &\to u_{j1}, \quad \text{strongly in } L_\rho^\kappa(\mathbb{R}^n).
\end{aligned} \tag{8.113}$$

Taking $e_{ji} = g_{ji}$ in (8.111) yields the following Cauchy problem for an ordinary differential equation with unknown g_{ji}^n:

$$\begin{aligned}
&(|g_{jit}^n(t)|^{\kappa-2} g_{jit}^n(t))_t + a g_{ji}^n(t) + \lambda_i(g_{ji}^n(t) + \omega g_{jit}^n(t)) \\
&= \lambda_i \int_0^t \mu_j(t-s) g_{ji}^n(s)\,ds + (f_j(u_1, u_2, \dots, u_m), g_{ji}).
\end{aligned} \tag{8.114}$$

The problems (8.111)–(8.112) have solutions

$$(g_{1in}, g_{2in}, \dots, g_{\min})_{i=1,n} \in (H^3[0,T])^m,$$

and by using the embedding

$$H^m[0,T] \to C^m[0,T],$$

we deduce that the solution

$$(g_{1in}, g_{2in}, \dots, g_{\min})_{i=1,n} \in (C^2[0,T])^4.$$

In turn, this gives a unique

$$(z_{1n}, z_{2n}, \dots, z_{mn}),$$

defined by (8.110) and satisfying (8.111).

We return now to (8.96), and find

$$\mathsf{T} : (u_1, u_2, \dots, u_m) \mapsto (z_1, z_2, \dots, z_m)$$

from X_T^m to X_T^m. We can now show that T is a contraction mapping in an appropriate subset of X_T^m for a small $T > 0$. Then T has a fixed point,

$$\mathsf{T}(u_1, u_2, \dots, u_m) = (u_1, u_2, \dots, u_m),$$

which gives a unique solution in X_T^m. □

Let us introduce the potential energy

$$J : \mathscr{H}^m \to \mathbb{R},$$

defined by

$$J(u_1, u_2, \dots, u_m) = \sum_{j=1}^{m} \left(1 - \int_0^t \mu_j(s)\, ds \right) \|u_j\|_{\mathscr{H}}^2 + (\mu_j \circ u_j). \tag{8.115}$$

The modified energy is defined by

$$\begin{aligned} \mathscr{E}(t) = {} & \frac{\kappa - 1}{\kappa} \sum_{j=1}^{m} \|\partial_t u_j\|_{L_\rho^\kappa}^\kappa + \frac{1}{2} J(u_1, u_2, \dots, u_m) \\ & - \int_{\mathbb{R}^n} \rho(x) F(u_1, u_2, \dots, u_m) dx. \end{aligned} \tag{8.116}$$

Theorem 8.6 (Global existence). *Let* (1.58), (8.97)–(8.102) *hold. Under* (8.107), (8.108) *and for sufficiently small*

$$(u_{10}, u_{11}), (u_{20}, u_{21}), \dots, (u_{m0}, u_{m1}) \in \mathscr{H} \times L_\rho^\kappa(\mathbb{R}^n),$$

problem (8.96) *admits a unique global solution*

$$(u_1, u_2, \dots, u_m),$$

such that

$$(u_1, u_2, \dots, u_m) \in X^m, \tag{8.117}$$

where

$$X \equiv C([0, +\infty); \mathscr{H}) \cap C^1([0, +\infty); L_\rho^\kappa(\mathbb{R}^n)).$$

Lemma 8.10. *For*

$$(u_1, u_2, \dots, u_m) \in X_T^m,$$

the functional $\mathscr{E}(t)$ associated with problem (8.96) is a decreasing energy.

Proof. For

$$0 \le t_1 < t_2 \le T,$$

we have

$$\begin{aligned}
&\mathscr{E}(t_2) - \mathscr{E}(t_1)\\
&= \int_{t_1}^{t_2} \frac{d}{dt} E(t)\, dt\\
&= -\sum_{j=1}^{m} \int_{t_1}^{t_2} \left(a\|\partial_t u_j\|_{L_\rho^2}^2 + \omega\|\partial_t u_j\|_{\mathscr{H}}^2 + \frac{1}{2}\mu_j(t)\|u_j\|_{\mathscr{H}}^2 - \frac{1}{2}(\mu_j' \circ u_j) \right) dt\\
&\le 0,
\end{aligned}$$

owing to (8.99)–(8.102). □

By (8.108), we get

$$(v, v)_* = \omega \int_{\mathbb{R}^n} |\nabla_x v|^2\, dx + a \int_{\mathbb{R}^n} \rho v^2\, dx \ge (\omega\lambda_1 + a) \int_{\mathbb{R}^n} \rho v^2\, dx \ge 0. \tag{8.118}$$

To show the global solution, we need to use

$$\begin{aligned}
\mathscr{E}(t) &\ge \frac{1}{2} J(u_1, u_2, \dots, u_m) - \int_{\mathbb{R}^n} \rho(x) F(u_1, u_2, \dots, u_m) dx\\
&\ge \frac{1}{2} J(u_1, u_2, \dots, u_m) - \frac{1}{p+1} \left\| \sum_{j=1}^{m} u_j \right\|_{L_\rho^{(p+1)}}^{(p+1)}\\
&\quad - \frac{2}{p+1} \left(\left\| \sum_{j=1}^{m-1} u_j u_{j+1} \right\|_{L_\rho^{(p+1)/2}}^{(p+1)/2} + \|u_m u_1\|_{L_\rho^{(p+1)/2}}^{(p+1)/2} \right)\\
&\ge \frac{1}{2} J(u_1, u_2, \dots, u_m) - \frac{\eta}{p+1} \left[\sum_{j=1}^{m} \rho_j \|u_j\|_{\mathscr{H}}^2 \right]^{(p+1)/2}\\
&\ge \frac{1}{2} J(u_1, u_2, \dots, u_m) - \frac{\eta}{p+1} (J(u_1, u_2, \dots, u_m))^{(p+1)/2}\\
&= G(\beta).
\end{aligned} \tag{8.119}$$

Here,

$$\beta^2 = J(u_1, u_2, \dots, u_m),$$

for $t \in [0, T)$, where

$$G(\xi) = \frac{1}{2}\xi^2 - \frac{\eta}{p+1}\xi^{(p+1)}.$$

Noting that $\mathcal{E}_0 = G(\lambda_0)$, given in (8.105). Then

$$\begin{cases} G(\xi) \geq 0 & \text{in } \xi \in [0, \lambda_0], \\ G(\xi) < 0 & \text{in } \xi > \lambda_0. \end{cases} \tag{8.120}$$

Moreover,

$$\lim_{\xi \to +\infty} G(\xi) \to -\infty.$$

Then we have the following.

Lemma 8.11. *Let*

$$0 \leq \mathcal{E}(0) < \mathcal{E}_0.$$

(i) *If* $\sum_{j=1}^m \|u_{j0}\|^2_{\mathcal{H}} < \lambda_0^2$, *then*

$$J(u_1, u_2, \dots, u_m) < \lambda_0^2, \quad \forall t \in [0, T).$$

(ii) *If* $\sum_{j=1}^m \|u_{j0}\|^2_{\mathcal{H}} > \lambda_0^2$, *then*

$$\sum_{j=1}^m \|u_j\|^2_{\mathcal{H}} > \lambda_1^2, \quad \forall t \in [0, T), \lambda_1 > \lambda_0.$$

Proof. Since

$$0 \leq \mathcal{E}(0) < \mathcal{E}_0 = G(\lambda_0),$$

there exist ξ_1 and ξ_2, such that

$$G(\xi_1) = G(\xi_2) = \mathcal{E}(0),$$

with

$$0 < \xi_1 < \lambda_0 < \xi_2.$$

The case (i). By (8.119), we have

$$G(J(u_{10}, u_{20}, \dots u_{m0})) \le \mathscr{E}(0) = G(\xi_1),$$

which implies that

$$J(u_{10}, u_{20}, \dots u_{m0}) \le \xi_1^2.$$

Then we claim that

$$J(u_1, u_2, \dots, u_m) \le \xi_1^2, \quad \forall t \in [0, T).$$

Moreover, there exists $t_0 \in (0, T)$ such that

$$\xi_1^2 < J(u_1(t_0), u_2(t_0), \dots, u_m(t_0)) < \xi_2^2.$$

Then

$$G(J(u_1(t_0), u_2(t_0), \dots, u_m(t_0)) > \mathscr{E}(0) \ge \mathscr{E}(t_0),$$

by Lemma 8.10, which contradicts (8.119). Hence, we have

$$J(u_1, u_2, \dots, u_m) \le \xi_1^2 < \lambda_0^2, \quad \forall t \in [0, T).$$

The case (ii). We can now show that

$$\sum_{j=1}^{m} \|u_{j0}\|_{\mathscr{H}}^2 \ge \xi_2^2,$$

and that

$$\sum_{j=1}^{m} \|u_j\|_{\mathscr{H}}^2 \ge \xi_2^2 > \lambda_0^2,$$

in the same way as (i). □

Proof of Theorem 8.6. Let

$$(u_0, u_1), (u_{20}, u_{21}), \dots, (u_{m0}, u_{m1}) \in \mathscr{H} \times L_\rho^\kappa(\mathbb{R}^n),$$

satisfy both

$$0 \le \mathscr{E}(0) < \mathscr{E}_0,$$

and

$$\sum_{j=1}^{m} \|u_{j0}\|_{\mathscr{H}}^2 < \lambda_0^2.$$

By Lemma 8.10 and Lemma 8.11, we have

$$\begin{aligned}
&\frac{2(\kappa-1)}{\kappa} \sum_{j=1}^{m} \|\partial_t u_j\|_{L_\rho^\kappa}^\kappa + \sum_{j=1}^{m} \rho_j \|u_j\|_{\mathscr{H}}^2 \\
&\leq \frac{2(\kappa-1)}{\kappa} \sum_{j=1}^{m} \|\partial_t u_j\|_{L_\rho^\kappa}^\kappa + \sum_{j=1}^{m} \left[\left(1 - \int_0^t \mu_j(s)\, ds \right) \|u_j\|_{\mathscr{H}}^2 + (\mu_j \circ u_j) \right] \\
&\leq 2\mathscr{E}(t) + \frac{2\eta}{p+1} \left(\sum_{j=1}^{m} \rho_j \|u_j\|_{\mathscr{H}}^2 \right)^{(p+1)/2} \\
&\leq 2\mathscr{E}(0) + \frac{2\eta}{p+1} (J(u_1, u_2, \ldots, u_m))^{(p+1)/2} \\
&\leq 2\mathscr{E}_0 + \frac{2\eta}{p+1} \lambda_0^{p+1} \\
&= \eta^{-2/(p-1)}.
\end{aligned} \tag{8.121}$$

The proof is completed. □

8.3.2 Nonclassical decay of solution

The nonclassical decay rate for solution is given in the next theorem.

Theorem 8.7 (Decay of solution). *Let* (1.58), (8.97)–(8.102) *hold. Under conditions* (8.107), (8.108) *and*

$$\gamma = \eta \left(\frac{2(p+1)}{p\ \ 1} \mathscr{E}(0) \right)^{(p-1)/2} < 1, \tag{8.122}$$

there exists $t_0 > 0$ depending only on μ_j, a, ω, λ_1 and $H'(0)$ such that

$$0 \leq \mathscr{E}(t) < \mathscr{E}(t_0) \exp\left(- \int_{t_0}^{t} \frac{\mu(s)}{1 - \mu_0(t)} \right), \tag{8.123}$$

holds for all $t > t_0$.

Owing to the positivity of μ in (8.100), we have

$$0 \leq \mathscr{E}(t) < \mathscr{E}(t_0) \exp\left(- \int_{t_0}^{t} \mu(s)\, ds \right),$$

for a single wave equation. Condition (8.102) is imposed to lead

$$(\mu + \nu\mu) \circ u;$$

here, $\nu \in \mathbb{R}$.

Let

$$\Lambda(u_1, u_2, \ldots, u_m) = \frac{1}{2}\sum_{j=1}^{m}\left[\left(1 - \int_0^t \mu_j(s)\,ds\right)\|u_j\|_{\mathcal{H}}^2 + (\mu_j \circ u_j)\right] - \int_{\mathbb{R}^n} \rho(x)F(u_1, u_2, \ldots, u_m)dx, \tag{8.124}$$

$$\Pi(u_1, u_2, \ldots, u_m) = \sum_{j=1}^{m}\left[\left(1 - \int_0^t \mu_j(s)\,ds\right)\|u_j\|_{\mathcal{H}}^2 + (\mu_j \circ u_j)\right] - (p+1)\int_{\mathbb{R}^n} \rho(x)F(u_1, u_2, \ldots, u_m)dx. \tag{8.125}$$

Lemma 8.12. *Let* $(u_1, u_2, \ldots, u_m)$ *be the solution of problem* (8.96). *If*

$$\sum_{j=1}^{m}\|u_{j0}\|_{\mathcal{H}}^2 - (p+1)\int_{\mathbb{R}^n} \rho(x)F(u_1, u_2, \ldots, u_m)dx > 0. \tag{8.126}$$

Then under condition (8.122), *the functional*

$$\Pi(u_1, u_2, \ldots, u_m) > 0, \quad \forall t > 0.$$

Proof. By (8.126) and continuity, there exists a time $t_1 > 0$ such that

$$\Pi(u_1, u_2, \ldots, u_m) \geq 0, \quad \forall t < t_1.$$

Let

$$Y = \{(u_1, u_2, \ldots, u_m) \mid \Pi(u_1(t_0), u_2(t_0), \ldots, u_m(t_0)) = 0, \Pi(u_1, u_2, \ldots, u_m) > 0, \forall t \in [0, t_0)\}. \tag{8.127}$$

Then, by (8.124), we have for all $(u_1, u_2, \ldots, u_m) \in Y$,

$$\begin{aligned}\Lambda(u_1, u_2, \ldots, u_m) &= \frac{p-1}{2(p+1)}\sum_{j=1}^{m}\left(1 - \int_0^t \mu_j(s)\,ds\right)\|u_j\|_{\mathcal{H}}^2 + \frac{p-1}{2(p+1)}\sum_{j=1}^{m}(\mu_j \circ u_j)\\ &\quad + \frac{1}{p+1}\Pi(u_1, u_2, \ldots, u_m)\end{aligned}$$

$$\geq \frac{p-1}{2(p+1)} \sum_{j=1}^{m} [\rho_j \|u_j\|_{\mathcal{H}}^2 + (\mu_j \circ u_j)].$$

Owing to (8.116), it follows for $(u_1, u_2, \dots, u_m) \in Y$,

$$\begin{aligned} \rho_j \|u_j\|_{\mathcal{H}}^2 &\leq \frac{2(p+1)}{p-1} \Lambda(u_1, u_2, \dots, u_m) \\ &\leq \frac{2(p+1)}{p-1} \mathcal{E}(t) \leq \frac{2(p+1)}{p-1} \mathcal{E}(0). \end{aligned} \tag{8.128}$$

By (8.104), (8.122), we have

$$\begin{aligned} &(p+1) \int_{\mathbb{R}^n} F(u_1(t_0), u_2(t_0), \dots, u_m(t_0)) \\ &\leq \eta \sum_{j=1}^{m} (\rho_j \|u_j(t_0)\|_{\mathcal{H}}^2)^{(p+1)/2} \\ &\leq \eta \left(\frac{2(p+1)}{p-1} E(0) \right)^{(p-1)/2} \sum_{j=1}^{m} \rho_j \|u_j(t_0)\|_{\mathcal{H}}^2 \\ &\leq \gamma \sum_{j=1}^{m} \rho_j \|u_j(t_0)\|_{\mathcal{H}}^2 \\ &< \sum_{j=1}^{m} \left(1 - \int_0^{t_0} \mu_j(s) ds \right) \|u_j(t_0)\|_{\mathcal{H}}^2 \\ &< \sum_{j=1}^{m} \left(1 - \int_0^{t_0} \mu_j(s) ds \right) \|u_j(t_0)\|_{\mathcal{H}}^2 \\ &\quad + \sum_{j=1}^{m} (\mu_j \circ u_j(t_0)); \end{aligned} \tag{8.129}$$

hence,

$$\Pi(u_1(t_0), u_2(t_0), \dots, u_m(t_0)) > 0,$$

on Y, which contradicts the definition of Y since

$$\Pi(u_1(t_0), u_2(t_0), \dots, u_m(t_0)) = 0.$$

Thus,

$$\Pi(u_1, u_2, \dots, u_m) > 0, \quad \forall t > 0. \qquad \square$$

We are now in position to prove the decay rate.

Proof of Theorem 8.7. By (8.104) and (8.128), we have for $t \geq 0$,

$$0 < \sum_{j=1}^{m} \rho_j \|u_j\|_{\mathcal{H}}^2 \leq \frac{2(p+1)}{p-1}\mathcal{E}(t). \tag{8.130}$$

Let

$$I(t) = \frac{\mu(t)}{1-\mu_0(t)},$$

where μ and μ_0 defined in (8.100) and (8.101).

Noting that

$$\lim_{t\to+\infty} \mu(t) = 0,$$

by (8.99)–(8.101), we have

$$\lim_{t\to+\infty} I(t) = 0, \quad I(t) > 0, \ \forall t \geq 0.$$

Then we take $t_0 > 0$ such that

$$0 < \frac{2(\kappa-1)}{\kappa} I(t) < \min\{2(\omega\lambda_1 + a), \chi'(0)\}, \tag{8.131}$$

with (8.102) for all $t > t_0$. Due to (8.116), we have

$$\begin{aligned}\mathcal{E}(t) &\leq \frac{(\kappa-1)}{\kappa}\sum_{j=1}^{m} \|\partial_t u_j\|_{L_\rho^\kappa}^\kappa + \frac{1}{2}\sum_{j=1}^{m}(\mu_j \circ u_j) + \frac{1}{2}\sum_{j=1}^{m}\left(1 - \int_0^t \mu_j(s)\,ds\right)\|u_j\|_{\mathcal{H}}^2 \\ &\leq \frac{(\kappa-1)}{\kappa}\sum_{j=1}^{m} \|\partial_t u_j\|_{L_\rho^\kappa}^\kappa + \frac{1}{2}\sum_{j=1}^{m}(\mu_j \circ u_j) + \frac{1}{2}(1-\mu_0(t))\sum_{j=1}^{m}\|u_j\|_{\mathcal{H}}^2.\end{aligned}$$

Then by definition of $I(t)$, we have

$$I(t)\mathcal{E}(t) \leq \frac{(\kappa-1)}{\kappa} I(t)\sum_{j=1}^{m} \|\partial_t u_j\|_{L_\rho^\kappa}^\kappa + \frac{1}{2}\mu(t)\sum_{j=1}^{m}\|u_j\|_{\mathcal{H}}^2 + \frac{1}{2}I(t)\sum_{j=1}^{m}(\mu_j \circ u_j),$$

and Lemma 8.10. We have for all $t_1, t_2 \geq 0$,

$$\begin{aligned}&\mathcal{E}(t_2) - \mathcal{E}(t_1) \\ &\quad\leq -\int_{t_1}^{t_2}\left(a\sum_{j=1}^{m}\|\partial_t u_j\|_{L_\rho^2}^2 + \omega\sum_{j=1}^{m}\|\partial_t u_j\|_{\mathcal{H}}^2 + \frac{1}{2}\mu(t)\sum_{j=1}^{m}\|u_j\|_{\mathcal{H}}^2\right)dt\end{aligned}$$

$$+\int_{t_1}^{t_2}\frac{1}{2}\sum_{j=1}^{m}(\mu_j'\circ u_j)\,dt.$$

Then

$$\mathscr{E}'(t)\le-(\omega\lambda_1+a)\sum_{j=1}^{m}\|\partial_t u_j\|_{L^2_\rho}^2-\frac{1}{2}\mu(t)\sum_{j=1}^{m}\|u_j\|_{\mathscr{H}}^2+\frac{1}{2}\sum_{j=1}^{m}(\mu_j'\circ u_j).$$

Then, by (8.131), $\forall t\ge t_0$, we get

$$\begin{aligned}
&\mathscr{E}'(t)+I(t)\mathscr{E}(t)\\
&\quad\le\left\{\frac{(\kappa-1)}{\kappa}I(t)-(\omega\lambda_1+a)\right\}\sum_{j=1}^{m}\|\partial_t u_j\|_{L^2_\rho}^2\\
&\qquad+\frac{1}{2}\sum_{j=1}^{m}(\mu_j'\circ u_j)+\frac{1}{2}I(t)\sum_{j=1}^{m}(\mu_j\circ u_j)\\
&\quad\le\frac{1}{2}\sum_{j=1}^{m}\int_0^t\{\mu_j'(t-\tau)+I(t)\mu_j(t-\tau)\}\|u_j(t)-u_j(\tau)\|_{\mathscr{H}}^2\,d\tau\\
&\quad\le\frac{1}{2}\sum_{j=1}^{m}\int_0^t\{\mu_j'(\tau)+I(t)\mu_j(\tau)\}\|u_j(t)-u_j(t-\tau)\|_{\mathscr{H}}^2\,d\tau\\
&\quad\le\frac{1}{2}\sum_{j=1}^{m}\int_0^t\{-\chi(\mu_j(\tau))+\chi'(0)\mu_j(\tau)\}\|u_j(t)-u_j(t-\tau)\|_{\mathscr{H}}^2\,d\tau\\
&\quad\le 0.
\end{aligned}$$

By the convexity of χ and (8.102), we have

$$\chi(\xi)\ge\chi(0)+\chi'(0)\xi=\chi'(0)\xi.$$

Then

$$\mathscr{E}(t)\le\mathscr{E}(t_0)\exp\left(-\int_{t_0}^{t}I(s)ds\right),$$

which completes the proof. □

9 Damped wave problems with memory term in Fourier spaces

Let F denote the Fourier transform in $L^2(\mathbb{R}^n)$ defined as follows:

$$F[f](\xi) = \widehat{f}(\xi) = (2\pi)^{-n/2} \int_{\mathbb{R}^n} \exp(-ix.\xi) f(x)dx. \tag{9.1}$$

Here, ξ is the variable associated with the Fourier transform, where

$$i = \sqrt{-1}, \quad x.\xi = \sum_{i=1}^{n} x_i \xi_i,$$

and denote its inverse transform by F^{-1}. The operator $-\Delta$ is defined by

$$-\Delta v(x) = F^{-1}(|\xi|^2 F(v)(\xi))(x), \quad v \in H^2(\mathbb{R}^n), x \in \mathbb{R}^n.$$

9.1 1D-nonlinear wave equations with more regularity

In $\mathbb{R}$, we address well-posedness of the Cauchy problem for the nonlinear wave equation

$$\frac{\partial^2 u}{\partial t^2} - u_{xx} = \pm u^p, \tag{9.2}$$

$$u(x,0) = u_0(x), \quad \frac{\partial u}{\partial t}(x,0) = u_1(x), \tag{9.3}$$

in a one-space dimension; here, $p \geq 2$ is a positive integer. We search the solutions u whose Fourier transform $\hat{u}$ with respect to x is supported in $\mathbb{R}$. Let

$$H_\Gamma^s(\mathbb{R}) = \{u \in H_\Gamma^s(\mathbb{R}) : \operatorname{supp} \hat{u} \subset [0,\infty)\}.$$

We denote by $\Gamma \subset \mathbb{R}$ a closed half-ray, which may be assumed to be the half-line $[0,\infty)$. Suppose that $u \in H_\Gamma^s(\mathbb{R})$. Then

$$u^p \in H_\Gamma^s(\mathbb{R}),$$

in the following cases:

$$\begin{aligned} &p = 2, \quad \frac{1}{2} < s < \infty, \\ &p \geq 3, \quad \frac{1}{2} - \frac{1}{2p-4} < s < \infty. \end{aligned} \tag{9.4}$$

https://doi.org/10.1515/9783111633787-009

In all cases,

$$\|u^p\|_{H^{s-1}} \le C\|u\|^p_{H^s}, \quad C > 0.$$

Theorem 9.1. *Assume that p and s satisfy* (9.3). *Let*

$$u_0 \in H^s_\Gamma(\mathbb{R}), \quad u_1 \in H^{s-1}_\Gamma(\mathbb{R}).$$

Then the systems (9.2)–(9.3) *admit a unique distributional solution*

$$u \in C([-T,T], H^s_\Gamma(\mathbb{R})) \cap C^1((-T,T], H^{s-1}_\Gamma(\mathbb{R})), \quad T > 0.$$

Further, the map $(u_0, u_1) \to u$ *is locally Lipschitz continuous.*

Proof. Let

$$E(t) = F^{-1}\left(\frac{\sin t|\xi|}{|\xi|}\right),$$

and

$$\mathscr{M}u(t) = \frac{d}{dt}E(t) * u_0 + E(t) * u_1 + \int_0^t E(t-\tau) * u^p(\tau)d\tau.$$

The goal is to construct a fixed-point $u = \mathscr{M}u$ in the ball

$$B_T = \left\{u \in C([-T,T], H^s_\Gamma(\mathbb{R})) : \sup_{t\in[-T,T]}\left\|u - \frac{d}{dt}E * u_0 - E * u_1\right\|_{H^s(\mathbb{R})} \le 1\right\}, \quad \mathbb{T} > 0.$$

We have for all $t \ge 0$,

$$\widehat{\mathscr{M}u}(\xi,t) = \cos(t|\xi|)\widehat{u}_0(\xi) + \frac{\sin(t|\xi|)}{|\xi|}\widehat{u}_1(\xi) + \int_0^t \frac{\sin((t-\tau)|\xi|)}{|\xi|}\widehat{u}^{*p}(\xi,\tau)d\tau,$$

from where

$$\|Mu\|_{H^s(\mathbb{R})} \le \|u_0\|_{H^s(\mathbb{R})} + \|u_1\|_{H^{s-1}(\mathbb{R})} + \int_0^t \|u^p\|_{H^{s-1}(\mathbb{R})}d\tau.$$

The case

$$p = 2, \quad s > -\frac{1}{2}.$$

Let

$$v \in H_\Gamma^s(\mathbb{R}),$$

then

$$v^2 \in H_\Gamma^\sigma(\mathbb{R}),$$

for

$$\sigma < -\frac{1}{2} + 2s,$$

(say $\sigma = -\frac{1}{2} + 2s - \epsilon$). We have

$$-\frac{1}{2} + 2s - \epsilon > s - 1, \quad \epsilon > 0,$$

and then

$$v^2 \in H_\Gamma^{s-1}(\mathbb{R}),$$

and

$$\|v^2\|_{H^{s-1}(\mathbb{R})} \leq C\|v\|^2_{H^s(\mathbb{R})}.$$

Similarly, if

$$v, w \in H_\Gamma^\sigma(\mathbb{R}),$$

then

$$v^2 - w^2 \in H_\Gamma^{s-1}(\mathbb{R}),$$

and

$$\|v^2 - w^2\|_{H^{s-1}(\mathbb{R})} \leq C\|v - w\|^2_{H^s(\mathbb{R})}\|v + w\|^2_{H^s(\mathbb{R})}.$$

It follows that:
1. We have

$$u \in C([-T, T], H_\Gamma^s(\mathbb{R})),$$

   which implies

$$u^2 \in C([-T, T], H_\Gamma^{s-1}(\mathbb{R})).$$

   Then

$$v = u(t+h), \quad w = u.$$

2. We have

$$\mathcal{M} : B_T \to B_T.$$

This follows from the estimate

$$\|\mathcal{M}u\|_{H^s(\mathbb{R})} \le \|u_0\|_{H^s(\mathbb{R})} + \|v_0\|_{H^s(\mathbb{R})} + C\int_0^t \|u\|^p_{H^s(\mathbb{R})} d\tau.$$

3. $\mathcal{M}$ is a contraction on B_T for small T. This follows

$$\|\mathcal{M}u - \mathcal{M}v\|_{H^s(\mathbb{R})} \le C\int_0^t \|u(\tau)-v(\tau)\|_{H^s(\mathbb{R})} \sup_{\tau\in[0,T]} \|u(\tau)-v(\tau)\|_{H^s(\mathbb{R})} d\tau.$$

Existence and uniqueness follow. Also,

$$u = \mathcal{M}u,$$

and

$$u \in C([-T,T], H^{s-1}_\Gamma(\mathbb{R})),$$

implies that

$$u \in C^1([-T,T], H^{s-1}_\Gamma(\mathbb{R})).$$

Finally, Gronwall's inequality gives local Lipschitz continuity. For $p \ge 3$, a similar argument works using factorization of $v^p w^p$ and Proposition 1.1. □

9.2 Damped wave equation with weak-memory

Let us consider the weak-viscoelastic case in the following problem:

$$\begin{aligned} &\frac{\partial^2 u}{\partial t^2} - \Delta u - \Delta\frac{\partial u}{\partial t} + \alpha(t)\int_0^t \mu(t-s)\Delta u(s,x)ds = 0, \\ &u(0,x) = u_0(x) \in H^1(\mathbb{R}^n), \quad \frac{\partial u}{\partial t}(0,x) = u_1(x) \in L^2(\mathbb{R}^n), \quad x \in \mathbb{R}^n, \end{aligned} \tag{9.5}$$

where $n \ge 2$. The function μ is assumed to be as in (1.40), (1.42) and (1.43). The energy of u at time t is given by

$$E(t) = \frac{1}{2}\left\|\frac{\partial u}{\partial t}\right\|_2^2 + \frac{1}{2}\left(1 - \alpha(t)\int_0^t \mu(s)ds\right)\|\nabla u\|_2^2 + \frac{1}{2}\alpha(t)(\mu \circ \nabla u), \tag{9.6}$$

and the following energy functional law holds:

$$\begin{aligned} E'(t) &= \frac{1}{2}\alpha(t)(\mu' \circ \nabla u) - \frac{1}{2}\alpha(t)\mu(t)\|\nabla u\|_2^2 - \left\|\nabla\frac{\partial u}{\partial t}\right\|_2^2 \\ &\quad + \frac{1}{2}\alpha'(t)(\mu \circ \nabla u) - \frac{1}{2}\alpha'(t)\int_0^t \mu(s)ds\|\nabla u\|_2^2. \end{aligned} \tag{9.7}$$

Lemma 9.1. [1] *For any two functions,*

$$\mu \in C^1(\mathbb{R}), \quad v \in W^{1,2}(0,T),$$

it holds that

$$\begin{aligned} &\operatorname{Re}\left\{\alpha(t)\int_0^t \mu(t-s)v(s)ds\frac{\partial v}{\partial t}\right\} \\ &= -\frac{1}{2}\alpha(t)\mu(t)|v(t)|^2 + \frac{1}{2}\alpha(t)(\mu' \circ v)(t) \\ &\quad - \frac{1}{2}\frac{d}{dt}\alpha(t)(\mu \circ v)(t) + \frac{1}{2}\frac{d}{dt}\alpha(t)\int_0^t \mu(s)ds|v(t)|^2 \\ &\quad + \frac{1}{2}\alpha'(t)(\mu \circ v)(t) - \frac{1}{2}\alpha'(t)\int_0^t \mu(s)ds|v|^2 \end{aligned} \tag{9.8}$$

and

$$\left|\int_0^t \mu(t-s)(v(s)-v(t))ds\right|^2 \le \int_0^t |\mu(s)|ds\int_0^t |\mu|(t-s)|v(t)-v(s)|^2 ds.$$

Finally, we give the definition of *weak solutions* for the problem (9.5).

Definition 9.1. A weak solution of (9.5) is u such that

– $u \in C([0,T);H^1(\mathbb{R}^n))$, $\frac{\partial u}{\partial t} \in C^1([0,T);L^2(\mathbb{R}^n))$

---

1 M. N. Priyanjana, J. E. M R. Dharmawardane and J. E. M Rivera. Decay property for second-order hyperbolic systems of viscoelastic materials. J. Math. Anal. Appl. 366 (2010), 621–635.

- For all $v \in C_0^\infty([0,T] \times \mathbb{R}^n)$, u satisfies the generalized formulae:

$$\begin{aligned} 0 = & \int_0^T \left(\frac{\partial^2 u}{\partial t^2}, v\right) ds + \int_0^T \int_{\mathbb{R}^n} \nabla u \nabla v dx ds + \int_0^T \int_{\mathbb{R}^n} \nabla \frac{\partial u}{\partial t} \nabla v dx ds \\ & - \int_0^T \int_{\mathbb{R}^n} \alpha(t) \int_0^s \mu(s-\tau) \nabla u(\tau) d\tau \nabla v(s) dx ds, \end{aligned} \tag{9.9}$$

- u satisfies the initial conditions

$$u_0(x) \in H^1(\mathbb{R}^n), \quad u_1(x) \in L^2(\mathbb{R}^n).$$

Let us set

$$\hat{u}(t,\xi) = F(u(t,\cdot))(\xi).$$

9.2.1 Solution decays time asymptotically to zero

Theorem 9.2. *Assume u is the solution of* (9.5), *then the next general exponential estimate satisfies in the Fourier space*

$$E(t) \le W \exp\left(-\omega \int_0^t \alpha(s)\beta(s) ds\right), \quad \forall t \ge 0, \tag{9.10}$$

for some positive constants W, ω.

Proof. We take the Fourier transform of both sides of (9.5). Then one has the reduced equation for $\xi \in \mathbb{R}^n$, $t \in \mathbb{R}_*^+$:

$$\begin{aligned} & \frac{\partial^2 \hat{u}}{\partial t^2}(t,\xi) + |\xi|^2 \hat{u}(t,\xi) - |\xi|^2 \alpha(t) \int_0^t \mu(t-s) \hat{u}(s,\xi) ds + |\xi|^2 \frac{\partial \hat{u}}{\partial t}(t,\xi) = 0, \\ & \hat{u}(0,\xi) = \hat{u}_0(\xi) \in H^1(\mathbb{R}^n), \quad \frac{\partial \hat{u}}{\partial t}(0,\xi) = \hat{u}_1(\xi) \in L^2(\mathbb{R}^n). \end{aligned} \tag{9.11}$$

We apply the multiplier techniques in Fourier space in order to obtain useful estimates and prepare some functional associated with the nature of our problem to introduce appropriate Lyapunov functions.

First, to derive the equality for the physical energy, we multiply both sides of (9.11) by

$$\overline{\frac{\partial \hat{u}}{\partial t}}.$$

Then, taking the real part of the resulting identities, we obtain

$$E_1(t) = \frac{1}{2}\left|\frac{\partial \widehat{u}}{\partial t}\right|^2 + \frac{1}{2}|\xi|^2\left(1 - \alpha(t)\int_0^t \mu(s)ds\right)|\widehat{u}|^2 + \frac{1}{2}|\xi|^2\alpha(t)(\mu \circ \widehat{u})(t)$$

and

$$\begin{aligned} e_1(t) = {} & \frac{1}{2}|\xi|^2\left(\alpha(t)\mu(t)|\widehat{u}|^2 - \alpha(t)(\mu' \circ \widehat{u})(t) + 2\left|\frac{\partial \widehat{u}}{\partial t}\right|^2\right) \\ & + \frac{1}{2}|\xi|^2\left(\alpha'(t)(\mu \circ \widehat{u})(t) - \alpha'(t)\int_0^t \mu(s)ds|\widehat{u}|^2\right). \end{aligned} \tag{9.12}$$

Then

$$\frac{d}{dt}E_1(t) + e_1(t) = 0. \tag{9.13}$$

Second, the existence of the memory term forces us to make the first modification of the energy by multiplying (9.11) by

$$\left(-\frac{d}{dt}\left(\alpha(t)\int_0^t \mu(t-s)\bar{\widehat{u}}(s)ds\right)\right),$$

and taking the real part, we have that

$$\begin{aligned} 0 = {} & -\operatorname{Re}\left\{\frac{\partial^2 \widehat{u}}{\partial t^2}(t,\xi)\frac{d}{dt}\left(\alpha(t)\int_0^t \mu(t-s)\bar{\widehat{u}}(s)ds\right)\right\} \\ & - \operatorname{Re}\left\{|\xi|^2\widehat{u}\frac{d}{dt}\left(\alpha(t)\int_0^t \mu(t-s)\bar{\widehat{u}}(s)ds\right)\right\} \\ & + \frac{1}{2}|\xi|^2\frac{d}{dt}\left(\left|\alpha(t)\int_0^t \mu(t-s)\widehat{u}(s)ds\right|^2\right) \\ & - \operatorname{Re}\left\{|\xi|^2\frac{\partial \widehat{u}}{\partial t}\frac{d}{dt}\left(\alpha(t)\int_0^t \mu(t-s)\bar{\widehat{u}}(s)ds\right)\right\}. \end{aligned} \tag{9.14}$$

Since

$$\begin{aligned}\frac{d}{dt}\left(\alpha(t)\int_0^t \mu(t-s)\bar{\hat{u}}(s)ds\right) &= \alpha'(t)\int_0^t \mu(t-s)\bar{\hat{u}}(s)ds + \alpha(t)\frac{d}{dt}\left(\int_0^t \mu(t-s)\bar{\hat{u}}(s)ds\right)\\ &= \alpha'(t)\int_0^t \mu(t-s)\bar{\hat{u}}(s)ds + \alpha(t)\mu_0\bar{\hat{u}} + \alpha(t)\int_0^t \mu'(t-s)\bar{\hat{u}}(s)ds. \end{aligned}\tag{9.15}$$

The first term in (9.14) takes the form

$$\begin{aligned}-\operatorname{Re}&\left\{\frac{\partial^2\hat{u}}{\partial t^2}(t,\xi)\frac{d}{dt}\left(\alpha(t)\int_0^t \mu(t-s)\bar{\hat{u}}(s)ds\right)\right\}\\ &= -\operatorname{Re}\frac{\partial}{\partial t}\left\{\frac{\partial\hat{u}}{\partial t}\frac{d}{dt}\left(\alpha(t)\int_0^t \mu(t-s)\bar{\hat{u}}(s)ds\right)\right\}\\ &\quad + \operatorname{Re}\left\{\frac{\partial\hat{u}}{\partial t}\frac{d^2}{dt^2}\left(\alpha(t)\int_0^t \mu(t-s)\bar{\hat{u}}(s)ds\right)\right\}\\ &= -\operatorname{Re}\frac{\partial}{\partial t}\left\{\frac{\partial\hat{u}}{\partial t}\frac{d}{dt}\left(\alpha(t)\int_0^t \mu(t-s)\bar{\hat{u}}(s)ds\right)\right\} + \alpha(t)\mu_0\left|\frac{\partial\hat{u}}{\partial t}\right|^2\\ &\quad + \operatorname{Re}\left\{\frac{\partial\hat{u}}{\partial t}\left(\alpha(t)\frac{d}{dt}\left(\int_0^t \mu'(t-s)\bar{\hat{u}}(s)ds\right) + \alpha'(t)\int_0^t \mu(t-s)\bar{\hat{u}}(s)ds\right)\right\}.\end{aligned}$$

Denote by

$$E_2(t) = \frac{1}{2}\left(|\xi|^2\left|\alpha(t)\int_0^t \mu(t-s)\hat{u}(s)ds\right|^2 - \operatorname{Re}\left\{\frac{\partial\hat{u}}{\partial t}\frac{d}{dt}\left(\alpha(t)\int_0^t \mu(t-s)\bar{\hat{u}}(s)ds\right)\right\}\right),$$

and

$$\begin{aligned}e_2(t) &= \alpha(t)\mu_0\left|\frac{\partial\hat{u}}{\partial t}\right|^2 - \operatorname{Re}\left\{|\xi|^2\frac{\partial\hat{u}}{\partial t}\frac{d}{dt}\left(\alpha(t)\int_0^t \mu(t-s)\bar{\hat{u}}(s)ds\right)\right\}\\ &\quad + \operatorname{Re}\left\{\alpha'(t)\frac{\partial\hat{u}}{\partial t}\int_0^t \mu(t-s)\bar{\hat{u}}(s)ds\right\},\\ R_2(t) &= -\operatorname{Re}\left\{|\xi|^2\hat{u}\frac{d}{dt}\left(\alpha(t)\int_0^t \mu(t-s)\bar{\hat{u}}(s)ds\right)\right\}\end{aligned}$$

$$+ \operatorname{Re}\left\{\frac{\partial \hat{u}}{\partial t}\frac{d}{dt}\left(\alpha(t)\int_0^t \mu'(t-s)\bar{\hat{u}}(s)ds\right)\right\}.$$

Then

$$\frac{d}{dt}E_2(t) + e_2(t) + R_2(t) = 0. \tag{9.16}$$

Next, to make the second modification of the energy, which corresponds to the strong damping, we multiply (9.11) by $\bar{\hat{u}}$ and taking the real part, we have

$$\begin{aligned} 0 = {} & \frac{\partial}{\partial t}\left(\operatorname{Re}\left\{\frac{\partial \hat{u}}{\partial t}\bar{\hat{u}}\right\}\right) - \left|\frac{\partial \hat{u}}{\partial t}\right|^2 + |\xi|^2|\hat{u}|^2 \\ & - |\xi|^2 \operatorname{Re}\left\{\alpha(t)\int_0^t \mu(t-s)\hat{u}(s)\bar{\hat{u}}(t)ds\right\} + \frac{1}{2}|\xi|^2\frac{\partial}{\partial t}(|\hat{u}|^2), \end{aligned}$$

and using results in Lemma 9.1, we get

$$\begin{aligned} 0 = {} & \frac{\partial}{\partial t}\left(\operatorname{Re}\left\{\frac{\partial \hat{u}}{\partial t}\bar{\hat{u}}\right\}\right) - \left|\frac{\partial \hat{u}}{\partial t}\right|^2 + |\xi|^2|\hat{u}|^2 + \frac{1}{2}|\xi|^2\frac{\partial}{\partial t}(|\hat{u}|^2) \\ & - |\xi|^2\left(\alpha(t)\int_0^t \mu(s)ds|\hat{u}|^2 + \operatorname{Re}\left\{\alpha(t)\int_0^t \mu(t-s)(\hat{u}(s)-\hat{u}(t))\bar{\hat{u}}(s)ds\right\}\right). \end{aligned}$$

Denote

$$E_3(t) = \operatorname{Re}\left\{\frac{\partial \hat{u}}{\partial t}\bar{\hat{u}}\right\} + \frac{1}{2}|\xi|^2|\hat{u}|^2,$$

and

$$e_3(t) = |\xi|^2\left(1 - \alpha(t)\int_0^t \mu(s)ds\right)|\hat{u}|^2,$$

$$R_3(t) = -\left|\frac{\partial \hat{u}}{\partial t}\right|^2 - \operatorname{Re}\left\{\alpha(t)\int_0^t \mu(t-s)(\hat{u}(s)-\hat{u}(t))\bar{\hat{u}}(s)ds\right\}.$$

Then

$$\frac{d}{dt}E_3(t) + e_3(t) + R_3(t) = 0. \tag{9.17}$$

Let us define for some constants $\varepsilon_1, \varepsilon_2 > 0$ to be chosen later

$$E_4(t) = E_1(t) + \varepsilon_1\alpha(t)E_2(t) + \varepsilon_2\alpha(t)E_3(t)$$

$$= \frac{1}{2}\left\{\left|\frac{\partial \hat{u}}{\partial t}\right|^2 + |\xi|^2\left(1 - \alpha(t)\int_0^t \mu(s)ds\right)|\hat{u}|^2 + |\xi|^2\alpha(t)(\mu \circ \hat{u})(t)\right\}$$
$$+ \frac{\varepsilon_1\alpha(t)}{2}\left(|\xi|^2\left|\alpha(t)\int_0^t \mu(t-s)\hat{u}(s)ds\right|^2 - \mathrm{Re}\left\{\frac{\partial \hat{u}}{\partial t}\frac{d}{dt}\left(\alpha(t)\int_0^t \mu(t-s)\bar{\hat{u}}(s)ds\right)\right\}\right)$$
$$+ \varepsilon_2\alpha(t)\left(\mathrm{Re}\left\{\frac{\partial \hat{u}}{\partial t}\bar{\hat{u}}\right\} + \frac{1}{2}|\xi|^2|\hat{u}|^2\right)$$

and

$$e_4(t) = e_1(t) + \varepsilon_1\alpha(t)e_2(t) + \varepsilon_2\alpha(t)e_3(t)$$
$$= \frac{|\xi|^2}{2}\alpha(t)\left(\mu(t)|\hat{u}|^2 - (\mu' \circ \hat{u})(t) + 2\alpha^{-1}(t)\left|\frac{\partial \hat{u}}{\partial t}\right|^2\right)$$
$$+ \frac{|\xi|^2}{2}\alpha'(t)\left((\mu \circ \hat{u})(t) - \frac{\partial \hat{u}}{\partial t}\int_0^t \mu(s)ds|\hat{u}|^2\right)$$
$$+ \varepsilon_1\alpha(t)\left(\alpha(t)\mu_0\left|\frac{\partial \hat{u}}{\partial t}\right|^2 - \mathrm{Re}\left\{|\xi|^2\frac{\partial \hat{u}}{\partial t}\frac{d}{dt}\left(\alpha(t)\int_0^t \mu(t-s)\bar{\hat{u}}(s)ds\right)\right\}\right)$$
$$+ \varepsilon_1\alpha(t)\left(\mathrm{Re}\left\{\alpha'(t)\int_0^t \mu(t-s)\bar{\hat{u}}(s)ds\right\}\right)$$
$$+ \varepsilon_2|\xi|^2\alpha(t)\left(1 - \alpha(t)\int_0^t \mu(s)ds\right)|\hat{u}|^2$$

and

$$R_4(t) = \varepsilon_1\alpha(t)R_2(t) + \varepsilon_2\alpha(t)R_3(t)$$
$$= \varepsilon_1\alpha(t)\left(-\mathrm{Re}\left\{|\xi|^2\hat{u}\frac{d}{dt}\left(\alpha(t)\int_0^t \mu(t-s)\bar{\hat{u}}(s)ds\right)\right\}\right)$$
$$+ \varepsilon_1\alpha(t)\left(\mathrm{Re}\left\{\frac{\partial \hat{u}}{\partial t}\frac{d}{dt}\left(\alpha(t)\int_0^t \mu'(t-s)\bar{\hat{u}}(s)ds\right)\right\}\right)$$
$$+ \varepsilon_2\alpha(t)\left(-\left|\frac{\partial \hat{u}}{\partial t}\right|^2 - \mathrm{Re}\left\{\alpha(t)\int_0^t \mu(t-s)(\hat{u}(s) - \hat{u}(t))\bar{\hat{u}}(s)ds\right\}\right).$$

At this point, we introduce the Lyapunov functions as

$$L_1(t) = \left\{\left|\frac{\partial \hat{u}}{\partial t}\right|^2 + k|\xi|^2|\hat{u}|^2 + |\xi|^2\alpha(t)(\mu \circ \hat{u})(t)\right\} \tag{9.18}$$

and

$$L_2(t) = \alpha(t)\mu(t)|\widehat{u}|^2 + \alpha(t)\beta(t)(\mu \circ \widehat{u})(t). \tag{9.19}$$

It is easy to verify that there exists positive constants $c_1(\mu_0)$, $c_2(\mu_0)$ such that

$$c_1 L_1(t) \le E_1(t) \le c_2 L_1(t), \quad \forall t > 0. \tag{9.20}$$

Thanks to Hölder and Young's inequalities, one gets for some constant c_3,

$$\left|\varepsilon_1 E_2(t) + \varepsilon_2 E_3(t)\right| \le c_3 L_1(t),$$

which means that

$$L_1(t) \sim E(t).$$

Using again (1.43), Hölder and Young's inequalities and assumptions on μ to obtain

$$\begin{aligned}
\left|R_4(t)\right| &= \varepsilon_1 \alpha(t) R_2(t) + \varepsilon_2 \alpha(t) R_3(t) \\
&\le \varepsilon_1 \alpha(t) \operatorname{Re}\left\{ |\xi|^2 \widehat{u} \alpha(t) \frac{d}{dt}\left(\int_0^t \mu(t-s) \overline{\widehat{u}}(s) ds \right) \right\} \\
&\quad + \varepsilon_1 \alpha(t) \operatorname{Re}\left\{ \frac{\partial \widehat{u}}{\partial t} \alpha(t) \frac{d}{dt}\left(\int_0^t \mu'(t-s) \overline{\widehat{u}}(s) ds \right) \right\} \\
&\quad + \varepsilon_2 \alpha(t) \left(\left| \frac{\partial \widehat{u}}{\partial t} \right|^2 + \operatorname{Re}\left\{ \alpha(t) \int_0^t \mu(t-s)(\widehat{u}(s) - \widehat{u}(t)) \overline{\widehat{u}}(s) ds \right\} \right) \\
&\le \varepsilon_1 \alpha(t) \left| \frac{\partial \widehat{u}}{\partial t} \right|^2 + c_4 \varepsilon_1 \alpha(t) |\xi|^2 |\widehat{u}|^2 + c_5 \varepsilon_1 |\xi|^2 L_2(t) \\
&\quad + \varepsilon_2 \alpha(t) \left[\left| \frac{\partial \widehat{u}}{\partial t} \right|^2 + c_6 |\xi|^2 (\lambda |\widehat{u}|^2 + c_\lambda \alpha(t)(\mu \circ \widehat{u})(t)) \right] \\
&\le (\varepsilon_1 + \varepsilon_2) \alpha(t) \left| \frac{\partial \widehat{u}}{\partial t} \right|^2 + (c_4 \varepsilon_1 + \varepsilon_2 c_6 \lambda) \alpha(t) |\xi|^2 |\widehat{u}|^2 + (c_5 \varepsilon_1 + c_\lambda \varepsilon_2) |\xi|^2 L_2(t).
\end{aligned}$$

Since

$$L_2(t) \le c_3\, e_1(t),$$

one can easily check that there exists positive constants

$$\varepsilon_1, \varepsilon_2, \lambda, c_4, c_5, c_6,$$

such that

$$|R_4(t)| \le ce_4(t), \quad c > 0. \tag{9.21}$$

By (9.13), (9.16) and (9.17), we get

$$\frac{d}{dt}E_4(t) = \frac{d}{dt}E_1(t) + \varepsilon_1\alpha(t)\frac{d}{dt}E_2(t) + \varepsilon_2\alpha(t)\frac{d}{dt}E_3(t) + \varepsilon_1\alpha'(t)E_2(t) + \varepsilon_2\alpha'(t)E_3(t).$$

We use

$$\lim_{t\to\infty}\frac{\alpha'(t)}{\alpha(t)} = 0,$$

by (1.41)–(1.43) to choose $t_1 > t_0$ and since $e_4(t) \ge cE_4(t)$, then (9.21) gives for some positive constant N,

$$\frac{d}{dt}E_4(t) \le -N\alpha(t)E_4(t) + c\alpha(t)(\mu \circ \hat{u})(t). \tag{9.22}$$

Multiplying (9.22) by $\beta(t)$ and using (1.43), (9.19), we obtain

$$\begin{aligned}
\beta(t)\frac{d}{dt}E_4(t) &\le -N\beta(t)\alpha(t)E_4(t) + c\beta(t)\alpha(t)(\mu \circ \hat{u})(t)\\
&\le -N\beta(t)\alpha(t)E_4(t) - c\alpha(t)(\mu' \circ \hat{u})(t)\\
&\le -N\beta(t)\alpha(t)E_4(t) - c|\xi|^2\alpha'(t)\int_0^t \mu(s)ds|\hat{u}|^2 - 2c\frac{d}{dt}E_4(t), \quad \forall t > t_1.
\end{aligned} \tag{9.23}$$

Since $\beta'(t) \le 0$, we set

$$L(s) = (\beta(s) + 2c)E_4(s),$$

which is equivalent to $E_4(t)$, then

$$\begin{aligned}
\frac{d}{dt}L(t) &\le -N\beta(t)\alpha(t)E_4(t) - c|\xi|^2\alpha'(t)\int_0^t \mu(s)ds|\hat{u}|^2\\
&\le -\beta(t)\alpha(t)\left[N - \frac{2\alpha'(t)}{k\beta(t)\alpha(t)}\int_0^t \mu(s)ds\right]E_4(t), \quad \forall t > t_1.
\end{aligned} \tag{9.24}$$

By (1.43), we can choose $t_2 > t_1$ such that

$$\begin{aligned}
\frac{d}{dt}L(t) &\le -c\beta(t)\alpha(t)E_4(t)\\
&\le -c\beta(t)\alpha(t)L(t), \quad \forall t > t_2.
\end{aligned} \tag{9.25}$$

Integrating (9.25) over $[t_2, t]$ using equivalence between Lyapunov function and the energy function, it yields that

$$E(t) \leq W \exp\left(-\omega \int_0^t \alpha(s)\beta(s)ds\right), \quad W, \omega > 0. \qquad \square$$

9.3 Coupled system of wave equations

We consider a linear evolution system in the weak viscoelastic of the form:

$$\begin{cases} \frac{\partial^2 u}{\partial t^2} - \Delta(u + \frac{\partial u}{\partial t} - \alpha_1(t) \int_0^t \mu(t-s)u(s,x)ds) + \gamma v = 0, \\ \frac{\partial^2 v}{\partial t^2} - \Delta(v + \frac{\partial v}{\partial t} - \alpha_2(t) \int_0^t h(t-s)v(s,x)ds) + \gamma u = 0, \\ u(0,x) = u_0(x) \in H^1(\mathbb{R}^n), \quad \frac{\partial u}{\partial t}(0,x) = u_1(x) \in L^2(\mathbb{R}^n), \\ v(0,x) = v_0(x) \in H^1(\mathbb{R}^n), \quad \frac{\partial v}{\partial t}(0,x) = v_1(x) \in L^2\mathbb{R}^n), \end{cases} \tag{9.26}$$

where $n \geq 2$, $\gamma \neq 0$. The energy of (u, v) at time t is given by

$$\begin{aligned} E(t) = {} & \frac{1}{2}\left(\left\|\frac{\partial u}{\partial t}\right\|_2^2 + \left\|\frac{\partial v}{\partial t}\right\|_2^2\right) + \frac{1}{2}\left(1 - \alpha_1(t)\int_0^t \mu(s)ds\right)\|\nabla u\|_2^2 + \frac{1}{2}\alpha_1(t)(\mu \circ \nabla u) \\ & + \frac{1}{2}\left(1 - \alpha_2(t)\int_0^t h(s)ds\right)\|\nabla v\|_2^2 + \frac{1}{2}\alpha_2(t)(h \circ \nabla v) + \gamma \int_{\mathbb{R}^n} uvdx, \end{aligned} \tag{9.27}$$

and the following energy functional law holds:

$$\begin{aligned} E'(t) \leq {} & \frac{1}{2}\alpha_1(t)(\mu' \circ \nabla u) - \frac{1}{2}\alpha_1(t)\mu(t)\|\nabla u\|_2^2 + \frac{1}{2}\alpha_1'(t)(\mu \circ \nabla u) - \frac{1}{2}\alpha_1'(t)\int_0^t \mu(s)ds\|\nabla u\|_2^2 \\ & + \frac{1}{2}\alpha_2(t)(h' \circ \nabla v) - \frac{1}{2}\alpha_2(t)h(t)\|\nabla v\|_2^2 \\ & + \frac{1}{2}\alpha_2'(t)(h \circ \nabla v) - \frac{1}{2}\alpha_2'(t)\int_0^t h(s)ds\|\nabla v\|_2^2. \end{aligned} \tag{9.28}$$

We also consider the assumption on μ and h as in (1.41)–(1.43) and give the definition of *weak solutions* for the problem (9.26).

Definition 9.2. A weak solution of (9.26) is (u, v) such that
- $(u, v) \in (C([0,T); H^1(\mathbb{R}^n)))^2$, $(\frac{\partial u}{\partial t}, \frac{\partial v}{\partial t}) \in (C^1([0,T); L^2(\mathbb{R}^n)))^2$

- For all $w_1, w_2 \in C_0^\infty([0,T] \times \mathbb{R}^n)$, (u,v) satisfies the generalized formulae:

$$
\begin{aligned}
0 &= \int_0^T \left(\frac{\partial^2 u}{\partial t^2}, w_1 \right)_{L^2} ds + \int_0^T \int_{\mathbb{R}^n} \nabla u \nabla w_1 dx ds + \int_0^T \int_{\mathbb{R}^n} \nabla \frac{\partial u}{\partial t} \nabla w_1 dx ds \\
&\quad - \int_0^T \int_{\mathbb{R}^n} a_1(t) \int_0^s \mu(s-\tau) \nabla u(\tau) d\tau \nabla w_1(s) dx ds + \gamma \int_0^T \int_{\mathbb{R}^n} v w_1 dx ds, \\
0 &= \int_0^T \left(\frac{\partial^2 v}{\partial t^2}, w_2 \right)_{L^2} ds + \int_0^T \int_{\mathbb{R}^n} \nabla v \nabla w_2 dx ds + \int_0^T \int_{\mathbb{R}^n} \nabla \frac{\partial v}{\partial t} \nabla w_2 dx ds \\
&\quad - \int_0^T \int_{\mathbb{R}^n} a_2(t) \int_0^s h(s-\tau) \nabla v(\tau) d\tau \nabla w_2(s) dx ds + \gamma \int_0^T \int_{\mathbb{R}^n} u w_2 dx ds,
\end{aligned}
\tag{9.29}
$$

- (u,v) satisfies the initial conditions $(u_0(x), v_0(x)) \in (H^1(\mathbb{R}^n))^2$, $(u_1(x), v_1(x)) \in (L^2(\mathbb{R}^n))^2$.

We can now state and prove the asymptotic behavior of the solution of (9.26). Throughout this paper, let us set

$$\hat{u}(t,\xi) = F(u(t,\cdot))(\xi),$$

and

$$\hat{v}(t,\xi) = F(v(t,\cdot))(\xi).$$

9.3.1 Decay rate of solution

Theorem 9.3. *Assume (u,v) is the solution of (9.26), then the next general exponential estimate satisfies in the Fourier space*

$$E(\xi,t) \le W \exp\left(-\omega \int_0^t a(s)\beta(s) ds \right) E(\xi,0), \quad \forall t \ge 0, \tag{9.30}$$

for some positive constants W, ω, where

$$\beta(t) = \min\{\beta_1(t), \beta_2(t)\}.$$

Proof. We take the Fourier transform of both sides of (9.26). Then one has the reduced equation for $\xi \in \mathbb{R}^n$, $t \in \mathbb{R}_*^+$:

$$\begin{cases} \frac{\partial^2\hat{u}}{\partial t^2}(t,\xi) + |\xi|^2(\hat{u}(t,\xi) - \alpha_1(t)\int_0^t \mu(t-s)\hat{u}(s,\xi)ds + \frac{\partial\hat{u}}{\partial t}(t,\xi)) + \gamma\hat{v}(t,\xi) = 0, \\ \frac{\partial^2\hat{v}}{\partial t^2}(t,\xi) + |\xi|^2(\hat{v}(t,\xi) - \alpha_2(t)\int_0^t h(t-s)\hat{v}(s,\xi)ds + \frac{\partial\hat{v}}{\partial t}(t,\xi)) + \gamma\hat{u}(t,\xi) = 0, \\ (\hat{u}(0,\xi),\hat{v}(0,\xi)) = (\hat{u}_0(\xi),\hat{v}_0(\xi)) \in (H^1(\mathbb{R}^n))^2, \\ (\frac{\partial\hat{u}}{\partial t}(0,\xi), \frac{\partial\hat{v}}{\partial t}(0,\xi)) = (\hat{u}_1(\xi),\hat{v}_1(\xi)) \in (L^2(\mathbb{R}^n))^2. \end{cases} \tag{9.31}$$

For the proof, we shall need to proceed in three steps. First, to derive the equality for the physical energy, we multiply the first equation of (9.31) by

$$\overline{\frac{\partial\hat{u}}{\partial t}},$$

and the second by

$$\overline{\frac{\partial\hat{v}}{\partial t}}.$$

Then, summing ant taking the real part of the resulting identities, we obtain

$$\begin{aligned} E_1(t) &= \frac{1}{2}\left(\left|\frac{\partial\hat{u}}{\partial t}\right|^2 + \left|\frac{\partial\hat{v}}{\partial t}\right|^2\right) + \frac{1}{2}|\xi|^2\left(\left(1-\alpha_1(t)\int_0^t \mu(s)ds\right)|\hat{u}|^2 + \alpha_1(t)(\mu\circ\hat{u})(t)\right) \\ &\quad + \frac{1}{2}|\xi|^2\left(\left(1-\alpha_2(t)\int_0^t h(s)ds\right)|\hat{u}|^2 + \alpha_2(t)(h\circ\hat{u})(t)\right) + \gamma|\hat{u}\hat{v}| \end{aligned}$$

and

$$\begin{aligned} e_1(t) &= \frac{1}{2}|\xi|^2\Bigg(\alpha_1(t)\mu(t)|\hat{u}|^2 - \alpha_1(t)(\mu'\circ\hat{u})(t) \\ &\quad + 2\left|\frac{\partial\hat{u}}{\partial t}\right|^2 + \alpha_1'(t)(\mu\circ\hat{u})(t) - \alpha_1'(t)\int_0^t \mu(s)ds|\hat{u}|^2\Bigg) \\ &\quad + \frac{1}{2}|\xi|^2\Bigg(\alpha_2(t)h(t)|\hat{u}|^2 - \alpha_2(t)(h'\circ\hat{u})(t) \\ &\quad + 2\left|\frac{\partial\hat{u}}{\partial t}\right|^2 + \alpha_2'(t)(h\circ\hat{u})(t) - \alpha_2'(t)\int_0^t h(s)ds|\hat{u}|^2\Bigg). \end{aligned}$$

Then

$$\frac{d}{dt}E_1(t) + e_1(t) = 0. \tag{9.32}$$

Second, the existence of the memory terms forces us to make the first modification of the energy by multiplying $(9.11)_1$ by

$$\left(-\frac{d}{dt}\left(\alpha_1(t)\int_0^t \mu(t-s)\overline{\hat{u}}(s)ds\right)\right),$$

and $(9.11)_2$ by

$$\left(-\frac{d}{dt}\left(\alpha_2(t)\int_0^t h(t-s)\overline{\hat{v}}(s)ds\right)\right).$$

Summing and taking the real part, we have that

$$\begin{aligned}
0 = &-\operatorname{Re}\left\{\frac{\partial^2\hat{u}}{\partial t^2}\frac{d}{dt}\left(\alpha_1(t)\int_0^t \mu(t-s)\overline{\hat{u}}(s)ds\right)\right\}\\
&-\operatorname{Re}\left\{\frac{\partial^2\hat{v}}{\partial t^2}\frac{d}{dt}\left(\alpha_2(t)\int_0^t h(t-s)\overline{\hat{v}}(s)ds\right)\right\}\\
&-\operatorname{Re}\left\{|\xi|^2\hat{u}\frac{d}{dt}\left(\alpha_1(t)\int_0^t \mu(t-s)\overline{\hat{u}}(s)ds\right)+|\xi|^2\hat{v}\frac{d}{dt}\left(\alpha_2(t)\int_0^t h(t-s)\overline{\hat{v}}(s)ds\right)\right\}\\
&+\frac{1}{2}|\xi|^2\frac{d}{dt}\left(\left(\left|\alpha_1(t)\int_0^t \mu(t-s)\hat{u}(s)ds\right|^2\right)+\left(\left|\alpha_2(t)\int_0^t h(t-s)\hat{v}(s)ds\right|^2\right)\right)\\
&-|\xi|^2\operatorname{Re}\left\{\frac{\partial\hat{u}}{\partial t}\frac{d}{dt}\left(\alpha_1(t)\int_0^t \mu(t-s)\overline{\hat{u}}(s)ds\right)\frac{\partial\hat{v}}{\partial t}\frac{d}{dt}\left(\alpha_2(t)\int_0^t h(t-s)\overline{\hat{v}}(s)ds\right)\right\}\\
&+\gamma\operatorname{Re}\left\{\hat{v}\frac{d}{dt}\left(\alpha_1(t)\int_0^t \mu(t-s)\overline{\hat{u}}(s)ds\right)+\hat{u}\frac{d}{dt}\left(\alpha_2(t)\int_0^t h(t-s)\overline{\hat{v}}(s)ds\right)\right\}.
\end{aligned}\tag{9.33}$$

Since, for $\alpha = \alpha_1, \alpha_2, f = g, h$,

$$\begin{aligned}
&\frac{d}{dt}\left(\alpha(t)\int_0^t f(t-s)\overline{\hat{u}}(s)ds\right)\\
&\quad = \alpha'(t)\int_0^t f(t-s)\overline{\hat{u}}(s)ds + \alpha(t)\frac{d}{dt}\left(\int_0^t f(t-s)\overline{\hat{u}}(s)ds\right)\\
&\quad = \alpha'(t)\int_0^t f(t-s)\overline{\hat{u}}(s)ds + \alpha(t)f_0\overline{\hat{u}} + \alpha(t)\int_0^t f'(t-s)\overline{\hat{u}}(s)ds.
\end{aligned}\tag{9.34}$$

The $(9.33)_1$ takes the form

$$\begin{aligned}
&-\operatorname{Re}\left\{\frac{\partial^2\hat{u}}{\partial t^2}\frac{d}{dt}\left(\alpha_1(t)\int_0^t\mu(t-s)\bar{\hat{u}}(s)ds\right)\right\}\\
&=-\operatorname{Re}\frac{\partial}{\partial t}\left\{\frac{\partial\hat{u}}{\partial t}\frac{d}{dt}\left(\alpha_1(t)\int_0^t\mu(t-s)\bar{\hat{u}}(s)ds\right)\right\}\\
&\quad+\operatorname{Re}\left\{\frac{\partial\hat{u}}{\partial t}\frac{d^2}{dt^2}\left(\alpha_1(t)\int_0^t\mu(t-s)\bar{\hat{u}}(s)ds\right)\right\}\\
&=-\operatorname{Re}\frac{\partial}{\partial t}\left\{\frac{\partial\hat{u}}{\partial t}\frac{d}{dt}\left(\alpha_1(t)\int_0^t\mu(t-s)\bar{\hat{u}}(s)ds\right)\right\}+\alpha_1(t)\mu_0\left|\frac{\partial\hat{u}}{\partial t}\right|^2\\
&\quad+\operatorname{Re}\left\{\frac{\partial\hat{u}}{\partial t}\left(\alpha_1(t)\frac{d}{dt}\left(\int_0^t\mu'(t-s)\bar{\hat{u}}(s)ds\right)+\alpha_1'(t)\int_0^t\mu(t-s)\bar{\hat{u}}(s)ds\right)\right\}.
\end{aligned}$$

Similar to obtain

$$\begin{aligned}
&-\operatorname{Re}\left\{\frac{\partial^2\hat{v}}{\partial t^2}\frac{d}{dt}\left(\alpha_2(t)\int_0^t h(t-s)\bar{\hat{v}}(s)ds\right)\right\}\\
&=-\operatorname{Re}\frac{\partial}{\partial t}\left\{\frac{\partial\hat{v}}{\partial t}\frac{d}{dt}\left(\alpha_2(t)\int_0^t h(t-s)\bar{\hat{v}}(s)ds\right)\right\}+\alpha_2(t)h_0\left|\frac{\partial\hat{v}}{\partial t}\right|^2\\
&\quad+\operatorname{Re}\left\{\frac{\partial\hat{v}}{\partial t}\left(\alpha_2(t)\frac{d}{dt}\left(\int_0^t h'(t-s)\bar{\hat{v}}(s)ds\right)+\alpha_2'(t)\int_0^t h(t-s)\bar{\hat{v}}(s)ds\right)\right\}.
\end{aligned}$$

Denote by

$$\begin{aligned}
E_2(t)=\frac{1}{2}|\xi|^2&\left(\left|\alpha_1(t)\int_0^t\mu(t-s)\hat{u}(s)ds\right|^2+\left|\alpha_2(t)\int_0^t h(t-s)\hat{v}(s)ds\right|^2\right)\\
&-\operatorname{Re}\left\{\frac{\partial\hat{u}}{\partial t}\frac{d}{dt}\left(\alpha_1(t)\int_0^t\mu(t-s)\bar{\hat{u}}(s)ds\right)\right\}\\
&-\operatorname{Re}\left\{\frac{\partial\hat{v}}{\partial t}\frac{d}{dt}\left(\alpha_2(t)\int_0^t h(t-s)\bar{\hat{v}}(s)ds\right)\right\},
\end{aligned}$$

and

$$e_2(t)=\alpha_1(t)\mu_0\left|\frac{\partial\hat{u}}{\partial t}\right|^2+\alpha_2(t)h_0\left|\frac{\partial\hat{v}}{\partial t}\right|^2$$

$$- |\xi|^2 \operatorname{Re}\left\{ \frac{\partial \widehat{u}}{\partial t} \frac{d}{dt}\left(\alpha_1(t) \int_0^t \mu(t-s)\overline{\widehat{u}}(s)ds \right) + \frac{\partial \widehat{v}}{\partial t} \frac{d}{dt}\left(\alpha_2(t) \int_0^t h(t-s)\overline{\widehat{v}}(s)ds \right) \right\}$$
$$+ \operatorname{Re}\left\{ \alpha_1'(t) \frac{\partial \widehat{u}}{\partial t} \int_0^t \mu(t-s)\overline{\widehat{u}}(s)ds + \alpha_2'(t) \frac{\partial \widehat{v}}{\partial t} \int_0^t h(t-s)\overline{\widehat{v}}(s)ds \right\}$$

and

$$R_2(t) = -|\xi|^2 \operatorname{Re}\left\{ \widehat{u} \frac{d}{dt}\left(\alpha_1(t) \int_0^t \mu(t-s)\overline{\widehat{u}}(s)ds \right) + \widehat{v} \frac{d}{dt}\left(\alpha_2(t) \int_0^t h(t-s)\overline{\widehat{v}}(s)ds \right) \right\}$$
$$+ \operatorname{Re}\left\{ \frac{\partial \widehat{u}}{\partial t} \frac{d}{dt}\left(\alpha_1(t) \int_0^t \mu'(t-s)\overline{\widehat{u}}(s)ds \right) \right\}$$
$$+ \operatorname{Re}\left\{ \frac{\partial \widehat{v}}{\partial t} \frac{d}{dt}\left(\alpha_2(t) \int_0^t h'(t-s)\overline{\widehat{v}}(s)ds \right) \right\}$$
$$+ \gamma \operatorname{Re}\left\{ \widehat{v} \frac{d}{dt}\left(\alpha_1(t) \int_0^t \mu(t-s)\overline{\widehat{u}}(s)ds \right) + \widehat{u} \frac{d}{dt}\left(\alpha_2(t) \int_0^t h(t-s)\overline{\widehat{v}}(s)ds \right) \right\}.$$

Then

$$\frac{d}{dt} E_2(t) + e_2(t) + R_2(t) = 0. \tag{9.35}$$

Next, to make the second modification of the energy, which corresponds to the strong damping, we multiply the first equation of (9.11) by $\overline{\widehat{u}}$, the second by $\overline{\widehat{v}}$. Summing and taking the real part, we have

$$0 = \frac{\partial}{\partial t}\left(\operatorname{Re}\left\{ \frac{\partial \widehat{u}}{\partial t} \overline{\widehat{u}} \right\} \right) + \frac{\partial}{\partial t}\left(\operatorname{Re}\left\{ \frac{\partial \widehat{v}}{\partial t} \overline{\widehat{v}} \right\} \right) - \left| \frac{\partial \widehat{u}}{\partial t} \right|^2 - \left| \frac{\partial \widehat{v}}{\partial t} \right|^2 + |\xi|^2 |\widehat{u}|^2 + |\xi|^2 |\widehat{v}|^2$$
$$- |\xi|^2 \operatorname{Re}\left\{ \alpha_1(t) \int_0^t \mu(t-s)\widehat{u}(s)\overline{\widehat{u}}(t)ds \right\} - |\xi|^2 \operatorname{Re}\left\{ \alpha_2(t) \int_0^t h(t-s)\widehat{v}(s)\overline{\widehat{v}}(t)ds \right\}$$
$$+ \frac{1}{2}|\xi|^2 \frac{\partial}{\partial t}(|\widehat{u}|^2) + \frac{1}{2}|\xi|^2 \frac{\partial}{\partial t}(|\widehat{v}|^2) + \gamma(\widehat{v}\overline{\widehat{u}} + \widehat{u}\overline{\widehat{v}}),$$

using results in Lemma 9.1, we get

$$0 = \frac{\partial}{\partial t}\left(\operatorname{Re}\left\{ \frac{\partial \widehat{u}}{\partial t} \overline{\widehat{u}} \right\} \right) + \frac{\partial}{\partial t}\left(\operatorname{Re}\left\{ \frac{\partial \widehat{v}}{\partial t} \overline{\widehat{v}} \right\} \right) + \frac{1}{2}|\xi|^2 \frac{\partial}{\partial t}(|\widehat{u}|^2) + \frac{1}{2}|\xi|^2 \frac{\partial}{\partial t}(|\widehat{v}|^2)$$
$$- \left| \frac{\partial \widehat{u}}{\partial t} \right|^2 - \left| \frac{\partial \widehat{v}}{\partial t} \right|^2 + |\xi|^2 \left(1 - \alpha_1(t) \int_0^t \mu(s)ds \right) |\widehat{u}|^2 + |\xi|^2 \left(1 - \alpha_2(t) \int_0^t h(s)ds \right) |\widehat{v}|^2$$

$$-|\xi|^2\left(\mathrm{Re}\left\{\alpha_1(t)\int_0^t \mu(t-s)(\widehat{u}(s)-\widehat{u}(t))\overline{\widehat{u}}(s)ds\right\}\right.$$
$$\left.-\mathrm{Re}\left\{\alpha_2(t)\int_0^t h(t-s)(\widehat{v}(s)-\widehat{v}(t))\overline{\widehat{v}}(s)ds\right\}\right).$$

Denote

$$E_3(t)=\mathrm{Re}\left\{\frac{\partial\widehat{u}}{\partial t}\overline{\widehat{u}}\right\}+\mathrm{Re}\left\{\frac{\partial\widehat{v}}{\partial t}\overline{\widehat{v}}\right\}+\frac{1}{2}|\xi|^2|\widehat{u}|^2+\frac{1}{2}|\xi|^2|\widehat{v}|^2,$$

and

$$e_3(t)=|\xi|^2\left(1-\alpha_1(t)\int_0^t \mu(s)ds\right)|\widehat{u}|^2+|\xi|^2\left(1-\alpha_2(t)\int_0^t h(s)ds\right)|\widehat{v}|^2,$$

$$R_3(t)=-\left|\frac{\partial\widehat{u}}{\partial t}\right|^2-\mathrm{Re}\left\{\alpha_1(t)\int_0^t \mu(t-s)(\widehat{u}(s)-\widehat{u}(t))\overline{\widehat{u}}(s)ds\right\}$$
$$-\left|\frac{\partial\widehat{v}}{\partial t}\right|^2-\mathrm{Re}\left\{\alpha_2(t)\int_0^t h(t-s)(\widehat{v}(s)-\widehat{v}(t))\overline{\widehat{v}}(s)ds\right\}$$
$$+\gamma(\widehat{v}\overline{\widehat{u}}+\widehat{u}\overline{\widehat{v}}).$$

Then

$$\frac{d}{dt}E_3(t)+e_3(t)+R_3(t)=0. \tag{9.36}$$

Let us define for some constants $\varepsilon_1,\varepsilon_2>0$ to be chosen later

$$E_4(t)=E_1(t)+\varepsilon_1\alpha(t)E_2(t)+\varepsilon_2\alpha(t)E_3(t)$$
$$=\frac{1}{2}\left(\left|\frac{\partial\widehat{u}}{\partial t}\right|^2+\left|\frac{\partial\widehat{v}}{\partial t}\right|^2\right)+\frac{1}{2}|\xi|^2\left(1-\alpha_1(t)\int_0^t \mu(s)ds\right)|\widehat{u}|^2+\frac{1}{2}|\xi|^2\alpha_1(t)(\mu\circ\widehat{u})(t)$$
$$+\frac{1}{2}|\xi|^2\left(1-\alpha_2(t)\int_0^t h(s)ds\right)|\widehat{u}|^2+\frac{1}{2}|\xi|^2\alpha_2(t)(h\circ\widehat{u})(t)+\gamma|\widehat{u}\widehat{v}|$$
$$+\frac{\varepsilon_1\alpha(t)}{2}|\xi|^2\left(\left|\alpha_1(t)\int_0^t \mu(t-s)\widehat{u}(s)ds\right|^2+\left|\alpha_2(t)\int_0^t h(t-s)\widehat{v}(s)ds\right|^2\right)$$
$$-\frac{\varepsilon_1\alpha(t)}{2}\mathrm{Re}\left\{\frac{\partial\widehat{u}}{\partial t}\frac{d}{dt}\left(\alpha_1(t)\int_0^t \mu(t-s)\overline{\widehat{u}}(s)ds\right)\right\}$$

$$- \frac{\varepsilon_1 \alpha(t)}{2} \operatorname{Re}\left\{ \frac{\partial \widehat{v}}{\partial t} \frac{d}{dt}\left(\alpha_2(t) \int_0^t h(t-s)\overline{\widehat{v}}(s)ds \right)\right\}$$

$$+ \varepsilon_2 \alpha(t)\left(\operatorname{Re}\left\{ \frac{\partial \widehat{u}}{\partial t}\overline{\widehat{u}} \right\} + \operatorname{Re}\left\{ \frac{\partial \widehat{v}}{\partial t}\overline{\widehat{v}} \right\} + \frac{1}{2}|\xi|^2|\widehat{u}|^2 + \frac{1}{2}|\xi|^2|\widehat{v}|^2 \right)$$

and

$$\begin{aligned}
e_4(t) &= e_1(t) + \varepsilon_1 \alpha(t) e_2(t) + \varepsilon_2 \alpha(t) e_3(t) \\
&= \frac{1}{2}|\xi|^2\Bigg(\alpha_1(t)\mu(t)|\widehat{u}|^2 - \alpha_1(t)(\mu' \circ \widehat{u})(t) \\
&\quad + 2\left|\frac{\partial \widehat{u}}{\partial t}\right|^2 + \alpha_1'(t)(\mu \circ \widehat{u})(t) - \alpha_1'(t) \int_0^t \mu(s)ds|\widehat{u}|^2 \Bigg) \\
&\quad + \frac{1}{2}|\xi|^2\Bigg(\alpha_2(t)h(t)|\widehat{u}|^2 - \alpha_2(t)(h' \circ \widehat{u})(t) \\
&\quad + 2\left|\frac{\partial \widehat{u}}{\partial t}\right|^2 + \alpha_2'(t)(h \circ \widehat{u})(t) - \alpha_2'(t) \int_0^t h(s)ds|\widehat{u}|^2 \Bigg) \\
&\quad + \varepsilon_1 \alpha(t)\left(\alpha_1(t)\mu_0\left|\frac{\partial \widehat{u}}{\partial t}\right|^2 + \alpha_2(t)h_0\left|\frac{\partial \widehat{v}}{\partial t}\right|^2 \right) \\
&\quad - \varepsilon_1 \alpha(t)|\xi|^2 \operatorname{Re}\left\{ \frac{\partial \widehat{u}}{\partial t}\frac{d}{dt}\left(\alpha_1(t) \int_0^t \mu(t-s)\overline{\widehat{u}}(s)ds \right) + \frac{\partial \widehat{v}}{\partial t}\frac{d}{dt}\left(\alpha_2(t) \int_0^t h(t-s)\overline{\widehat{v}}(s)ds \right) \right\} \\
&\quad + \varepsilon_1 \alpha(t) \operatorname{Re}\left\{ \alpha_1'(t)\frac{\partial \widehat{u}}{\partial t} \int_0^t \mu(t-s)\overline{\widehat{u}}(s)ds \right\} \\
&\quad + \varepsilon_1 \alpha(t) \operatorname{Re}\left\{ \alpha_2'(t)\frac{\partial \widehat{v}}{\partial t} \int_0^t h(t-s)\overline{\widehat{v}}(s)ds \right\} \\
&\quad + \varepsilon_2 |\xi|^2 \alpha(t)\left(\left(1 - \alpha_1(t) \int_0^t \mu(s)ds\right)|\widehat{u}|^2 + \left(1 - \alpha_2(t) \int_0^t h(s)ds\right)|\widehat{v}|^2 \right)
\end{aligned} \tag{9.37}$$

and

$$\begin{aligned}
R_4(t) &= \varepsilon_1 \alpha(t) R_2(t) + \varepsilon_2 \alpha(t) R_3(t) \\
&= -\varepsilon_1 \alpha(t) \operatorname{Re}\left\{ |\xi|^2 \widehat{u} \frac{d}{dt}\left(\alpha_1(t) \int_0^t \mu(t-s)\overline{\widehat{u}}(s)ds \right)\right\} \\
&\quad - \varepsilon_1 \alpha(t) \operatorname{Re}\left\{ |\xi|^2 \widehat{v} \frac{d}{dt}\left(\alpha_2(t) \int_0^t h(t-s)\overline{\widehat{v}}(s)ds \right)\right\}
\end{aligned}$$

$$+\varepsilon_1\alpha(t)\operatorname{Re}\left\{\frac{\partial\hat{u}}{\partial t}\frac{d}{dt}\left(\alpha_1(t)\int_0^t \mu'(t-s)\bar{\hat{u}}(s)ds\right)\right\}$$

$$+\varepsilon_1\alpha(t)\operatorname{Re}\left\{\frac{\partial\hat{v}}{\partial t}\frac{d}{dt}\left(\alpha_2(t)\int_0^t h'(t-s)\bar{\hat{v}}(s)ds\right)\right\}$$

$$+\gamma\varepsilon_1\alpha(t)\operatorname{Re}\left\{\hat{v}\frac{d}{dt}\left(\alpha_1(t)\int_0^t \mu(t-s)\bar{\hat{u}}(s)ds\right)+\hat{u}\frac{d}{dt}\left(\alpha_2(t)\int_0^t h(t-s)\bar{\hat{v}}(s)ds\right)\right\}$$

$$-\varepsilon_2\alpha(t)\left|\frac{\partial\hat{u}}{\partial t}\right|^2-\varepsilon_2\alpha(t)\operatorname{Re}\left\{\alpha_1(t)\int_0^t \mu(t-s)(\hat{u}(s)-\hat{u}(t))\bar{\hat{u}}(s)ds\right\}$$

$$-\varepsilon_2\alpha(t)\left|\frac{\partial\hat{v}}{\partial t}\right|^2-\varepsilon_2\alpha(t)\operatorname{Re}\left\{\alpha_2(t)\int_0^t h(t-s)(\hat{v}(s)-\hat{v}(t))\bar{\hat{v}}(s)ds\right\}$$

$$+\gamma\varepsilon_2\alpha(t)(\hat{v}\bar{\hat{u}}+\hat{u}\bar{\hat{v}}). \tag{9.38}$$

At this point, we introduce the Lyapunov functions as

$$L_1(t)=\left\{\left|\frac{\partial\hat{u}}{\partial t}\right|^2+\left|\frac{\partial\hat{v}}{\partial t}\right|^2 + |\xi|^2(k_1|\hat{u}|^2+k_2|\hat{v}|^2+\alpha(t)((\mu\circ\hat{u})(t)+(h\circ\hat{v})(t))\right\} \tag{9.39}$$

and

$$L_2(t)=\alpha(t)(\mu(t)|\hat{u}|^2+h(t)|\hat{v}|^2)+\alpha(t)\beta(t)((\mu\circ\hat{u})(t)+(h\circ\hat{v})(t)). \tag{9.40}$$

It is easy to verify that there exists positive constants c_1, c_2 such that

$$c_1L_1(t)\le E_1(t)\le c_2L_1(t),\quad \forall t>0. \tag{9.41}$$

Thanks to Hölder and Young's inequalities, one gets for some constant c_3,

$$|\varepsilon_1E_2(t)+\varepsilon_2E_3(t)|\le c_3L_1(t),$$

which means that

$$L_1(t)\sim E(t).$$

Since $l\ge 2$, using again (1.43), Hölder and Young's inequalities and assumptions on g, h to obtain

$$|R_4(t)|\le\varepsilon_1\alpha(t)\operatorname{Re}\left\{|\xi|^2\hat{u}\frac{d}{dt}\left(\alpha(t)\int_0^t \mu(t-s)\bar{\hat{u}}(s)ds\right)\right\}$$

$$+\varepsilon_1\alpha(t)\,\mathrm{Re}\left\{|\xi|^2\hat{v}\frac{d}{dt}\left(\alpha(t)\int_0^t h(t-s)\bar{\hat{v}}(s)ds\right)\right\}$$

$$+\varepsilon_1\alpha(t)\,\mathrm{Re}\left\{\frac{\partial\hat{u}}{\partial t}\frac{d}{dt}\left(\alpha(t)\int_0^t \mu'(t-s)\bar{\hat{u}}(s)ds\right)\right\}$$

$$+\varepsilon_1\alpha(t)\,\mathrm{Re}\left\{\frac{\partial\hat{v}}{\partial t}\frac{d}{dt}\left(\alpha(t)\int_0^t h'(t-s)\bar{\hat{v}}(s)ds\right)\right\}$$

$$+\gamma\varepsilon_1\alpha(t)\,\mathrm{Re}\left\{\hat{v}\frac{d}{dt}\left(\alpha(t)\int_0^t \mu(t-s)\bar{\hat{u}}(s)ds\right)+\hat{u}\frac{d}{dt}\left(\alpha(t)\int_0^t h(t-s)\bar{\hat{v}}(s)ds\right)\right\}$$

$$+\varepsilon_2\alpha(t)\left|\frac{\partial\hat{u}}{\partial t}\right|^2+\varepsilon_2\alpha(t)\,\mathrm{Re}\left\{\alpha(t)\int_0^t \mu(t-s)(\hat{u}(s)-\hat{u}(t))\bar{\hat{u}}(s)ds\right\}$$

$$+\varepsilon_2\alpha(t)\left|\frac{\partial\hat{v}}{\partial t}\right|^2+\varepsilon_2\alpha(t)\,\mathrm{Re}\left\{\alpha(t)\int_0^t h(t-s)(\hat{v}(s)-\hat{v}(t))\bar{\hat{v}}(s)ds\right\}$$

$$+\gamma\varepsilon_2\alpha(t)(\hat{v}\bar{\hat{u}}+\hat{u}\bar{\hat{v}})$$

$$\leq\lambda(\varepsilon_1+\varepsilon_2)\alpha(t)\left(\left|\frac{\partial\hat{u}}{\partial t}\right|^2+\left|\frac{\partial\hat{v}}{\partial t}\right|^2\right)+\gamma c_\lambda(\varepsilon_1+\varepsilon_2)|\xi|^2L_2(t).$$

One can easily check that there exists positive constants ε_1, ε_2, λ, c_λ, γ such that

$$|R_4(t)|\leq ce_4(t),\quad c>0. \tag{9.42}$$

By (9.32), (9.35) and (9.36), we get

$$\frac{d}{dt}E_4(t)=\frac{d}{dt}E_1(t)+\varepsilon_1\alpha(t)\frac{d}{dt}E_2(t)+\varepsilon_2\alpha(t)\frac{d}{dt}E_3(t)+\varepsilon_1\alpha'(t)E_2(t)+\varepsilon_2\alpha'(t)E_3(t).$$

We use

$$\lim_{t\to\infty}\frac{\alpha'(t)}{\alpha(t)}=0,$$

by (1.41)–(1.43) to choose $t_1>0$ and since

$$e_4(t)\geq cE_4(t),$$

then (9.42) gives for some positive constant N,

$$\frac{d}{dt}E_4(t)\leq -N\alpha(t)E_4(t)+c\alpha(t)((\mu\circ\hat{u})(t)+(h\circ\hat{v})(t)). \tag{9.43}$$

Multiplying (9.43) by $\beta(t)$ and using (1.43), (9.40), we obtain

$$\begin{aligned}
\beta(t)\frac{d}{dt}E_4(t) &\le -N\beta(t)\alpha(t)E_4(t) + c\beta(t)\alpha(t)((\mu \circ \widehat{u})(t) + (h \circ \widehat{v})(t)) \\
&\le -N\beta(t)\alpha(t)E_4(t) - c\alpha(t)((\mu' \circ \widehat{u})(t) + (h' \circ \widehat{v})(t)) \\
&\le -N\beta(t)\alpha(t)E_4(t) - c|\xi|^2\alpha'(t)\int_0^t \mu(s)ds|\widehat{u}|^2 \\
&\quad - c|\xi|^2\alpha'(t)\int_0^t h(s)ds|\widehat{v}|^2 - 2c\frac{d}{dt}E_4(t), \quad \forall t > t_1. \qquad (9.44)
\end{aligned}$$

Since

$$\beta'(t) \le 0,$$

we set

$$L(s) = (\beta(s) + 2c)E_4(s),$$

which is equivalent to $E_4(t)$, then

$$\begin{aligned}
\frac{d}{dt}L(t) &\le -N\beta(t)\alpha(t)E_4(t) - c|\xi|^2\alpha'(t)\left(\int_0^t \mu(s)ds|\widehat{u}|^2 + \int_0^t h(s)ds|\widehat{v}|^2\right) \\
&\le -\beta(t)\alpha(t)\left[N - \frac{2\alpha'(t)}{k\beta(t)\alpha(t)}\int_0^t f(s)ds\right]E_4(t), \quad \forall t > t_1, \qquad (9.45)
\end{aligned}$$

where

$$f(t) = \max\{\mu(t), h(t)\}.$$

By (1.43), we have

$$\lim_{t\to\infty} = \frac{\alpha'(t)}{\beta(t)\alpha(t)} = 0,$$

then we can choose $t_2 > t_1$ such that

$$\begin{aligned}
\frac{d}{dt}L(t) &\le -c\beta(t)\alpha(t)E_4(t) \\
&\le -c\beta(t)\alpha(t)L(t), \quad \forall t > t_2. \qquad (9.46)
\end{aligned}$$

Integrating (9.46) over $[t_2, t]$ using equivalence between Lyapunov function and the energy function, it yields that

$$E(\xi, t) \le W\exp\left(-\omega\int_0^t \alpha(s)\beta(s)ds\right)E(\xi, 0), \quad W, \omega > 0. \qquad \square$$

10 Degenerate evolution equations in $\mathbb{R}^n$

10.1 Wave equation with structural δ-evolution

We study the decay rate of solution to the Cauchy problem for structural damped δ-evolution with memory term in Fourier spaces

$$\frac{\partial^2 u}{\partial t^2} + (-\Delta)^\delta u - \int_0^t \mu(t-s)(-\Delta)^\delta u(s)ds = 0, \quad x \in \mathbb{R}^n, t > 0, \tag{10.1}$$

with the initial conditions

$$u(x,0) = u_0(x), \quad \frac{\partial u}{\partial t}(x,0) = u_1(x) \quad \text{and} \quad \delta > 1. \tag{10.2}$$

We assume that the function μ satisfies (1.35)–(1.40). The following lemma, which is useful to obtain estimate of solutions in the Fourier space.

Lemma 10.1. *For any $\mu \in C^1(\mathbb{R}^+)$ and any $h \in H^1(\mathbb{R}^+)$, we have*

$$\begin{aligned} \operatorname{Re}(\mu * h)(t)\bar{h}_t(t) = &-\frac{1}{2}\mu(t)|h(t)|^2 + \frac{1}{2}(\mu' \circ h)(t) \\ &- \frac{1}{2}\frac{d}{dt}\left\{(\mu \circ h)(t) - \left(\int_0^t \mu(\tau)d\tau\right)|h(t)|^2\right\}, \end{aligned} \tag{10.3}$$

and

$$\left|\int_0^t \mu(t-s)(h(s)-h(t))ds\right|^2 \le \int_0^t |\mu(s)|ds \int_0^t |\mu|(t-s)|h(t)-h(s)|^2 ds.$$

The following lemma is very useful in the sequel.

Lemma 10.2. [1] *Assume that $\beta(t) > 0$ for all $t \ge 0$. Then we have*

$$\left\| |\xi|^l \exp\left\{-c|\xi|^2 \int_0^t \beta(\tau)d\tau\right\}\right\|_{L^p} \le C\left(1 + \int_0^t \beta(\tau)d\tau\right)^{-\frac{l}{2}-\frac{N}{2p}},$$

where C is a positive constant.

1 J. Wirth. Asymptotic properties of solutions to wave equations with time-dependent dissipation. PhD thesis, TU Bergakademie Freiberg, 2004.

https://doi.org/10.1515/9783111633787-010

After Fourier transformation of (10.1)–(10.2), we get the following system:

$$\begin{cases} \hat{u}_{tt} + |\xi|^{2\delta}(\hat{u} - \int_0^t \mu(t-s)\hat{u}(s)ds) = 0, & \xi \in \mathbb{R}^n, t \geq 0, \\ \hat{u}(\xi, 0) = \hat{u}_0(x), \quad \hat{u}_t(\xi, 0) = \hat{u}_1(x), & \xi \in \mathbb{R}^n. \end{cases} \tag{10.4}$$

The energy associated to (10.4) is given as

$$\hat{E}(\xi, t) = \frac{1}{2}\{|\hat{u}_t|^2 + |\xi|^{2\delta}|\hat{u}|^2\}(\xi, t). \tag{10.5}$$

Lemma 10.3. *The modified energy associated to* (10.4) *as follows:*

$$\hat{\mathscr{E}}(\xi, t) = \frac{1}{2}\left\{|\hat{u}_t|^2 + \left(1 - \int_0^t \mu(s)ds\right)|\xi|^{2\delta}|\hat{u}|^2 + |\xi|^{2\delta}(\mu \circ \hat{u})\right\}(\xi, t), \tag{10.6}$$

and the modified energy $\hat{\mathscr{E}}(\xi, t)$ *is nonincreasing and satisfies for all* $t \geq 0$,

$$\frac{d\hat{\mathscr{E}}(\xi, t)}{dt} = |\xi|^{2\delta}(\mu' \circ \hat{u})(\xi, t) - |\xi|^{2\delta}\mu(t)|\hat{u}(\xi, t)|^2 \leq 0. \tag{10.7}$$

Proof. Multiplying the equation in (10.4) by $\bar{\hat{u}}_t$ and taking the real part, we get

$$\frac{1}{2}\frac{d}{dt}\{|\hat{u}_t(\xi, t)|^2 + |\xi|^{2\delta}|\hat{u}(\xi, t)|^2\} = \operatorname{Re}\left\{|\xi|^{2\delta}\bar{\hat{u}}_t \int_0^t \mu(s)\hat{u}(\xi, t-s)ds\right\}. \tag{10.8}$$

Using Lemma 10.1, we easily see that

$$\begin{aligned} &\operatorname{Re}\left\{|\xi|^{2\delta}\bar{\hat{u}}_t \int_0^t \mu(s)\hat{u}(\xi, t-s)ds\right\} \\ &= -\frac{1}{2}|\xi|^{2\delta}\mu(t)|\hat{u}(\xi, t)|^2 + \frac{1}{2}|\xi|^{2\delta}(\mu' \circ \hat{u})(\xi, t) \\ &\quad - |\xi|^{2\delta}\frac{1}{2}\frac{d}{dt}\left\{(\mu \circ \hat{u})(\xi, t) - \left(\int_0^t \mu(\tau)d\tau\right)|\hat{u}(\xi, t)|^2\right\}. \end{aligned}$$

Substituting this last equality into (10.8), the identity (10.7) then follows.

On the other hand, using (1.35), there exists a positive constant $c_1 > 0$ such that, for all $t \geq 0$ and for all $\xi \in \mathbb{R}^n$, we have

$$\hat{E}(\xi, t) \leq c_1\hat{\mathscr{E}}(\xi, t). \tag{10.9}$$

□

We will prove the following exponential stability result.

Proposition 10.1. *Let* $\hat{u}(\xi,t)$ *be the solution of* (10.4). *Then there exist two positive* C, β_0 *such that*

$$\hat{E}(\xi,t) \le \hat{E}(\xi,0)\exp\left\{-\frac{\beta_0|\xi|^{2\delta}}{1+|\xi|^{2\delta}}\int_0^t \alpha(s)ds\right\}, \quad \forall t \ge t_0 > 0. \tag{10.10}$$

To prove Proposition 10.1, the key point is to apply the multiplier techniques in Fourier spaces in order to obtain a useful estimates and prepare some functional associated with the nature of our problem to introduce an appropriate Lyapunov functions.

Proof. Multiplying (10.4) by $\bar{\hat{u}}$ and taking the real part, we get

$$\begin{aligned}
&\frac{d}{dt}\operatorname{Re}(\hat{u}_t\bar{\hat{u}}) + \left(1 - \int_0^t \mu(s)ds\right)|\xi|^{2\delta}|\hat{u}|^2 - |\hat{u}_t|^2 \\
&\quad + |\xi|^{2\delta}\operatorname{Re}\left\{\bar{\hat{u}}(t)\int_0^t \mu(t-s)(\hat{u}(s)-\hat{u}(t))ds\right\} = 0.
\end{aligned}$$

Applying Young's inequality, we obtain, for any $\varepsilon_1 > 0$,

$$\begin{aligned}
&\frac{d}{dt}\operatorname{Re}(\hat{u}_t\bar{\hat{u}}) + \left(1 - \varepsilon_1 - \int_0^t \mu(s)ds\right)|\xi|^{2\delta}|\hat{u}|^2 - |\hat{u}_t|^2 \\
&\quad \le \frac{1}{4\varepsilon_1}|\xi|^{2\delta}(1-k)(\mu\circ\hat{u})(\xi,t).
\end{aligned} \tag{10.11}$$

Next, the existence of the memory term forces us to make the first modification of the energy by multiplying (10.4) by

$$\int_0^t \mu(t-s)(\bar{\hat{u}}(s) - \bar{\hat{u}}(t))ds,$$

and taking the real part, we have that

$$\begin{aligned}
&-\frac{d}{dt}\operatorname{Re}\left(\hat{u}_t\int_0^t \mu(t-s)(\bar{\hat{u}}(s)-\bar{\hat{u}}(t))ds\right) + \operatorname{Re}\left(\hat{u}_t\int_0^t \mu'(t-s)(\bar{\hat{u}}(s)-\bar{\hat{u}}(t))ds\right) \\
&\quad + |\hat{u}_t|^2\int_0^t \mu(s)ds - |\xi|^{2\delta}\operatorname{Re}\left(\hat{u}(t)\int_0^t \mu(t-s)(\bar{\hat{u}}(s)-\bar{\hat{u}}(t))ds\right) \\
&\quad + |\xi|^{2\delta}\operatorname{Re}\left\{\left(\hat{u}\int_0^t \mu(t-s)ds\right)\left(\int_0^t \mu(t-s)(\bar{\hat{u}}(s)-\bar{\hat{u}}(t))ds\right)\right\}
\end{aligned}$$

$$-|\xi|^{2\delta}\left|\int_0^t \mu(t-s)(\hat{u}(s)-\hat{u}(t))ds\right|^2 = 0.$$

Young's inequality gives, for any $\varepsilon_2 > 0$,

$$-\frac{d}{dt}\operatorname{Re}\left(\hat{u}_t\int_0^t \mu(t-s)(\bar{\hat{u}}(s)-\bar{\hat{u}}(t))ds\right) + |\hat{u}_t|^2\int_0^t \mu(s)ds \tag{10.12}$$
$$\leq \varepsilon_2|\hat{u}_t|^2 - \frac{\mu(0)}{4\varepsilon_2}\int_0^t \mu'(t-s)|\hat{u}(s)-\hat{u}(t)|^2 ds + \varepsilon_2(2-k)|\xi|^{2\delta}|\hat{u}|^2$$
$$+\left(|\xi|^{2\delta}\left\{\frac{(1-k)}{4\varepsilon_2} + (1-k)^2 + \frac{1}{4\varepsilon_2}\right\}\right)(\mu\circ\hat{u})(t), \quad \forall t \geq 0.$$

Then, for any $t \geq t_0$, the estimate (10.12) can be rewritten as

$$-\frac{d}{dt}\operatorname{Re}\left(\hat{u}_t\int_0^t \mu(t-s)(\bar{\hat{u}}(s)-\bar{\hat{u}}(t))ds\right) + \left(\int_0^{t_0}\mu(s)ds - \varepsilon_2\right)|\hat{u}_t|^2$$
$$\leq \varepsilon_2(2-k)|\xi|^{2\delta}|\hat{u}|^2 - \frac{\mu(0)}{4\varepsilon_2}(\mu'\circ\hat{u})(t) + |\xi|^{2\delta}\left\{\frac{(2-k)}{4\varepsilon_2} + (1-k)^2\right\}(\mu\circ\hat{u})(t). \tag{10.13}$$

Now, we define the functional $\mathscr{L}(\xi,t)$ as follows:

$$\mathscr{L}(\xi,t) = \frac{\mu_0}{2}\operatorname{Re}(\hat{u}_t\bar{\hat{u}}) - \operatorname{Re}\left(\hat{u}_t\int_0^t \mu(t-s)(\bar{\hat{u}}(s)-\bar{\hat{u}}(t))ds\right) \tag{10.14}$$

where

$$\mu_0 = \int_0^{t_0}\mu(s)ds.$$

Taking the derivative of $\mathscr{L}(\xi,t)$ with respect to t and exploiting the estimates (10.11) and (10.13), we arrive at

$$\frac{d}{dt}\mathscr{L}(\xi,t) + \frac{\mu_0}{2}(l-\varepsilon_1-\varepsilon_2(2-k)|\xi|^{2\delta}|\hat{u}|^2 + \left\{\frac{\mu_0}{2}-\varepsilon_2\right\}|\hat{u}_t|^2$$
$$\leq -c_2(\mu'\circ\hat{u})(t) + c_3|\xi|^{2\delta}(\mu\circ\hat{u})(t), \quad \forall t \geq 0, \tag{10.15}$$

where c_2 and c_3 are two positive constants depending on ε_1 and ε_2. In the last estimate, we have used the fact that

$$1 - \int_0^t \mu(s)ds \geq k,$$

and

$$\mu_0 = \int_0^{t_0} \mu(s)ds \le \int_0^{t} \mu(s)ds,$$

for any $t \ge t_0$.

Now, choosing $\varepsilon_1 < k$ in (10.15) and ε_2 small enough such that

$$\varepsilon_2 < \min\left(\frac{\mu_0}{2}, \frac{l - \varepsilon_1}{2 - k}\right);$$

hence, there exist $\lambda_i (i = 1, 2, 3)$ positive constants such that

$$\begin{aligned} &\frac{d}{dt}\mathscr{L}(\xi, t) + \lambda_1 |\xi|^{2\delta} |\hat{u}|^2 + \lambda_2 |\hat{u}_t|^2 + \lambda_3 |\xi|^{2\delta} (\mu \circ \hat{u})(t) \\ &\quad \le -c_2(\mu' \circ \hat{u})(t) + (c_3 + \lambda_3)|\xi|^{2\delta}(\mu \circ \hat{u})(t), \quad \forall t \ge 0. \end{aligned} \tag{10.16}$$

Define the Lyapunov functional $F(\xi, t)$ as follows:

$$F(\xi, t) = \eta(1 + |\xi|^{2\delta})\mathscr{E}(\xi, t) + |\xi|^{2\delta}\mathscr{L}(\xi, t), \tag{10.17}$$

for a large positive constant η that has to be chosen later. It is straightforward to see that for η large enough, we can find two positive constants β_1 and β_2 such that

$$\beta_1(1 + |\xi|^{2\delta})\mathscr{E}(\xi, t) \le F(\xi, t) \le \beta_2(1 + |\xi|^{2\delta})\mathscr{E}(\xi, t), \quad \forall t \ge 0. \tag{10.18}$$

On the other hand, taking the derivative of $F(\xi, t)$ with respect to t and using (10.7) and (10.16), we have

$$\begin{aligned} &\frac{d}{dt}F(\xi, t) + \lambda_1 |\xi|^{4\delta} |\hat{u}|^2 + \lambda_2 |\xi|^{2\delta} |\hat{u}_t|^2 + \lambda_3 |\xi|^{4\delta} (\mu \circ \hat{u})(t) \\ &\quad \le (\eta - c_2)|\xi|^{2\delta}(\mu' \circ \hat{u})(t) + (c_3 + \lambda_3)|\xi|^{4\delta}(\mu \circ \hat{u})(t), \quad \forall t \ge 0. \end{aligned} \tag{10.19}$$

At this point, we choose η large enough such that (10.18) holds, $\eta - c_2 > 0$. Consequently, (10.19) yields

$$\frac{d}{dt}F(\xi, t) + \lambda_0 |\xi|^{2\delta}\mathscr{E}(\xi, t) \le c_4 |\xi|^{4\delta}(\mu \circ \hat{u})(t), \quad \forall t \ge t_0, \tag{10.20}$$

for some $\lambda_0, c_4 > 0$.

Multiplying (10.20) by $\alpha(t)$, we get

$$\begin{aligned} \alpha(t)\frac{d}{dt}F(\xi, t) + \lambda_0 \alpha(t)|\xi|^{2\delta}\mathscr{E}(\xi, t) &\le c_4 \alpha(t)|\xi|^{4\delta}(\mu \circ \hat{u})(t) \\ &\le c_4 |\xi|^{4\delta}(-g' \circ \hat{u})(t), \quad \forall t \ge t_0. \end{aligned}$$

Exploiting (10.7), we have

$$a(t)\frac{d}{dt}F(\xi,t)+\lambda_0 a(t)|\xi|^{2\delta}\mathscr{E}(\xi,t)\le -c_4|\xi|^{2\delta}\frac{d}{dt}\mathscr{E}(\xi,t),\quad \forall t\ge t_0, \tag{10.21}$$

which, in turns, gives

$$\frac{d}{dt}\{a(t)F(\xi,t)+c_4|\xi|^{2\delta}\mathscr{E}(\xi,t)\}-a'(t)F(\xi,t)+\lambda_0 a(t)|\xi|^{2\delta}\mathscr{E}(\xi,t)\le 0,\quad \forall t\ge t_0. \tag{10.22}$$

Recalling that $a'(t)<0$ and setting

$$K(\xi,t)=a(t)F(\xi,t)+c_4|\xi|^{2\delta}\mathscr{E}(\xi,t), \tag{10.23}$$

we get

$$\frac{d}{dt}K(\xi,t)+\lambda_0 a(t)|\xi|^{2\delta}\mathscr{E}(\xi,t)\le 0,\quad \forall t\ge t_0. \tag{10.24}$$

On the other hand, and since $a(t)$ is bounded, we deduce that

$$K(\xi,t)\sim(1+|\xi|^{2\delta})\mathscr{E}(\xi,t). \tag{10.25}$$

Consequently, these last two estimates lead to

$$\frac{d}{dt}K(\xi,t)+\beta_0\frac{|\xi|^{2\delta}}{1+|\xi|^{2\delta}}a(t)K(\xi,t)\le 0,\quad \forall t\ge t_0, \tag{10.26}$$

for some $\beta_0>0$. Integrating (10.26) with respect to t yields

$$K(\xi,t)\le K(\xi,t_0)\exp\left\{-\frac{\beta_0|\xi|^{2\delta}}{1+|\xi|^{2\delta}}\int_{t_0}^{t}a(t)ds\right\}. \tag{10.27}$$

Thus, using (10.25), we obtain

$$\begin{aligned}\mathscr{E}(\xi,t)&\le C\mathscr{E}(\xi,t_0)\exp\left\{-\frac{\beta_0|\xi|^{2\delta}}{1+|\xi|^{2\delta}}\int_{t_0}^{t}a(t)ds\right\}\\&\le C\mathscr{E}(\xi,0)\exp\left\{-\frac{\beta_0|\xi|^{2\delta}}{1+|\xi|^{2\delta}}\int_{0}^{t}a(t)ds\right\}.\end{aligned} \tag{10.28}$$

Finally, the estimate (10.10) holds by combining (1) and (10.28). This completes the proof of Proposition 10.1. □

Theorem 10.1. *Let σ be a nonnegative integer. Assume that*

$$U_0 = (u_1, u_0')^T \in H^\sigma(\mathbb{R}^n) \cap L^1(\mathbb{R}^n).$$

Then the solution

$$U = \left(\frac{\partial u}{\partial t}, u_x\right)^T,$$

of problem (10.1) *satisfies, for all $t \geq 0$, the following decay estimates:*

$$\|\partial_x^k U(t)\|_{L^2} \leq C\left(1 + \int_0^t \alpha(s)ds\right)^{-\frac{\delta k}{2}-\frac{N}{4}} \|U_0\|_{L^1} + Ce^{-c\int_0^t \alpha(s)ds}\|\partial_x^k U_0\|_{L^2}, \tag{10.29}$$

where C and c are two positive constants and $k \leq \sigma$.

Proof. Applying the Plancherel theorem, observing that $|\widehat{U}(\xi,t)|^2$ and $\widehat{E}(\xi,t)$ are equivalent and making use of (10.10), we obtain

$$\begin{aligned}
\|\partial_x^k U(t)\|_2^2 &= \int_{\mathbb{R}^n} |\xi|^{2k}|\widehat{U}(\xi,t)|^2 d\xi \\
&\leq C\int_{\mathbb{R}^n} |\xi|^{2k} e^{-\frac{\beta_0|\xi|^{2\delta}}{1+|\xi|^{2\delta}}\int_0^t \alpha(s)ds}|\widehat{U}(\xi,0)|^2 d\xi \\
&= C\int_{|\xi|\leq 1} |\xi|^{2k} e^{-\frac{\beta_0|\xi|^{2\delta}}{1+|\xi|^{2\delta}}\int_0^t \alpha(s)ds}|\widehat{U}(\xi,0)|^2 d\xi \\
&\quad + C\int_{|\xi|\geq 1} |\xi|^{2k} e^{-\frac{\beta_0|\xi|^{2\delta}}{1+|\xi|^{2\delta}}\int_0^t \alpha(s)ds}|\widehat{U}(\xi,0)|^2 d\xi \\
&= I_1 + I_2.
\end{aligned} \tag{10.30}$$

Looking carefully to the form of

$$\rho(\xi) = \frac{|\xi|^{2\delta}}{1+|\xi|^{2\delta}},$$

we see that

$$\rho(\xi) \geq \frac{1}{2}|\xi|^{2\delta},$$

for

$$|\xi| \leq 1,$$

and

$$\rho(\xi) \geq \frac{1}{2} \quad \text{for } |\xi| \geq 1.$$

Consequently, we infer that

$$I_1 \leq C\|\widehat{U}_0\|^2_{L_\infty} \int\limits_{|\xi|\leq 1} |\xi|^{2k} e^{-\frac{1}{2}|\xi|^2\delta \int_0^t \alpha(s)ds} d\xi \leq C\left(1 + \int\limits_0^t \alpha(\tau)d\tau\right)^{-k\delta-\frac{n}{2}} \|U_0\|^2_{L^1}, \tag{10.31}$$

where we have used Lemma 10.2.

On the other hand,

$$I_2 \leq Ce^{-\frac{1}{2}\int_0^t \alpha(s)ds} \int\limits_{|\xi|\geq 1} |\xi|^{2k} \left|\widehat{U}_0(\xi)\right|^2 d\xi \leq Ce^{-\frac{1}{2}\int_0^t \alpha(s)ds} \|\partial_x^k U_0\|^2_{L^2}. \tag{10.32}$$

Collecting (10.31) and (10.32), then the estimate (10.29) holds. This completes the proof of Theorem 10.1. □

10.2 Plate equation with weak-memory

Let us consider the weak viscoelastic case in the following problem:

$$\begin{aligned} &\frac{\partial^2 u}{\partial t^2} + \Delta^2 u + \alpha(t)\int\limits_0^t \mu(t-s)\Delta u(s,x)ds - \Delta\frac{\partial u}{\partial t} = 0, \\ &u(0,x) = u_0(x) \in H^1(\mathbb{R}^n), \quad \frac{\partial u}{\partial t}(0,x) = u_1(x) \in L^2(\mathbb{R}^n), \end{aligned} \tag{10.33}$$

where $n \geq 2$. The function μ satisfy (1.41), (1.42) and (1.43). Let us set

$$\hat{u}(t,\xi) = F(u(t,\cdot))(\xi).$$

Theorem 10.2. *Assume u is the solution of* (10.33), *then the next general exponential estimate satisfies in the Fourier space*

$$E(t) \leq W\exp\left(-\omega\int\limits_0^t \alpha(s)\beta(s)ds\right), \quad \forall t \geq 0, \tag{10.34}$$

for some positive constants W, ω.

Proof. We take the Fourier transform of both sides of (10.33). Then one has the reduced equation for $\xi \in \mathbb{R}^n, t \in \mathbb{R}^+_*$:

$$
\begin{aligned}
&\frac{\partial^2\widehat{u}}{\partial t^2}(t,\xi)+|\xi|^4\widehat{u}(t,\xi)-|\xi|^2\alpha(t)\int_0^t\mu(t-s)\widehat{u}(s,\xi)ds+|\xi|^2\frac{\partial\widehat{u}}{\partial t}(t,\xi)=0,\\
&\widehat{u}(0,\xi)=\widehat{u}_0(\xi)\in H^1(\mathbb{R}^n),\quad \frac{\partial\widehat{u}}{\partial t}(0,\xi)=\widehat{u}_1(\xi)\in L^2(\mathbb{R}^n).
\end{aligned}
\tag{10.35}
$$

We apply the multiplier techniques in Fourier space in order to obtain useful estimates and prepare some functional associated with the nature of our problem to introduce appropriate Lyapunov functions.

First, to derive the equality for the physical energy, we multiply both sides of (10.35) by $\overline{\frac{\partial\widehat{u}}{\partial t}}$. We denote

$$
E_1(t)=\frac{1}{2}\left|\frac{\partial\widehat{u}}{\partial t}\right|^2+\frac{1}{2}|\xi|^2\left(|\xi|^2-\alpha(t)\int_0^t\mu(s)ds\right)|\widehat{u}|^2+\frac{1}{2}|\xi|^2\alpha(t)(\mu\circ\widehat{u})(t)
$$

and

$$
\begin{aligned}
e_1(t)=&\frac{1}{2}|\xi|^2\left(\alpha(t)\mu(t)|\widehat{u}|^2-\alpha(t)(\mu'\circ\widehat{u})(t)+2\left|\frac{\partial\widehat{u}}{\partial t}\right|^2\right)\\
&+\frac{1}{2}|\xi|^2\left(\alpha'(t)(\mu\circ\widehat{u})(t)-\alpha'(t)\int_0^t\mu(s)ds|\widehat{u}|^2\right).
\end{aligned}
$$

Then, taking the real part of the resulting identities and by Lemma 9.1, we obtain

$$
\frac{d}{dt}E_1(t)+e_1(t)=0. \tag{10.36}
$$

Second, the existence of the memory term forces us to make the first modification of the energy by multiplying (10.35) by

$$
\left(-\frac{d}{dt}\left(\alpha(t)\int_0^t\mu(t-s)\overline{\widehat{u}}(s)ds\right)\right),
$$

and taking the real part, we have that

$$
\begin{aligned}
0=&-\operatorname{Re}\left\{\frac{\partial^2\widehat{u}}{\partial t^2}\frac{d}{dt}\left(\alpha(t)\int_0^t\mu(t-s)\overline{\widehat{u}}(s)ds\right)\right\}\\
&-\operatorname{Re}\left\{|\xi|^4\widehat{u}\frac{d}{dt}\left(\alpha(t)\int_0^t\mu(t-s)\overline{\widehat{u}}(s)ds\right)\right\}\\
&+\frac{1}{2}|\xi|^2\frac{d}{dt}\left(\left|\alpha(t)\int_0^t\mu(t-s)\widehat{u}(s)ds\right|^2\right)
\end{aligned}
$$

$$- \operatorname{Re}\left\{ |\xi|^2 \frac{\partial \hat{u}}{\partial t} \frac{d}{dt}\left(\alpha(t) \int_0^t \mu(t-s)\bar{\hat{u}}(s)ds \right) \right\}. \tag{10.37}$$

Since

$$\begin{aligned}
&\frac{d}{dt}\left(\alpha(t) \int_0^t \mu(t-s)\bar{\hat{u}}(s)ds \right) \\
&\quad = \alpha'(t) \int_0^t \mu(t-s)\bar{\hat{u}}(s)ds + \alpha(t)\frac{d}{dt}\left(\int_0^t \mu(t-s)\bar{\hat{u}}(s)ds \right) \\
&\quad = \alpha'(t) \int_0^t \mu(t-s)\bar{\hat{u}}(s)ds + \alpha(t)\mu_0\bar{\hat{u}} + \alpha(t) \int_0^t \mu'(t-s)\bar{\hat{u}}(s)ds.
\end{aligned} \tag{10.38}$$

The first term in (10.37) takes the form

$$\begin{aligned}
&- \operatorname{Re}\left\{ \frac{\partial^2 \hat{u}}{\partial t^2} \frac{d}{dt}\left(\alpha(t) \int_0^t \mu(t-s)\bar{\hat{u}}(s)ds \right) \right\} \\
&\quad = - \operatorname{Re} \frac{\partial}{\partial t}\left\{ \frac{\partial \hat{u}}{\partial t} \frac{d}{dt}\left(\alpha(t) \int_0^t \mu(t-s)\bar{\hat{u}}(s)ds \right) \right\} \\
&\qquad + \operatorname{Re}\left\{ \frac{\partial \hat{u}}{\partial t} \frac{d^2}{dt^2}\left(\alpha(t) \int_0^t \mu(t-s)\bar{\hat{u}}(s)ds \right) \right\} \\
&\quad = - \operatorname{Re} \frac{\partial}{\partial t}\left\{ \frac{\partial \hat{u}}{\partial t} \frac{d}{dt}\left(\alpha(t) \int_0^t \mu(t-s)\bar{\hat{u}}(s)ds \right) \right\} + \alpha(t)\mu_0\left| \frac{\partial \hat{u}}{\partial t} \right|^2 \\
&\qquad + \operatorname{Re}\left\{ \frac{\partial \hat{u}}{\partial t}\left(\alpha(t)\frac{d}{dt}\left(\int_0^t \mu'(t-s)\bar{\hat{u}}(s)ds \right) + \alpha'(t) \int_0^t \mu(t-s)\bar{\hat{u}}(s)ds \right) \right\}.
\end{aligned}$$

Denote by

$$E_2(t) = \frac{1}{2}\left(|\xi|^2 \left| \alpha(t) \int_0^t \mu(t-s)\hat{u}(s)ds \right|^2 - \operatorname{Re}\left\{ \frac{\partial \hat{u}}{\partial t} \frac{d}{dt}\left(\alpha(t) \int_0^t \mu(t-s)\bar{\hat{u}}(s)ds \right) \right\} \right),$$

and

$$e_2(t) = \alpha(t)\mu_0\left| \frac{\partial \hat{u}}{\partial t} \right|^2 - \operatorname{Re}\left\{ |\xi|^2 \frac{\partial \hat{u}}{\partial t} \frac{d}{dt}\left(\alpha(t) \int_0^t \mu(t-s)\bar{\hat{u}}(s)ds \right) \right\}$$

$$+\operatorname{Re}\left\{\alpha'(t)\frac{\partial\hat{u}}{\partial t}\int_0^t \mu(t-s)\bar{\hat{u}}(s)ds\right\},$$

$$R_2(t) = -\operatorname{Re}\left\{|\xi|^4\hat{u}\frac{d}{dt}\left(\alpha(t)\int_0^t \mu(t-s)\bar{\hat{u}}(s)ds\right)\right\}$$

$$+\operatorname{Re}\left\{\frac{\partial\hat{u}}{\partial t}\frac{d}{dt}\left(\alpha(t)\int_0^t \mu'(t-s)\bar{\hat{u}}(s)ds\right)\right\}.$$

Then

$$\frac{d}{dt}E_2(t) + e_2(t) + R_2(t) = 0. \tag{10.39}$$

Next, to make the second modification of the energy, which corresponds to the strong damping, we multiply (10.35) by $\bar{\hat{u}}$ and taking the real part, we have

$$0 = \frac{\partial}{\partial t}\left(\operatorname{Re}\left\{\frac{\partial\hat{u}}{\partial t}\bar{\hat{u}}\right\}\right) - \left|\frac{\partial\hat{u}}{\partial t}\right|^2 + |\xi|^4|\hat{u}|^2$$
$$-|\xi|^2\operatorname{Re}\left\{\alpha(t)\int_0^t \mu(t-s)\hat{u}(s)\bar{\hat{u}}(t)ds\right\} + \frac{1}{2}|\xi|^2\frac{\partial}{\partial t}(|\hat{u}|^2),$$

using results in Lemma 9.1, we get

$$0 = \frac{\partial}{\partial t}\left(\operatorname{Re}\left\{\frac{\partial\hat{u}}{\partial t}\bar{\hat{u}}\right\}\right) - \left|\frac{\partial\hat{u}}{\partial t}\right|^2 + |\xi|^4|\hat{u}|^2 + \frac{1}{2}|\xi|^2\frac{\partial}{\partial t}(|\hat{u}|^2)$$
$$-|\xi|^2\left(\alpha(t)\int_0^t \mu(s)ds|\hat{u}|^2 - \operatorname{Re}\left\{\alpha(t)\int_0^t \mu(t-s)(\hat{u}(s)-\hat{u}(t))\bar{\hat{u}}(s)ds\right\}\right).$$

Denote

$$E_3(t) = \operatorname{Re}\left\{\frac{\partial\hat{u}}{\partial t}\bar{\hat{u}}\right\} + \frac{1}{2}|\xi|^2|\hat{u}|^2,$$

and

$$e_3(t) = |\xi|^2\left(|\xi|^2 - \alpha(t)\int_0^t \mu(s)ds\right)|\hat{u}|^2,$$

$$R_3(t) = -\left|\frac{\partial\hat{u}}{\partial t}\right|^2 - \operatorname{Re}\left\{\alpha(t)\int_0^t \mu(t-s)(\hat{u}(s)-\hat{u}(t))\bar{\hat{u}}(s)ds\right\}.$$

Then

$$\frac{d}{dt}E_3(t) + e_3(t) + R_3(t) = 0. \tag{10.40}$$

Let us define for some constants $\varepsilon_1, \varepsilon_2 > 0$ to be chosen later

$$\begin{aligned}
E_4(t) &= E_1(t) + \varepsilon_1 \alpha(t) E_2(t) + \varepsilon_2 \alpha(t) E_3(t) \\
&= \frac{1}{2}\left\{ \frac{2(1)}{2}\left|\frac{\partial \hat{u}}{\partial t}\right|^2 + |\xi|^2\left(|\xi|^2 - \alpha(t)\int_0^t \mu(s)ds \right)|\hat{u}|^2 + |\xi|^2\alpha(t)(\mu \circ \hat{u})(t) \right\} \\
&\quad + \frac{\varepsilon_1 \alpha(t)}{2}\left(|\xi|^2\left|\alpha(t)\int_0^t \mu(t-s)\hat{u}(s)ds\right|^2 - \mathrm{Re}\left\{ \frac{\partial \hat{u}}{\partial t}\frac{d}{dt}\left(\alpha(t)\int_0^t \mu(t-s)\bar{\hat{u}}(s)ds \right) \right\} \right) \\
&\quad + \varepsilon_2 \alpha(t)\left(\mathrm{Re}\left\{ \frac{\partial \hat{u}}{\partial t}\bar{\hat{u}} \right\} + \frac{1}{2}|\xi|^2|\hat{u}|^2 \right)
\end{aligned}$$

and

$$\begin{aligned}
e_4(t) &= e_1(t) + \varepsilon_1 \alpha(t) e_2(t) + \varepsilon_2 \alpha(t) e_3(t) \\
&= \frac{|\xi|^2}{2}\alpha(t)\left(\mu(t)|\hat{u}|^2 - (\mu' \circ \hat{u})(t) + 2\alpha^{-1}(t)\left|\frac{\partial \hat{u}}{\partial t}\right|^2 \right) \\
&\quad + \frac{|\xi|^2}{2}\alpha'(t)\left((\mu \circ \hat{u})(t) - \frac{\partial \hat{u}}{\partial t}\int_0^t \mu(s)ds|\hat{u}|^2 \right) \\
&\quad + \varepsilon_1 \alpha(t)\left(\alpha(t)\mu_0\left|\frac{\partial \hat{u}}{\partial t}\right|^2 - \mathrm{Re}\left\{ |\xi|^2\frac{\partial \hat{u}}{\partial t}\frac{d}{dt}\left(\alpha(t)\int_0^t \mu(t-s)\bar{\hat{u}}(s)ds \right) \right\} \right) \\
&\quad + \varepsilon_1 \alpha(t)\left(\mathrm{Re}\left\{ \alpha'(t)\int_0^t \mu(t-s)\bar{\hat{u}}(s)ds \right\} \right) \\
&\quad + \varepsilon_2 |\xi|^2 \alpha(t)\left(|\xi|^2 - \alpha(t)\int_0^t \mu(s)ds \right)|\hat{u}|^2
\end{aligned}$$

and

$$\begin{aligned}
R_4(t) &= \varepsilon_1 \alpha(t) R_2(t) + \varepsilon_2 \alpha(t) R_3(t) \\
&= \varepsilon_1 \alpha(t)\left(-\mathrm{Re}\left\{ |\xi|^4 \hat{u}\frac{d}{dt}\left(\alpha(t)\int_0^t \mu(t-s)\bar{\hat{u}}(s)ds \right) \right\} \right) \\
&\quad + \varepsilon_1 \alpha(t)\left(\mathrm{Re}\left\{ \frac{\partial \hat{u}}{\partial t}\frac{d}{dt}\left(\alpha(t)\int_0^t \mu'(t-s)\bar{\hat{u}}(s)ds \right) \right\} \right)
\end{aligned}$$

$$+\varepsilon_2\alpha(t)\left(-\left|\frac{\partial\hat{u}}{\partial t}\right|^2-\operatorname{Re}\left\{\alpha(t)\int_0^t\mu(t-s)(\hat{u}(s)-\hat{u}(t))\bar{\hat{u}}(s)ds\right\}\right).$$

At this point, we introduce the Lyapunov functions as

$$L_1(t)=\left\{\left|\frac{\partial\hat{u}}{\partial t}\right|^2+k|\xi|^2|\hat{u}|^2+|\xi|^2\alpha(t)(\mu\circ\hat{u})(t)\right\} \tag{10.41}$$

and

$$L_2(t)=\alpha(t)\mu(t)|\hat{u}|^2+\alpha(t)\beta(t)(\mu\circ\hat{u})(t). \tag{10.42}$$

It is easy to verify that there exists positive constants $c_1(\mu_0)$, $c_2(\mu_0)$ such that

$$c_1L_1(t)\le E_1(t)\le c_2L_1(t),\quad \forall t>0. \tag{10.43}$$

Thanks to Hölder, Young's inequalities, one gets for some constant $c_3>0$,

$$\left|\varepsilon_1E_2(t)+\varepsilon_2E_3(t)\right|\le c_3L_1(t),$$

which means that

$$L_1(t)\sim E(t).$$

Using again (1.43), Hölder and Young's inequalities and assumptions on μ to obtain

$$\begin{aligned}
\left|R_4(t)\right| &= \varepsilon_1R_2(t)+\varepsilon_2R_3(t)\\
&\le \varepsilon_1\operatorname{Re}\left\{|\xi|^4\hat{u}\alpha(t)\frac{d}{dt}\left(\int_0^t\mu(t-s)\bar{\hat{u}}(s)ds\right)\right\}\\
&\quad+\varepsilon_1\operatorname{Re}\left\{\frac{\partial\hat{u}}{\partial t}\alpha(t)\frac{d}{dt}\left(\int_0^t\mu'(t-s)\bar{\hat{u}}(s)ds\right)\right\}\\
&\quad+\varepsilon_2\left(\left|\frac{\partial\hat{u}}{\partial t}\right|^2+\operatorname{Re}\left\{\alpha(t)\int_0^t\mu(t-s)(\hat{u}(s)-\hat{u}(t))\bar{\hat{u}}(s)ds\right\}\right)\\
&\le \varepsilon_1\left|\frac{\partial\hat{u}}{\partial t}\right|^2+c_4\varepsilon_1|\xi|^4|\hat{u}|^2+c_5\varepsilon_1|\xi|^2L_2(t)\\
&\quad+\varepsilon_2\left[\left|\frac{\partial\hat{u}}{\partial t}\right|^2+c_6|\xi|^2(\lambda|\hat{u}|^2+c_\lambda\alpha(t)(\mu\circ\hat{u})(t))\right]\\
&\le (\varepsilon_1+\varepsilon_2)\left|\frac{\partial\hat{u}}{\partial t}\right|^2+(c_4\varepsilon_1|\xi|^2+\varepsilon_2c_6\lambda)|\xi|^2|\hat{u}|^2+(c_5\varepsilon_1+c_\lambda\varepsilon_2)|\xi|^2L_2(t).
\end{aligned}$$

Since

$$L_2(t) \le c_3\, e_1(t),$$

one can easily check that there exists positive constants $\varepsilon_1, \varepsilon_2, \lambda, c_4, c_5, c_6$ such that

$$|R_4(t)| \le c e_4(t), \quad c > 0. \tag{10.44}$$

By (10.36), (10.39) and (10.40), we get

$$\frac{d}{dt}E_4(t) = \frac{d}{dt}E_1(t) + \varepsilon_1 \alpha(t)\frac{d}{dt}E_2(t) + \varepsilon_2 \alpha(t)\frac{d}{dt}E_3(t) + \varepsilon_1 \alpha'(t)E_2(t) + \varepsilon_2 \alpha'(t)E_3(t).$$

We use

$$\lim_{t\to\infty} \frac{\alpha'(t)}{\alpha(t)} = 0,$$

by (1.41)–(1.43) to choose $t_1 > t_0$ and since

$$e_4(t) \ge cE_4(t),$$

then (10.44) gives for some positive constant N

$$\frac{d}{dt}E_4(t) \le -N\alpha(t)E_4(t) + c\alpha(t)(\mu \circ \hat{u})(t). \tag{10.45}$$

Multiplying (10.45) by $\beta(t)$ and using (1.43), (10.42), we obtain

$$\begin{aligned}\beta(t)\frac{d}{dt}E_4(t) &\le -N\beta(t)\alpha(t)E_4(t) + c\beta(t)\alpha(t)(\mu \circ \hat{u})(t)\\ &\le -N\beta(t)\alpha(t)E_4(t) - c\alpha(t)(\mu' \circ \hat{u})(t) \qquad (10.46)\\ &\le -N\beta(t)\alpha(t)E_4(t) - c|\xi|^2\alpha'(t)\int_0^t \mu(s)ds|\hat{u}|^2 - 2c\frac{d}{dt}E_4(t), \quad \forall t > t_1.\end{aligned}$$

Since $\beta'(t) \le 0$, we set

$$L(s) = (\beta(s) + 2c)E_4(s),$$

which is equivalent to $E_4(t)$, then

$$\begin{aligned}\frac{d}{dt}L(t) &\le -N\beta(t)\alpha(t)E_4(t) - c|\xi|^2\alpha'(t)\int_0^t \mu(s)ds|\hat{u}|^2\\ &\le -\beta(t)\alpha(t)\left[N - \frac{2\alpha'(t)}{k\beta(t)\alpha(t)}\int_0^t \mu(s)ds\right]E_4(t), \quad \forall t > t_1. \qquad (10.47)\end{aligned}$$

By (1.43), we can choose $t_2 > t_1$ such that

$$\begin{aligned} \frac{d}{dt}L(t) &\leq -c\beta(t)\alpha(t)E_4(t) \\ &\leq -c\beta(t)\alpha(t)L(t), \quad \forall t > t_2. \end{aligned} \tag{10.48}$$

Integrating (10.48) over $[t_2, t]$ using equivalence between Lyapunov function and the energy function, it yields that

$$E_4(t) \leq W \exp\left(-\omega \int_0^t \alpha(s)\beta(s)ds\right), \quad W, \omega > 0. \qquad \square$$

11 Structural damped wave equations with nonlinear memory terms

Let us look at the next Cauchy problem

$$\begin{cases} \frac{\partial^2 u}{\partial t^2} - \Delta u + (t+1)^{r_1} \frac{\partial u}{\partial t} = \int_0^t (t-s)^{-\gamma} |v(s,\cdot)|^p ds, \\ \frac{\partial^2 v}{\partial t^2} - \Delta v + (t+1)^{r_2} \frac{\partial v}{\partial t} = \int_0^t (t-s)^{-\gamma} |u(s,\cdot)|^q ds, \\ u(t=0,x) = u_0(x); \quad \frac{\partial u}{\partial t}(t=0,x) = u_1(x), \\ v(t=0,x) = v_0(x); \quad \frac{\partial v}{\partial t}(t=0,x) = v_1(x), \end{cases} \tag{11.1}$$

where $n > 0$, $p, q \geq 1$, $r_1, r_2 \in (-1,1)$, $\gamma_1, \gamma_2 \in (0,1)$, $u_0(x)$, $u_1(x)$ and $v_0(x)$, $v_1(x)$ are given initial data.

11.1 Existence of unique solution

We consider (11.1) with $r_1 = r_2$ and $\gamma_1 = \gamma_2$. We present the scale $A_{m,\sigma}$ of Banach spaces with $\sigma > 0$ and $1 \leq m < 2$ as

$$A_{m,\sigma} = A_{m,\sigma}(\mathbb{R}^n) = (L^m(\mathbb{R}^n) \cap H^\sigma(\mathbb{R}^n)) \times (L^m(\mathbb{R}^n) \cap H^{\max\{0;\sigma-1\}}(\mathbb{R}^n)). \tag{11.2}$$

For $(\tau, t) \in [0,t] \times [0,\infty)$, let

$$B(t,\tau) = \int_\tau^t (s+1)^{-r} ds = \frac{1}{1-r}((t+1)^{1-r} - (\tau+1)^{1-r}).$$

In order to prove that small data solutions exist globally for nonlinear models, the following lemma will be used.

Lemma 11.1. [1] *Let $\gamma \in (0,1)$, $a \geq 0$ and $b > 1$ and $-1 < r < 1$. Then*

$$\int_0^t (s+1)^{-r}(1+B(t,s))^{-a} \int_0^s (s-\tau)^{-\gamma}(\tau+1)^{-b} d\tau ds$$

$$\leq C \begin{cases} (t+1)^{-\min\{a(1-r);\gamma\}} & \max\{a; r+\gamma\} > 1, \\ (t+1)^{-\min\{a(1-r);\gamma\}} \log(2+t) & \max\{a; r+\gamma\} = 1, \\ (t+1)^{1-a(1-r)-\gamma-r} & \max\{a; r+\gamma\} < 1, \end{cases}$$

with C a positive constant.

1 K. Tayeb Hadj and M. Reissing. Global well-posedness for effectively damped wave models with nonlinear memory. Commun. Pure Appl. Anal. 20 (5) (2021), 2039–2064.

https://doi.org/10.1515/9783111633787-011

Lemma 11.2. [2] *Let $\gamma \in (0,1)$, $a \geq 0$ and $b \in (0,1]$ and $-1 < r < 1$. Then*

$$\int_0^t (s+1)^{-r}(1+B(t,s))^{-a} \int_0^s (s-\tau)^{-\gamma}(\tau+1)^{-b} d\tau ds$$
$$\leq C \begin{cases} (t+1)^{-\min\{a(1-r);\gamma+b-1\}} & \max\{a; r+\gamma+b-1\} > 1, \\ (t+1)^{-\min\{a(1-r);\gamma+b-1\}} \log(2+t) & \max\{a; r+\gamma+b-1\} = 1, \\ (t+1)^{2-a(1-r)-b-\gamma-r} & \max\{a; r+\gamma+b-1\} < 1 \end{cases}$$

with C a positive constant.

Proposition 11.1. [3] *Let us consider the problem* (11.6) *with $u_1 \equiv 0$ and the data u_0 belongs to $A_{m,\sigma}$ with $\sigma \in (0,1)$ and $1 \leq m < 2$. The Sobolev solutions have decay estimates similar to those of Matsumura:*

$$\|u(t,\cdot)\|_{L^2} \leq C(t+1)^{-\frac{n}{2}(\frac{1}{m}-\frac{1}{2})(1-r)} \|u_0\|_{A_{m,\sigma}},$$
$$\|D^\sigma u(t,\cdot)\|_{L^2} \leq C(t+1)^{-\frac{n}{2}(\frac{1}{m}-\frac{1}{2})(1-r)-\frac{\sigma}{2}(1-r)} \|u_0\|_{A_{m,\sigma}}.$$

Proposition 11.2. [4] *Let us consider the problem* (11.6) *with $u_1 \equiv 0$ when the data u_0 belongs to $A_{m,\sigma}$ with $\sigma \geq 1$ and $1 \leq m < 2$. The Sobolev solutions have decay estimates similar to those of Matsumura:*

$$\|u(t,\cdot)\|_{L^2} \leq C(t+1)^{-\frac{n}{2}(\frac{1}{m}-\frac{1}{2})(1-r)} \|u_0\|_{A_{m,\sigma}},$$
$$\|D^\sigma u(t,\cdot)\|_{L^2} \leq C(t+1)^{-\frac{n}{2}(\frac{1}{m}-\frac{1}{2})(1-r)-\frac{\sigma}{2}(1-r)} \|u_0\|_{A_{m,\sigma}},$$
$$\left\|\frac{\partial u}{\partial t}(t,\cdot)\right\|_{L^2} \leq C(t+1)^{-\frac{n}{2}(\frac{1}{m}-\frac{1}{2})(1-r)-(1-r)} \|u_0\|_{A_{m,\sigma}}.$$
$$\left\|D^\sigma \frac{\partial u}{\partial t}(t,\cdot)\right\|_{L^2} \leq C(t+1)^{-\frac{n}{2}(\frac{1}{m}-\frac{1}{2})(1-r)-\frac{\sigma-1}{2}(1-r)-(1-r)} \|u_0\|_{A_{m,\sigma}}.$$

Proposition 11.3. [5] *Let us consider the Cauchy problem* (11.7) *with $h = h(\tau,u) \in L^m \cap L^2$ for some $1 \leq m < 2$. Then the Sobolev solutions $v = v(t,x)$ to* (11.7) *belong for $\sigma \in (0,1)$ to*

---

2 K. Tayeb Hadj and M. Reissing. Global well-posedness for effectively damped wave models with nonlinear memory. Commun. Pure Appl. Anal. 20 (5), (2021), 2039–2064.

3 M. D'Abbicco, S. Lucente and M. Reissig. Semilinear wave equations with effective damping. Chin. Ann. Math., Serie B 34 (2013), 345380.

4 J. Wirth. Asymptotic properties of solutions to wave equations with time-dependent dissipation. PhD thesis, TU Bergakademie Freiberg, 2004.

5 A. Djaouti and M. Reissig. Weakly coupled systems of semilinear effectively damped waves with time-dependent coefficient, different power nonlinearities and different regularity of the data. Nonlinear Anal. 175 (2018), 2855.

$$C([\tau,\infty),H^{\sigma}(\mathbb{R}^n)),$$

and satisfy the following estimates:

$$\|v(t,\cdot)\|_{L^2} \le C(\tau+1)^{-r}(1+B(\tau+1))^{-\frac{n}{2}(\frac{1}{m}-\frac{1}{2})}\|h(\tau,\cdot)\|_{L^m\cap L^2},$$
$$\|D^{\sigma}v(t,\cdot)\|_{L^2} \le C(\tau+1)^{-r}(1+B(\tau+1))^{-\frac{n}{2}(\frac{1}{m}-\frac{1}{2})-\frac{\sigma}{2}}\|h(\tau,\cdot)\|_{L^m\cap L^2}.$$

Proposition 11.4. *Let us consider the Cauchy problem* (11.7) *with*

$$h = h(\tau,u) \in L^m \cap H^{\sigma-1},$$

for some $1 \le m < 2$. *Then the Sobolev solutions* $v = v(t,x)$ *to* (11.7) *belong for* $\sigma \ge 1$ *to*

$$C([\tau,\infty),H^{\sigma}(\mathbb{R}^n)) \cap C^1([\tau,\infty),H^{\sigma-1}(\mathbb{R}^n)),$$

and satisfy the following estimates:

$$\|v(t,\cdot)\|_{L^2} \le C(\tau+1)^{-r}(1+B(\tau+1))^{-\frac{n}{2}(\frac{1}{m}-\frac{1}{2})}\|h(\tau,\cdot)\|_{L^m\cap H^{\sigma-1}},$$
$$\|D^{\sigma}v(t,\cdot)\|_{L^2} \le C(\tau+1)^{-r}(1+B(\tau+1))^{-\frac{n}{2}(\frac{1}{m}-\frac{1}{2})-\frac{\sigma}{2}}\|h(\tau,\cdot)\|_{L^m\cap H^{\sigma-1}},$$
$$\left\|\frac{\partial v}{\partial t}(t,\cdot)\right\|_{L^2} \le C(\tau+1)^{-r}(t+1)^{-r}(1+B(\tau+1))^{-\frac{n}{2}(\frac{1}{m}-\frac{1}{2})-1}\|h(\tau,\cdot)\|_{L^m\cap H^{\sigma-1}},$$
$$\left\|D^{\sigma-1}\frac{\partial v}{\partial t}(t,\cdot)\right\|_{L^2} \le C(\tau+1)^{-r}(t+1)^{-r}(1+B(\tau+1))^{-\frac{n}{2}(\frac{1}{m}-\frac{1}{2})-\frac{\sigma-1}{2}-1}\|h(\tau,\cdot)\|_{L^m\cap H^{\sigma-1}}.$$

Proposition 11.5. *Let* $1 < p, p_0, p_1 < \infty$ *and* $k \in [0,\sigma)$. *Then the following fractional Gagliardo–Nirenberg inequality holds for all* $u \in L^{p_0} \cap \dot{H}^{\sigma}_{p_1}$:

$$\|u\|_{\dot{H}^k_p} < \|u\|_{L^{p_0}}^{1-\theta}\|u\|_{\dot{H}^k_{p_1}}^{\theta} \quad \textit{for } \frac{k}{\sigma} \le \theta \le 1. \tag{11.3}$$

Corollary 11.1. *Let*

$$L^2 \cap \dot{H}^k_2.$$

Then the following inequality holds:

$$\|u\|_{L^p} < \|u\|_{L^2}^{1-\rho_k(q)}\|u\|_{\dot{H}^k_2}^{\rho_k(q)} \quad \textit{for } \rho_k(q) = \frac{n}{k}\left(\frac{1}{2}-\frac{1}{q}\right) \tag{11.4}$$

for any $k \in (0,\frac{n}{2})$ *and any* q *such that* $2 \le q \le \frac{2n}{n-2k}$.

11.1.1 Existence and uniqueness results

For $r = r_1 = r_2$ and $\gamma = \gamma_1 = \gamma_2$, let us assume that

$$(u_0, u_1), (v_0, v_1) \in A_{m,\sigma},$$

with $\sigma > 0$ and $1 \leq m < 2$. Introduce the parameters

$$p_m(\gamma, n, r) = \frac{1 + \frac{n}{2m}(1 - r)}{\frac{n}{2m}(1 - r) + r + \gamma - 1}, \tag{11.5}$$

for $r \in (-1, 1)$ and $\gamma \in (0, 1)$. In our study, we will discus a lot of cases. We start with the first one

$$r + \gamma < 1.$$

Theorem 11.1. *Assume that $r \in (-1, 1)$ and $\gamma \in (0, 1)$ and*

$$r + \gamma < 1,$$

moreover,

$$\frac{2}{m} < k < \frac{n}{2 - 2\sigma}, \quad \frac{2}{m} < k < \infty, \quad \text{if } n \leq 2\sigma,$$

with

$$k \in \{p, q\}, \quad n \geq 1, \quad 1 \leq m < 2,$$

and $\sigma \in (0, 1)$. Moreover, we suppose

$$\frac{n}{2m}(1 - r) + r + \gamma - 1 > 0$$

and in case $n \geq 3$ we assume

$$\frac{n}{2}\left(\frac{1}{m} - \frac{1}{2}\right) + \frac{\sigma}{2} < 1.$$

Finally, let us suppose the condition

$$k > p_m(\gamma, n, r).$$

Then there exist a positive constant ε_0 such that for any initial data

$$(u_0, u_1), (v_0, v_1) \in A_{m,\sigma} = (L^m \cap H^\sigma) \times (L^m \cap L^2),$$

there is a uniquely determined global (in time) Sobolev solution

$$(u,v) \in \left(C([0,T],H^\sigma)\right)^2.$$

Proof. To begin with, let us introduce the concept of Sobolev solutions to the Cauchy problem (11.1). The Cauchy problem under study can be reduced by using Duhamels' principle due to the nonlinear term on the right-hand side to the study of the two following Cauchy problem for $r \in (-1,1)$:

$$\begin{cases} \frac{\partial^2 u}{\partial t^2} - \Delta u + (t+1)^r \frac{\partial u}{\partial t} = 0, \\ u(t=0,x) = u_0(x), \quad \frac{\partial u}{\partial t}(t=0,x) = u_1(x), \end{cases} \tag{11.6}$$

and the family of parameter-dependent Cauchy problems

$$\begin{cases} \frac{\partial^2 v}{\partial t^2} - \Delta v + (t+1)^r \frac{\partial v}{\partial t} = 0, \\ v(\tau,x) = 0; \quad \frac{\partial v}{\partial t}(\tau,x) = h(\tau,u) = \int_0^\tau (\tau - s)^{-\gamma} |u(s,x)|^q ds. \end{cases} \tag{11.7}$$

Now, we denote by $v = v(t,\tau,x)$ the solution of linear system (11.7) and by $u^{\text{lin}} = u^{\text{lin}}(t,x)$ the solution of (11.6). Then the solution u of the Cauchy problem (11.1) can be written as

$$u(t,x) = u^{\text{lin}}(t,x) + \int_0^t v(t,\tau,x)d\tau = u^{\text{lin}}(t,x) + u^{\text{nonl}}(t,x) \tag{11.8}$$

with

$$u^{\text{nonl}}(t,x) = \int_0^t v(t,\tau,x)d\tau.$$

Using Duhamels' principle, we can defined the solution of the Cauchy problem (11.1) formally as a solution of the fixed-point equation:

$$u = E_0(t,0,x) *_x u_0(x) + E_1(t,0,x) *_x u_1(x) + \int_0^t E_1(t,\tau,x) *_x h(\tau,u)d\tau.$$

Now, we define the solution space to (11.1) by

$$X(T) = \left(C([0,T],H^\sigma)\right)^2 \tag{11.9}$$

with a norm

$$\|(u,v)\|_{X(T)} = M(u) + M(v), \tag{11.10}$$

where

$$M(u) = \sup_{t\in[0,T]} \Big\{(t+1)^{\frac{n}{2}(\frac{1}{m}-\frac{1}{2})(1-r)+\gamma+r-1}\|u(t,\cdot)\|_{L^2} + (t+1)^{\frac{n}{2}(\frac{1}{m}-\frac{1}{2})(1-r)+\gamma-(1-\frac{\sigma}{2})(1-r)}\|D^\sigma u(t,\cdot)\|_{L^2}\Big\}.$$

By using Duhamel's principle, we are ready to introduce the operator N such that

$$\begin{aligned} N:(u,v)\in X(T) \to N(u,v) &= (N_1(u,v);N_2(u,v)) \\ &= (u^{\text{lin}}(t,x)+u^{\text{nonl}}(t,x); v^{\text{lin}}(t,x)+v^{\text{nonl}}(t,x)), \end{aligned}$$

where

$$\begin{cases} N_1(u,v) = E_0(t,0,x)*_x u_0(x) + E_1(t,0,x)*_x u_1(x) \\ \qquad + \int_0^t E_1(t,\tau,x)*_x \int_0^\tau (\tau-s)^{-\gamma}|v(s,x)|^q ds d\tau \\ N_2(u,v) = E_0(t,0,x)*_x v_0(x) + E_1(t,0,x)*_x v_1(x) \\ \qquad + \int_0^t E_1(t,\tau,x)*_x \int_0^\tau (\tau-s)^{-\gamma}|u(s,x)|^q ds d\tau. \end{cases} \tag{11.11}$$

Next, we are going to consider as solution of (11.1), the fixed points of the operator N. In other words, we should prove

$$\begin{aligned} \|N(u,v)\|_{X(T)} &< \|(u_0,u_1)\|_{\mathcal{A}_{m,\sigma}} + \|(v_0,v_1)\|_{\mathcal{A}_{m,\sigma}} \\ &\quad + T(\|(u,v)\|^p_{X(T)} + \|(u,v)\|^q_{X(T)}), \end{aligned} \tag{11.12}$$

$$\|N(u,v)-N(\bar{u},\bar{v})\|_{X(T)} < T\|(u,v)-(\bar{u},\bar{v})\|_{X(T)} \sum_{r=p,q} [\|(u,v)\|^{r-1}_{X(T)} + \|(\bar{u},\bar{v})\|^{r-1}_{X(T)}]. \tag{11.13}$$

First, it easy to see from (11.3) and (11.1) the estimates

$$M(u^{\text{lin}}) < \|(v_0,v_1)\|_{\mathcal{A}_{m,\sigma}},$$
$$M(v^{\text{lin}}) < \|(u_0,u_1)\|_{\mathcal{A}_{m,\sigma}}$$

for $\tau = 0$ and

$$h(0,u(x)) = u_1(x),$$

and

$$h(0,v(x)) = v_1(x).$$

Then it is enough to prove that

$$\begin{cases} M(u^{\text{nonl}}) < \|(u,v)\|^q_{X(T)}, \\ M(v^{\text{nonl}}) < \|(u,v)\|^p_{X(T)}. \end{cases} \tag{11.14}$$

Again we call the Proposition (11.3) to get

$$\begin{cases} \|u^{\text{nonl}}(t,\cdot)\|_{L^2} \\ \quad < \int_0^t (\tau+1)^{-r}(1+B(\tau+1))^{-\frac{n}{2}(\frac{1}{m}-\frac{1}{2})} \int_0^\tau (\tau-s)^{-\gamma} \||v(s,\cdot)|^p\|_{L^m\cap L^2} ds d\tau, \\ \|u^{\text{nonl}}(t,\cdot)\|_{L^2} \\ \quad < \int_0^t (\tau+1)^{-r}(1+B(\tau+1))^{-\frac{n}{2}(\frac{1}{m}-\frac{1}{2})} \int_0^\tau (\tau-s)^{-\gamma} \||u(s,\cdot)|^q\|_{L^m\cap L^2} ds d\tau. \end{cases} \tag{11.15}$$

Now we need to estimate the following norms:

$$\||v(s,\cdot)|^p\|_{L^m}, \quad \||v(s,\cdot)|^p\|_{L^2}, \quad \||u(s,\cdot)|^q\|_{L^m} \quad \text{and} \quad \||u(s,\cdot)|^q\|_{L^2}.$$

We call the Gagliardo–Nirenberg inequality (11.4) to obtain for $j \in m, 2$, the next inequalities

$$\begin{aligned} \|v(s,\cdot)\|_{L^{jp}}^p &< \|v(s,\cdot)\|_{L^2}^{p(1-\theta(jp))} \|v(s,\cdot)\|_{L^2}^{p\theta(jp)} \\ &< (s+1)^{p(1-\theta(jp))(1-\frac{n}{2}(\frac{1}{m}-\frac{1}{2})(1-r)-\gamma-r)} \\ &\quad \times (s+1)^{p\theta(jp)(-\frac{n}{2}(\frac{1}{m}-\frac{1}{2})(1-r)+(1-\frac{\sigma}{2})(1-r)-\gamma)} [M(v)]^p \\ &< (s+1)^{-\beta_j(p)} \|(u,v)\|_{X(T)}^p, \end{aligned} \tag{11.16}$$

and

$$\begin{aligned} \|u(s,\cdot)\|_{L^{jq}}^q &< \|u(s,\cdot)\|_{L^2}^{q(1-\theta(jq))} \|u(s,\cdot)\|_{L^2}^{q\theta(jq)} \\ &< (s+1)^{q(1-\theta(jq))(1-\frac{n}{2}(\frac{1}{m}-\frac{1}{2})(1-r)-\gamma-r)} \\ &\quad \times (s+1)^{q\theta(jq)(-\frac{n}{2}(\frac{1}{m}-\frac{1}{2})(1-r)+(1-\frac{\sigma}{2})(1-r)-\gamma)} [M(u)]^q \\ &< (s+1)^{-\beta_j(q)} \|(u,v)\|_{X(T)}^q, \end{aligned} \tag{11.17}$$

where

$$\theta(jk) = \frac{n}{\sigma}\left(\frac{1}{2} - \frac{1}{jk}\right) \in [0,1], \tag{11.18}$$

if and only if

$$\frac{2}{j} \le k \le \frac{2n}{(n-2\sigma)_+ j}.$$

That is

$$\begin{cases} \frac{2}{m} \le k \le \frac{2n}{(n-2\sigma)} & \text{if } n > 2\sigma, \\ \frac{2}{m} \le k & \text{if } n = 1 \text{ and } \sigma \in [\frac{1}{2}, 1), \end{cases}$$

and

$$\beta_j(k) = \left(\frac{n}{2m}(1-r) + \gamma + r - 1\right)k - \frac{n}{2j}(1-r), \tag{11.19}$$

with $k \in \{p, q\}$. Now, from (11.16) and (11.17), if we take $j = m$ and $j = 2$, we get

$$\begin{cases} \|v(s,\cdot)\|^p_{L^{mp}} < (s+1)^{-\beta(p)}\|(u,v)\|^p_{X(T)}, \\ \|u(s,\cdot)\|^q_{L^{mq}} < (s+1)^{-\beta(q)}\|(u,v)\|^q_{X(T)}, \end{cases} \tag{11.20}$$

and

$$\begin{cases} \|v(s,\cdot)\|^p_{L^{2p}} < (s+1)^{-\beta(p)-\frac{n}{2}(\frac{1}{m}-\frac{1}{2})(1-r)}\|(u,v)\|^p_{X(T)} \\ \qquad\qquad < (s+1)^{-\beta(p)}\|(u,v)\|^p_{X(T)}, \\ \|u(s,\cdot)\|^q_{L^{2q}} < (s+1)^{-\beta(q)-\frac{n}{2}(\frac{1}{m}-\frac{1}{2})(1-r)}\|(u,v)\|^q_{X(T)} \\ \qquad\qquad < (s+1)^{-\beta(q)}\|(u,v)\|^q_{X(T)}, \end{cases} \tag{11.21}$$

where $r < 1$ and $m < 2$. Then

$$-\beta(k) - \frac{n}{2}\left(\frac{1}{m} - \frac{1}{2}\right)(1-r) > 0,$$

where

$$\beta(k) = \beta_m(k) = \left(\frac{n}{2m}(1-r) + \gamma + r - 1\right)k - \frac{n}{2m}(1-r). \tag{11.22}$$

From (11.20) and (11.21), we get

$$\begin{cases} \|v(s,\cdot)\|^p_{L^{2p}} + \|v(s,\cdot)\|^p_{L^{mp}} < (s+1)^{-\beta(p)}\|(u,v)\|^p_{X(T)}, \\ \|u(s,\cdot)\|^q_{L^{2q}} + \|u(s,\cdot)\|^q_{L^{mq}} < (s+1)^{-\beta(q)}\|(u,v)\|^q_{X(T)}. \end{cases} \tag{11.23}$$

It is clear that $\beta(k) > 1$ if and only if $p_m(\gamma, n, r)$ is given in (11.5). We put (11.23) in (11.15) to obtain

$$\begin{cases} \|u^{\mathrm{nonl}}(t,\cdot)\|_{L^2} < J_n(t,p)M(v)^p < J_n(t,p)\|(u,v)\|^p_{X(T)}, \\ \|u^{\mathrm{nonl}}(t,\cdot)\|_{L^2} < J_n(t,q)M(u)^q < J_n(t,q)\|(u,v)\|^q_{X(T)}, \end{cases} \tag{11.24}$$

where

$$J_n(t,k) = \int_0^t (\tau+1)^{-r}\big(1 + B(\tau+1)\big)^{-\frac{n}{2}(\frac{1}{m}-\frac{1}{2})} \times \int_0^\tau (\tau-s)^{-\gamma}(s+1)^{-\beta(k)}\,ds\,d\tau. \tag{11.25}$$

Since

$$\beta(k) > 1, \quad \gamma + r < 1,$$

and

$$\frac{n}{2}\left(\frac{1}{m} - \frac{1}{2}\right),$$

using Lemma 11.1 we estimate $J_n(t,k)$ as follows:

$$J_n(t,k) < (t+1)^{1-\frac{n}{2}(\frac{1}{m}-\frac{1}{2})(1-r)-\gamma-r}. \tag{11.26}$$

Now, we put (11.24) and (11.26) to get

$$\begin{cases} \|u^{\text{nonl}}(t,\cdot)\|_{L^2} < (t+1)^{1-\frac{n}{2}(\frac{1}{m}-\frac{1}{2})(1-r)-\gamma-r}\|(u,v)\|^p_{X(T)}, \\ \|u^{\text{nonl}}(t,\cdot)\|_{L^2} < (t+1)^{1-\frac{n}{2}(\frac{1}{m}-\frac{1}{2})(1-r)-\gamma-r}\|(u,v)\|^q_{X(T)}. \end{cases} \tag{11.27}$$

Now, let us turn to

$$\|D^\sigma v^{\text{nonl}}(t,\cdot)\|_{L^2},$$

and

$$\|D^\sigma u^{\text{nonl}}(t,\cdot)\|_{L^2},$$

with same way we call the Proposition (11.3) and we get for $0 < \sigma < 1$ the estimate

$$\begin{cases} \|D^\sigma u^{\text{nonl}}(t,\cdot)\|_{L^2} \\ \quad < \int_0^t (\tau+1)^{-r}(1+B(\tau+1))^{-\frac{n}{2}(\frac{1}{m}-\frac{1}{2})-\frac{\sigma}{2}} \int_0^\tau (\tau-s)^{-\gamma}\||v(s,\cdot)|^p\|_{L^m\cap L^2}\,ds\,d\tau, \\ \|D^\sigma v^{\text{nonl}}(t,\cdot)\|_{L^2} \\ \quad < \int_0^t (\tau+1)^{-r}(1+B(\tau+1))^{-\frac{n}{2}(\frac{1}{m}-\frac{1}{2})-\frac{\sigma}{2}} \int_0^\tau (\tau-s)^{-\gamma}\||u(s,\cdot)|^q\|_{L^m\cap L^2}\,ds\,d\tau. \end{cases} \tag{11.28}$$

Using the estimate (11.23), we get

$$\begin{cases} \|D^\sigma u^{\text{nonl}}(t,\cdot)\|_{L^2} < J_{n,\sigma}(t,p)[M(v)]^p < J_{n,\sigma}(t,p)\|(u,v)\|^p_{X(T)}, \\ \|D^\sigma v^{\text{nonl}}(t,\cdot)\|_{L^2} < J_{n,\sigma}(t,q)[M(u)]^q < J_{n,\sigma}(t,q)\|(u,v)\|^q_{X(T)}, \end{cases} \tag{11.29}$$

with

$$J_{n,\sigma}(t,k) = \int_0^t (\tau+1)^{-r}(1+B(\tau+1))^{-\frac{n}{2}(\frac{1}{m}-\frac{1}{2})-\frac{\sigma}{2}} \int_0^\tau (\tau-s)^{-\gamma}(s+1)^{-\beta(k)}\,ds\,d\tau. \tag{11.30}$$

Since $\beta(k) > 1$, we can estimate $J_{n,\sigma}(t,k)$ after using (11.2) with

$$a = -\frac{n}{2}\left(\frac{1}{m} - \frac{1}{2}\right) - \frac{\sigma}{2},$$

as follows

$$J_{n,\sigma}(t,k) < (t+1)^{-\frac{n}{2}(\frac{1}{m}-\frac{1}{2})(1-r)+(1-\frac{\sigma}{2})(1-r)-\gamma}. \tag{11.31}$$

Now, from (11.31) and (11.29), we get the estimate

$$\begin{cases} \|D^\sigma u^{\mathrm{nonl}}(t,\cdot)\|_{L^2} < (t+1)^{-\frac{n}{2}(\frac{1}{m}-\frac{1}{2})(1-r)+(1-\frac{\sigma}{2})(1-r)-\gamma}\|(u,v)\|^p_{X(T)}, \\ \|D^\sigma v^{\mathrm{nonl}}(t,\cdot)\|_{L^2} < (t+1)^{-\frac{n}{2}(\frac{1}{m}-\frac{1}{2})(1-r)+(1-\frac{\sigma}{2})(1-r)-\gamma}\|(u,v)\|^q_{X(T)}. \end{cases} \tag{11.32}$$

Then we can conclude from (11.32) and (11.27) and the definition of $\|(u,v)\|_{X(T)}$ to get (11.14).

Now, we deal with the inequality (11.13). By definition of the operator N, we have

$$\begin{aligned} \|N(u,v) - N(\bar{u},\bar{v})\|_{X(T)} &= \|(N_1(u,v), N_2(u,v)) - (N_1(\bar{u},\bar{v}), N_2(\bar{u},\bar{v}))\|_{X(T)} \\ &= \|(N_1(u,v) - N_1(\bar{u},\bar{v}), N_2(u,v) - N_2(\bar{u},\bar{v}))\|_{X(T)} \\ &= M[N_1(u,v) - N_1(\bar{u},\bar{v})] + M[N_2(u,v) - N_2(\bar{u},\bar{v})]. \end{aligned}$$

Thanks to Proposition (11.3), we can see

$$\begin{cases} \|N_1(u,v) - N_1(\bar{u},\bar{v})\|_{L^2} < \int_0^t (\tau+1)^{-r}(1+B(\tau+1))^{-\frac{n}{2}(\frac{1}{m}-\frac{1}{2})} \\ \qquad\qquad \times \int_0^\tau (\tau-s)^{-\gamma}\||v(s,\cdot)|^p - |\bar{v}(s,\cdot)|^p\|_{L^m\cap L^2}\,ds\,d\tau, \\ \|N_2(u,v) - N_2(\bar{u},\bar{v})\|_{L^2} < \int_0^t (\tau+1)^{-r}(1+B(\tau+1))^{-\frac{n}{2}(\frac{1}{m}-\frac{1}{2})} \\ \qquad\qquad \times \int_0^\tau (\tau-s)^{-\gamma}\||u(s,\cdot)|^q - |\bar{u}(s,\cdot)|^q\|_{L^m\cap L^2}\,ds\,d\tau. \end{cases} \tag{11.33}$$

By the same way, we need to estimate

$$\big\||v(s,\cdot)|^p - |\bar{v}(s,\cdot)|^p\big\|_{L^j},$$

and

$$\big\||u(s,\cdot)|^q - |\bar{u}(s,\cdot)|^q\big\|_{L^j},$$

with $j \in \{m,2\}$.

Hölder's inequality gives

$$\begin{cases} \||v(s,\cdot)|^p - |\bar{v}(s,\cdot)|^p\|_{L^j} \\ \quad < \|v(s,\cdot) - \bar{v}(s,\cdot)\|_{L^{jp}}(\|v(s,\cdot)\|^{p-1}_{L^{jp}} + \|\bar{v}(s,\cdot)\|^{p-1}_{L^{jp}}), \\ \||u(s,\cdot)|^q - |\bar{u}(s,\cdot)|^q\|_{L^j} \\ \quad < \|u(s,\cdot) - \bar{u}(s,\cdot)\|_{L^{jq}}(\|u(s,\cdot)\|^{q-1}_{L^{jp}} + \|\bar{u}(s,\cdot)\|^{q-1}_{L^{jq}}), \end{cases} \tag{11.34}$$

which means that we should estimate now

$$\big\||v(s,\cdot)| - |\bar{v}(s,\cdot)|\big\|_{L^{jp}},$$
$$\|v(s,\cdot)\|_{L^{jp}}^{p-1},$$

and

$$\|\bar{v}(s,\cdot)\|_{L^{jp}}^{p-1}.$$

For this aim, let begin with

$$\|v(s,\cdot) - \bar{v}(s,\cdot)\|_{L^{jp}}.$$

With Gagliardo–Nirenberg inequality and (11.4), we can write

$$\begin{aligned}
&\|v(s,\cdot) - \bar{v}(s,\cdot)\|_{L^{jp}} \\
&\quad < \|v(s,\cdot) - \bar{v}(s,\cdot)\|_{L^2}^{(1-\theta(jp))} \|D^\sigma(v-\bar{v})(s,\cdot)\|_{L^2}^{\theta(jp)} \\
&\quad < (s+1)^{(1-\theta(jp))(1-\frac{n}{2}(\frac{1}{m}-\frac{1}{2})-r-\gamma)} \\
&\qquad \times (s+1)^{\theta(jp)(-\frac{n}{2}(\frac{1}{m}-\frac{1}{2})+(1-\frac{\sigma}{2})(1-r)-\gamma)}\big[M(v-\bar{v})\big] \\
&\quad < (s+1)^{1-\frac{n}{2}(\frac{1}{m}-\frac{1}{2})-r-\gamma-\frac{\sigma}{2}(1-r)\theta(jp)} \|(u,v)-(\bar{u},\bar{v})\|_{X(T)}.
\end{aligned} \tag{11.35}$$

By the same way, we can estimate $\|v(s,\cdot)\|_{L^{jp}}^{p-1}$ as follows:

$$\begin{aligned}
\|v(s,\cdot)\|_{L^{jp}}^{p-1} &< \|v(s,\cdot)\|_{L^2}^{(p-1)(1-\theta(jp))} \|D^\sigma v(s,\cdot)\|_{L^2}^{(p-1)\theta(jp)} \\
&< (s+1)^{(p-1)(1-\theta(jp))(1-\frac{n}{2}(\frac{1}{m}-\frac{1}{2})-r-\gamma)} \\
&\quad \times (s+1)^{(p-1)\theta(jp)(-\frac{n}{2}(\frac{1}{m}-\frac{1}{2})+(1-\frac{\sigma}{2})(1-r)-\gamma)}\big[M(v)\big]^{p-1} \\
&< (s+1)^{(p-1)(1-\frac{n}{2}(\frac{1}{m}-\frac{1}{2})-r-\gamma)-\frac{\sigma}{2}(p-1)(1-r)\theta(jp)} \|(u,v)\|_{X(T)}^{p-1},
\end{aligned} \tag{11.36}$$

and

$$\|\bar{v}(s,\cdot)\|_{L^{jp}}^{p-1} < (s+1)^{(p-1)(1-\frac{n}{2}(\frac{1}{m}-\frac{1}{2})-r-\gamma)-\frac{\sigma}{2}(p-1)(1-r)\theta(jp)} \|(\bar{u},\bar{v})\|_{X(T)}^{p-1}. \tag{11.37}$$

Finally, by (11.35), (11.36), (11.37) and (11.34) for $j \in \{m, 2\}$, we estimate

$$\begin{cases}
\||v(s,\cdot)|^p - |\bar{v}(s,\cdot)|^p\|_{L^j} < (s+1)^{-\beta_j(p)}\|(u,v)-(\bar{u},\bar{v})\|_{X(T)} \\
\qquad\qquad \times (\|(u,v)\|_{X(T)}^{p-1} + \|(\bar{u},\bar{v})\|_{X(T)}^{p-1}), \\
\||u(s,\cdot)|^q - |\bar{u}(s,\cdot)|^q\|_{L^j} < (s+1)^{-\beta_j(q)}\|(u,v)-(\bar{u},\bar{v})\|_{X(T)} \\
\qquad\qquad \times (\|(u,v)\|_{X(T)}^{q-1} + \|(\bar{u},\bar{v})\|_{X(T)}^{q-1}).
\end{cases} \tag{11.38}$$

Taking $j = m$ and $j = 2$ in (11.38), we get

$$\begin{cases} \||v(s,\cdot)|^p - |\bar{v}(s,\cdot)|^p\|_{L^m} < (s+1)^{-\beta_m(p)}\|(u,v) - (\bar{u},\bar{v})\|_{X(T)} \\ \qquad \times (\|(u,v)\|_{X(T)}^{p-1} + \|(\bar{u},\bar{v})\|_{X(T)}^{p-1}), \\ \||u(s,\cdot)|^q - |\bar{u}(s,\cdot)|^q\|_{L^m} < (s+1)^{-\beta_m(q)}\|(u,v) - (\bar{u},\bar{v})\|_{X(T)} \\ \qquad \times (\|(u,v)\|_{X(T)}^{q-1} + \|(\bar{u},\bar{v})\|_{X(T)}^{q-1}) \end{cases} \tag{11.39}$$

and

$$\begin{cases} \||v(s,\cdot)|^p - |\bar{v}(s,\cdot)|^p\|_{L^2} < (s+1)^{-\beta_2(p)}\|(u,v) - (\bar{u},\bar{v})\|_{X(T)} \\ \qquad \times (\|(u,v)\|_{X(T)}^{p-1} + \|(\bar{u},\bar{v})\|_{X(T)}^{p-1}), \\ \||u(s,\cdot)|^q - |\bar{u}(s,\cdot)|^q\|_{L^2} < (s+1)^{-\beta_2(q)}\|(u,v) - (\bar{u},\bar{v})\|_{X(T)} \\ \qquad \times (\|(u,v)\|_{X(T)}^{q-1} + \|(\bar{u},\bar{v})\|_{X(T)}^{q-1}), \end{cases} \tag{11.40}$$

where

$$\beta_2(k) = \beta_m(k) + \frac{n}{2}\Big(\frac{1}{m} - \frac{1}{2}\Big)(1-r), \quad k \in \{p,q\}. \tag{11.41}$$

We take (11.40) and (11.39) to get

$$\begin{cases} \||v(s,\cdot)|^p - |\bar{v}(s,\cdot)|^p\|_{L^2\cap L^m} < (s+1)^{-\beta(p)}\|(u,v) - (\bar{u},\bar{v})\|_{X(T)} \\ \qquad \times (\|(u,v)\|_{X(T)}^{p-1} + \|(\bar{u},\bar{v})\|_{X(T)}^{p-1}), \\ \||u(s,\cdot)|^q - |\bar{u}(s,\cdot)|^q\|_{L^2\cap L^m} < (s+1)^{-\beta(q)}\|(u,v) - (\bar{u},\bar{v})\|_{X(T)} \\ \qquad \times (\|(u,v)\|_{X(T)}^{q-1} + \|(\bar{u},\bar{v})\|_{X(T)}^{q-1}) \end{cases} \tag{11.42}$$

with $\beta(k)$ is given by (11.22). We put the estimate (11.42) in (11.33). Then we may write

$$\begin{cases} \|N_1(u,v) - N_1(\bar{u},\bar{v})\|_{L^2} < J_n(t,p)\|(u,v) - (\bar{u},\bar{v})\|_{X(T)} \\ \qquad \times (\|(u,v)\|_{X(T)}^{p-1} + \|(\bar{u},\bar{v})\|_{X(T)}^{p-1}), \\ \|N_2(u,v) - N_2(\bar{u},\bar{v})\|_{L^2} < J_n(t,q)\|(u,v) - (\bar{u},\bar{v})\|_{X(T)} \\ \qquad \times (\|(u,v)\|_{X(T)}^{q-1} + \|(\bar{u},\bar{v})\|_{X(T)}^{q-1}), \end{cases} \tag{11.43}$$

where $J_n(t,k)$ is defined by (11.25). Then, using (11.26), we get the estimate

$$\begin{aligned} \big\|N_1(u,v) - N_1(\bar{u},\bar{v})\big\|_{L^2} &< (t+1)^{1-\frac{n}{2}(\frac{1}{m}-\frac{1}{2})(1-r)-\gamma-r}\big\|(u,v) - (\bar{u},\bar{v})\big\|_{X(T)} \\ &\quad \times \big(\big\|(u,v)\big\|_{X(T)}^{p-1} + \big\|(\bar{u},\bar{v})\big\|_{X(T)}^{p-1}\big), \\ \big\|N_2(u,v) - N_2(\bar{u},\bar{v})\big\|_{L^2} &< (t+1)^{1-\frac{n}{2}(\frac{1}{m}-\frac{1}{2})(1-r)-\gamma-r}\big\|(u,v) - (\bar{u},\bar{v})\big\|_{X(T)} \\ &\quad \times \big(\big\|(u,v)\big\|_{X(T)}^{q-1} + \big\|(\bar{u},\bar{v})\big\|_{X(T)}^{q-1}\big). \end{aligned}$$

Next, we will proceed to estimate the terms

$$\|D^\sigma(N_1(u,v) - N_1(\bar{u},\bar{v}))\|_{L^2},$$

and

$$\|D^\sigma(N_2(u,v) - N_2(\bar{u},\bar{v}))\|_{L^2}.$$

Applying the results of (11.3), we get

$$\begin{cases} \|D^\sigma(N_1(u,v) - N_1(\bar{u},\bar{v}))\|_{L^2} < \int_0^t (\tau+1)^{-r}(1+B(\tau+1))^{-\frac{n}{2}(\frac{1}{m}-\frac{1}{2})-\frac{\sigma}{2}} \\ \qquad \times \int_0^\tau (\tau-s)^{-\gamma}\||v(s,\cdot)|^p - |\bar{v}(s,\cdot)|^p\|_{L^m\cap L^2} ds d\tau, \\ \|D^\sigma(N_2(u,v) - N_2(\bar{u},\bar{v}))\|_{L^2} < \int_0^t (\tau+1)^{-r}(1+B(\tau+1))^{-\frac{n}{2}(\frac{1}{m}-\frac{1}{2})-\frac{\sigma}{2}} \\ \qquad \times \int_0^\tau (\tau-s)^{-\gamma}\||u(s,\cdot)|^q - |\bar{u}(s,\cdot)|^q\|_{L^m\cap L^2} ds d\tau. \end{cases} \tag{11.44}$$

We put the estimate (11.42) in (11.44) and we obtain

$$\begin{cases} \|D^\sigma(N_1(u,v) - N_1(\bar{u},\bar{v}))\|_{L^2} < J_{n,\sigma}(t,p)\|(u,v) - (\bar{u},\bar{v})\|_{X(T)} \\ \qquad \times (\|(u,v)\|_{X(T)}^{p-1} + \|(\bar{u},\bar{v})\|_{X(T)}^{p-1}), \\ \|D^\sigma(N_2(u,v) - N_2(\bar{u},\bar{v}))\|_{L^2} < J_{n,\sigma}(t,q)\|(u,v) - (\bar{u},\bar{v})\|_{X(T)} \\ \qquad \times (\|(u,v)\|_{X(T)}^{q-1} + \|(\bar{u},\bar{v})\|_{X(T)}^{q-1}), \end{cases} \tag{11.45}$$

with $J_{n,\sigma}(t,k)$ is defined in (11.30).

Now, from the estimate (11.31), we find

$$\begin{cases} \|D^\sigma(N_1(u,v) - N_1(\bar{u},\bar{v}))\|_{L^2} < (t+1)^{-\frac{n}{2}(\frac{1}{m}-\frac{1}{2})(1-r)+(1-\frac{\sigma}{2})(1-r)-\gamma} \\ \qquad \times \|(u,v) - (\bar{u},\bar{v})\|_{X(T)}(\|(u,v)\|_{X(T)}^{p-1} + \|(\bar{u},\bar{v})\|_{X(T)}^{p-1}), \\ \|D^\sigma(N_2(u,v) - N_2(\bar{u},\bar{v}))\|_{L^2} < (t+1)^{-\frac{n}{2}(\frac{1}{m}-\frac{1}{2})(1-r)+(1-\frac{\sigma}{2})(1-r)-\gamma} \\ \qquad \times \|(u,v) - (\bar{u},\bar{v})\|_{X(T)}(\|(u,v)\|_{X(T)}^{q-1} + \|(\bar{u},\bar{v})\|_{X(T)}^{q-1}). \end{cases} \tag{11.46}$$

Ultimately, the desired inequality is derived from the provided estimates (11.44) and (11.44) and definition of the norm $X(T)$. This completes the proof. □

Now, we give the well-posedness result for $\gamma + r > 0$ in the next theorem.

Theorem 11.2. *Assume that $\gamma, r \in (0,1)$ and $\gamma + r > 0$. Moreover,*

$$\frac{2}{m} < k < \frac{n}{2-2\sigma}, \quad \frac{2}{m} < k < \infty, \quad \text{if } n \le 2\sigma,$$

with $k \in \{p,q\}$, $n \ge 1$, $1 \le m < 2$ and $\sigma \in (0,1)$. Then there exist a positive constant ε_0 such that for any initial data

$$(u_0,u_1),(v_0,v_1) \in \mathrm{A}_{m,\sigma},$$

one has

$$\begin{cases} \|(v_0, v_1)\|_{A_{m,\sigma}} < \varepsilon, \\ \|(u_0, u_1)\|_{A_{m,\sigma}} < \varepsilon \end{cases} \quad \text{for all } \varepsilon < \varepsilon_0.$$

Then there is a uniquely determined global (in time) Sobolev solution

$$(u, v) \in \left[C([0, \infty), H^\sigma)\right]^2,$$

to the Cauchy problem (11.1) in following cases:

1. $a_m + \frac{\sigma}{2} \le 1$, *where the exponent k satisfies the condition*

$$k > 1 + \frac{2m}{n(1-r)}, \quad k \in \{p, q\} \tag{11.47}$$

2. $a_m < 1 < a_m + \frac{\sigma}{2}$, *where the exponent k satisfies the condition*

$$k > \begin{cases} 1 + \frac{2m}{n(1-r)} & \text{if } (a_m + \frac{\sigma}{2})(1-r) \le \gamma, \\ 1 + \frac{1 - a_m(1-r) - \frac{n}{\sigma}(\frac{1}{m} - \frac{1}{2})(a_m(1-r) - \gamma)}{[a_m(1-r) - \frac{n}{2\sigma}(a_m(1-r) - \gamma)]_+} & \text{if } (a_m + \frac{\sigma}{2})(1-r) > \gamma \end{cases} \tag{11.48}$$

3. $a_m > 1$, *where the exponent k satisfies the condition*

$$k > \begin{cases} \frac{1}{\gamma} & \text{if } \gamma \le a_m(1-r), \\ 1 + \frac{1 - a_m(1-r) - \frac{n}{\sigma}(\frac{1}{m} - \frac{1}{2})(a_m(1-r) - \gamma)}{[a_m(1-r) - \frac{n}{2\sigma}(a_m(1-r) - \gamma)]_+} & \text{if } 0 < \gamma < (a_m + \frac{\sigma}{2})(1-r), \\ 1 + \frac{2m}{n(1-r)} & \text{if } (a_m + \frac{\sigma}{2})(1-r) \le \gamma. \end{cases} \tag{11.49}$$

Proof. The proof follows the same logic as Theorem (11.1). We will use the same Sobolev space $X(T)$ and norm for the solution. We will estimate the norms until we arrive at the integrals $J_{n,\sigma}(t, k)$ and $J_n(t, k)$. We need to estimate the integrals with help of Lemma 11.1 with $\sigma + r > 0$ and $\beta > 1$ and every time we fix a bound for the exponent p. Let us start with

$$a_m = \frac{n}{2}\left(\frac{1}{m} - \frac{1}{2}\right).$$

1. If $a_m + \frac{\sigma}{2} \le 1$, then $a_m \le 1$. So, we have

$$\max\{a_m, r + \gamma\} = r + \gamma > 1.$$

   Hence,

$$\min\{a_m(1-r), \gamma\} = a_m(1-r).$$

   Using Lemma 11.1, we get

$$J_n(t,k) < (t+1)^{-\frac{n}{2}(\frac{1}{m}-\frac{1}{2})(1-r)}.$$

From the other side, we have

$$\max\left\{a_m + \frac{\sigma}{2}, r+\gamma\right\} = r+\gamma > 1.$$

Then we have

$$\min\left\{\left(a_m + \frac{\sigma}{2}\right)(1-r), \gamma\right\} = \left(a_m + \frac{\sigma}{2}\right)(1-r).$$

Using Lemma 11.1 again, we can see the estimate:

$$J_{n,\sigma}(t,k) < (t+1)^{-\frac{n}{2}(\frac{1}{m}-\frac{1}{2})(1-r)-(1-\frac{\sigma}{2})(1-r)}.$$

Then $\beta(k) > 1$ and we can get the exponent

$$k > 1 + \frac{2m}{n(1-r)}, \quad k \in \{p,q\}.$$

2. If $a_m < 1 < a_m + \frac{\sigma}{2}$, we get the same estimate for $J_n(t,k)$ in the case $a_m < 1$. Now, we focus in the case $1 < a_m + \frac{\sigma}{2}$ to have

$$\max\left\{a_m + \frac{\sigma}{2}, r+\gamma\right\} > 1.$$

 – If $(a_m + \frac{\sigma}{2})(1-r) \geq \gamma$, then

$$\min\left\{\left(a_m + \frac{\sigma}{2}\right)(1-r), \gamma\right\} = \gamma.$$

 So, we have

$$J_{n,\sigma}(t,k) < (t+1)^{-\gamma}$$

 and the lower bound for the exponent $k \in \{p,q\}$ to get the same result as

$$k > 1 + \frac{1 - a_m(1-r) - \frac{n}{\sigma}(\frac{1}{m}-\frac{1}{2})(a_m(1-r)-\gamma)}{[a_m(1-r) - \frac{n}{2\sigma}(a_m(1-r)-\gamma)]_+}.$$

 – If $(a_m + \frac{\sigma}{2})(1-r) < \gamma$, then

$$\min\left\{\left(a_m + \frac{\sigma}{2}\right)(1-r), \gamma\right\} = \left(a_m + \frac{\sigma}{2}\right)(1-r),$$

 and

$$J_{n,\sigma}(t,k) < (t+1)^{-(a_m+\frac{\sigma}{2})(1-r)},$$

and the exponents k satisfies

$$k > 1 + \frac{2m}{n(1-r)}, \qquad k \in \{p,q\}.$$

3. If $a_m > 1$, we have

$$a_m + \frac{\sigma}{2} > a_m > 1.$$

It is clear that

$$\max\{a_m, r+\gamma\} > 1.$$

Then
- If $a_m(1-r) < \gamma$, then we have seen this case in the previous study.
- If $a_m(1-r) > \gamma$, then we can show easily $k > \frac{1}{\gamma}$ for $k \in \{p,q\}$. After this, we group the both of cases as

$$J_n(t,k) < (t+1)^{-\min\{a_m(1-r),\gamma\}}.$$

Now, let us turn to $J_{n,\sigma}(t,k)$. We have also

$$\max\{a_m, r+\gamma\} > 1.$$

So, we devise two cases:
- If $(a_m + \frac{\sigma}{2})(1-r) < \gamma$, which means that

$$\min\left\{\left(a_m + \frac{\sigma}{2}\right)(1-r), \gamma\right\} = \left(a_m + \frac{\sigma}{2}\right)(1-r),$$

  then

$$J_{n,\sigma}(t,k) < (t+1)^{-(a_m+\frac{\sigma}{2})(1-r)},$$

  and

$$k > 1 + \frac{2m}{n(1-r)}.$$

- If $(a_m + \frac{\sigma}{2})(1-r) > \gamma$, which means that

$$\min\left\{\left(a_m + \frac{\sigma}{2}\right)(1-r), \gamma\right\} = \gamma,$$

  then we have

$$J_{n,\sigma}(t,k) < (t+1)^{-\gamma},$$

and

$$k > 1 + \frac{1 - a_m(1-r) - \frac{n}{\sigma}(\frac{1}{m} - \frac{1}{2})(a_m(1-r) - \gamma)}{[a_m(1-r) - \frac{n}{2\sigma}(a_m(1-r) - \gamma)]_+}.$$

Here are some critical cases we did not study yet:

1. If $(a_m + \frac{\sigma}{2}) = 1$ in the second case (11.2), then the condition

   $$\left(a_m + \frac{\sigma}{2}\right)(1-r) > \gamma,$$

   has no meaning $(\gamma + r > 1)$. Hence, we take only the condition in the

   $$\left(a_m + \frac{\sigma}{2}\right)(1-r) < \gamma,$$

   which means the same condition with the first case (11.47).
2. If $a_m = 1$ in the second case, then the condition $\gamma \le a_m(1-r)$ has no meaning $(\gamma + r > 1)$. Consequently, we get the same condition in (11.48).
3. $\gamma = a_m(1-r)$ in the second condition of (11.49). Then it has the same meaning of the first condition. □

Finally, we will see the last case when $\gamma + r = 1$.

Theorem 11.3. *We assume the same conditions in* (11.2), *but in this time we suppose*

$$\gamma + r = 1.$$

Then there exists a positive constant ε_0 *such for any initial data:*

$$(u_0, u_1), (v_0, v_1) \in A_{m,\sigma},$$

satisfying

$$\begin{cases} \|(v_0, v_1)\|_{A_{m,\sigma}} < \varepsilon, \\ \|(u_0, u_1)\|_{A_{m,\sigma}} < \varepsilon \end{cases} \quad \text{for all } \varepsilon < \varepsilon_0.$$

Hence, there is a uniquely determined global (in time) Sobolev solution

$$(u, v) \in [C([0,\infty), H^\sigma)]^2,$$

to the Cauchy problem (11.1) *in following cases:*

1. $a_m + \frac{\sigma}{2} \le 1$, *where the exponent k satisfies the condition*

$$k > 1 + \frac{2m}{n(1-r)}, \quad k \in \{p, q\}, \tag{11.50}$$

2. $a_m \ge 1$, *where the exponent k satisfies the condition*

$$k > \frac{1}{\gamma}. \tag{11.51}$$

Proof. 1. If $a_m + \frac{\sigma}{2} \le 1$, then

$$\max\left\{a_m + \frac{\sigma}{2}, r + \gamma\right\} = 1,$$

then

$$\min\left\{\left(a_m + \frac{\sigma}{2}\right)(1-r), \gamma\right\} = \min\left\{\left(a_m + \frac{\sigma}{2}\right)(1-r), (1-r)\right\} = \left(a_m + \frac{\sigma}{2}\right)(1-r),$$

and

$$J_{n,\sigma}(t,k) < (t+1)^{-\frac{n}{2}(\frac{1}{m}-\frac{1}{2})(1-r)-\frac{\sigma}{2}(1-r)} \log(2+t).$$

Now, for $a_m < a_m + \frac{\sigma}{2} \le 1$, then $a_m < 1$ and we have

$$\min\{a_m(1-r), \gamma\} = \min\{a_m(1-r), (1-r)\} = a_m(1-r).$$

From here, we get

$$J_n(t,k) < (t+1)^{-\frac{n}{2}(\frac{1}{m}-\frac{1}{2})(1-r)} \log(2+t),$$

and the exponent

$$k > 1 + \frac{2m}{n(1-r)}.$$

2. For $a_m + \frac{\sigma}{2} > a_m \ge 1$, we devise two cases:
   - If $a_m > 1$, then we have

$$\max\{a_m, r + \gamma\} = a_m.$$

   Hence,

$$\min\{a_m(1-r), \gamma\} = \min\{a_m\gamma, \gamma\} = \gamma,$$

   and we deduce

$$J_n(t,k) < (t+1)^{-\gamma}\log(2+t).$$

- If $a_m = 1$, we obtain

$$\max\{a_m, r+\gamma\} = 1.$$

Then

$$J_n(t,k) < (t+1)^{-\gamma}\log(2+t).$$

We can reduce both cases as

$$J_n(t,k) < l(t)(t+1)^{-\gamma}, \quad l(t) = \begin{cases} 1 & : a_m > 1, \\ \log(2+t) & : a_m = 1. \end{cases}$$

Finally, we have

$$1 \le a_m < a_m + \frac{\sigma}{2}.$$

Then

$$\max\left\{a_m + \frac{\sigma}{2}, r+\gamma\right\} = a_m + \frac{\sigma}{2}.$$

So,

$$\min\left\{\left(a_m + \frac{\sigma}{2}\right)(1-r), \gamma\right\} = \min\left\{\left(a_m + \frac{\sigma}{2}\right)\gamma, \gamma\right\} = \gamma.$$

Thus, we obtain

$$J_{n,\sigma}(t,k) < (t+1)^{-\gamma},$$

with the exponent $k > \frac{1}{\gamma}$.
We complete the proof with some critical cases:
- If we put $a_m + \frac{\sigma}{2} = 1$ and $r + \gamma = 1$ in (11.5), then we obtain the condition (11.50).
- If put $a_m = 1$ in second case, then the condition (11.51) coincides with the first condition in (11.49). □

11.2 Global time nonexistence of solutions

Now, let us deal with our Cauchy problem (11.1) with $r_1 \neq r_2$ and $\gamma_1 \neq \gamma_2$. We mention that for $0 < \varrho < 1$ the fractional derivative operator $D^{\varrho}_{0|t}$ of order ϱ in the sense of Riemann–Liouville defined by [2]:

$$D^{\varrho}_{0|t}u = d_t J^{1-\varrho}_{0|t}u, \tag{11.52}$$

and $J^{1-\varrho}_{0|t}$ is the fractional integral of order $1-\varrho$ defined by

$$J^{1-\varrho}_{0|t}v(t) = \frac{1}{\Gamma(1-\varrho)}\int_0^t \frac{v(s)}{(t-s)^{\varrho}}ds, \quad \forall v \in C(\mathbb{R}). \tag{11.53}$$

Definition 11.1 (Weak solution). Let

$$T > 0, \quad \gamma_1, \gamma_2 \in (0,1), \quad r_1, r_2 \in (-1,1),$$

and

$$u_0, u_1, v_0, v_1 \in L^1_{\mathrm{loc}}(\mathbb{R}^n)$$

We say that (u, v) is a weak solution if

$$(u,v) \in L^q((0,T_{\max}), L^q_{\mathrm{loc}}(\mathbb{R}^n)) \times L^p((0,T_{\max}), L^p_{\mathrm{loc}}(\mathbb{R}^n)),$$

and satisfies

$$\begin{aligned}
\Gamma(\varrho_1)&\int_0^T\int_{\mathbb{R}^n} \varpi J^{\varrho_1}_{0|t}(|v|^p)dxdt + \int_{\mathbb{R}^n} u_1(x)\varpi(t=0,x)dx \\
&- \int_{\mathbb{R}^n} u_0(x)\frac{\partial \varpi}{\partial t}(t=0,x)dx + b_1(0)\int_{\mathbb{R}^n} u_0(x)\varpi(t=0,x)dx \\
&= \int_0^T\int_{\mathbb{R}^n} u\varpi_{tt}dxdt - \int_0^T\int_{\mathbb{R}^n} u\Delta\varpi dxdt \\
&- \int_0^T\int_{\mathbb{R}^n} ub_1(t)\frac{\partial \varpi}{\partial t}dxdt - \int_0^T\int_{\mathbb{R}^n} ub_1'(t)\varpi dxdt,
\end{aligned}$$

and

$$\begin{aligned}
\Gamma(\varrho_2)&\int_0^T\int_{\mathbb{R}^n} \psi J^{\varrho_2}_{0|t}(|u|^p)dxdt + \int_{\mathbb{R}^n} v_1(x)\psi(t=0,x)dx \\
&- \int_{\mathbb{R}^n} v_0(x)\frac{\partial \psi}{\partial t}(t=0,x)dx + b_2(0)\int_{\mathbb{R}^n} v_0(x)\psi(t=0,x)dx \\
&= \int_0^T\int_{\mathbb{R}^n} v\frac{\partial^2\psi}{\partial t^2}dxdt - \int_0^T\int_{\mathbb{R}^n} v\Delta\psi dxdt
\end{aligned}$$

$$-\int_0^T\int_{\mathbb{R}^n} vb_2(t)\frac{\partial\psi}{\partial t}dxdt - \int_0^T\int_{\mathbb{R}^n} vb_2'(t)\psi dxdt,$$

for all test functions

$$(\varpi,\psi) \in \left(C^2([0,T]\times\mathbb{R}^n)\right)^2,$$

such that

$$\varpi(T,\cdot) = \frac{\partial\varpi}{\partial t}(T,\cdot) = \psi(T,\cdot) = \frac{\partial\psi}{\partial t}(T,\cdot) = 0,$$

where $\varrho_1 = 1-\gamma_1$ and $\varrho_2 = 1-\gamma_2$.

Proposition 11.6 ([2]). *Let*

$$f, g \in C[0,T],$$

such that

$$D^{\varrho}_{0|t}f(t), D^{\varrho}_{t|T}\mu(t) \in C[0,T].$$

The integration by parts formula in fractional version is given by

$$\int_0^t f(s)D^{\varrho}_{s|T}\mu(s)ds = \int_0^t (D^{\varrho}_{0|s}f(s))\mu(s)ds, \tag{11.54}$$

for all $t \in (0,1)$.

Proposition 11.7. *For all*

$$v \in L^q(0,T), \quad q \geq 1$$

and $t \in (0,T)$*, we have*

$$(D^{\varrho}_{0|t}\circ J^{\varrho}_{0|t})(v) = v. \tag{11.55}$$

Moreover, for all

$$v \in C^q[0,T], \quad T > 0,$$

the following rule holds:

$$(-1)^l d^l_t D^{\varrho}_{t|T}v(t) = D^{\varrho+l}_{t|T}v(t), \quad l \in \mathbb{N}, \varrho \in (0,1). \tag{11.56}$$

Corollary 11.2. *Let $\nu \geq 1$ and $T > 0$. Let also, ϖ_2 be the function defined by* (11.65). *Then* $\forall 0 < \varrho < 1$, *we have*

$$D^{\varrho}_{t|T}\varpi_2(t) = \frac{\Gamma(\nu+1)}{\Gamma(\nu-\varrho+1)}T^{-\nu}(T-t)_+^{\nu-\varrho} = \frac{\Gamma(\nu+1)}{\Gamma(\nu-\varrho+1)}T^{-\varrho}\left(1-\frac{t}{T}\right)_+^{\nu-\varrho},$$

$$D^{\varrho+1}_{t|T}\varpi_2(t) = \frac{\Gamma(\nu+1)}{\Gamma(\nu-\varrho+1)}T^{-\nu}(T-t)_+^{\nu-\varrho-1} = \frac{\Gamma(\nu+1)}{\Gamma(\nu-\varrho+1)}T^{-\varrho-1}\left(1-\frac{t}{T}\right)_+^{\nu-\varrho-1},$$

$$D^{\varrho+2}_{t|T}\varpi_2(t) = \frac{\Gamma(\nu+1)}{\Gamma(\nu-\varrho+1)}T^{-\nu}(T-t)_+^{\nu-\varrho-2} = \frac{\Gamma(\nu+1)}{\Gamma(\nu-\varrho+1)}T^{-\varrho-2}\left(1-\frac{t}{T}\right)_+^{\nu-\varrho-2}.$$

Corollary 11.3. *Let $T > 0$ and ϖ_2 be the function defined by* (11.65). *Then, for all* $0 < \varrho < 1$, *we have*

$$D^{\varrho}_{t|T}\varpi_2(0) = CT^{-\varrho}, \quad D^{\varrho+1}_{t|T}\varpi_2(0) = CT^{-\varrho-1}, \quad D^{\varrho+2}_{t|T}\varpi_2(0) = CT^{-\varrho-2}.$$

11.2.1 Finite time blow-up result

Theorem 11.4. *Let $0 < \gamma_1, \gamma_2 < 1$ and $p, q \in (1, \infty)$ and $pq > 1$. Assume that*

$$(u_0, u_1), (v_0, v_1) \in H^1(\mathbb{R}^n) \times L^2(\mathbb{R}^n),$$

and satisfies the conditions

$$\int_{\mathbb{R}^n} u_0(x)dx > 0 \quad \text{and} \quad \int_{\mathbb{R}^n} u_1(x)dx > 0, \tag{11.57}$$

$$\int_{\mathbb{R}^n} v_0(x)dx > 0 \quad \text{and} \quad \int_{\mathbb{R}^n} v_1(x)dx > 0. \tag{11.58}$$

If q satisfies for $r_1 \in (-1, 0)$ or $r_1 \in (0, 1)$ with the condition

$$\frac{q}{q-1} > \inf_{d>0}\max\left\{\frac{\frac{nd}{2}+1}{1-\gamma_1+d}; \sqrt{\frac{\frac{nd}{2}+1}{(1-\gamma_1)(1-r_1)} + \left(\frac{1-\gamma_1-(1-r_1)\frac{nd}{2}}{2(1-\gamma_1)(1-r_1)}\right)^2} - \frac{1-\gamma_1-(1-r_1)\frac{nd}{2}}{2(1-\gamma_1)(1-r_1)}\right\}, \tag{11.59}$$

or for $r_1 = 0$, where

$$\frac{q}{q-1} > \frac{\frac{n}{2}+1}{2-\gamma_1}, \tag{11.60}$$

or if p satisfies for $r_2 \in (-1, 0)$ or $r_1 \in (0, 1)$ with the condition

$$\frac{p}{p-1} > \inf_{d>0} \max\left\{\frac{\frac{nd}{2}+1}{1-\gamma_2+d}; \sqrt{\frac{\frac{nd}{2}+1}{(1-\gamma_2)(1-r_2)} + \left(\frac{1-\gamma_2-(1-r_2)\frac{nd}{2}}{2(1-\gamma_2)(1-r_2)}\right)^2} - \frac{1-\gamma_2-(1-r_2)\frac{nd}{2}}{2(1-\gamma_2)(1-r_2)}\right\}, \tag{11.61}$$

or for $r_2 = 0$, *where*

$$\frac{p}{p-1} > \frac{\frac{n}{2}+1}{2-\gamma_2}, \tag{11.62}$$

then the weak solution in the sense of definition (11.1) *of the Cauchy problem* (11.1) *does not exist globally in time.*

Proof. We will use the proof by contradiction. Suppose that (u, v) is a nontrivial weak solution of (11.1). For that, let us define the following test functions:

$$\begin{aligned} \varpi &= D^{\varrho_1}_{t|T}(\tilde{\varpi}) \\ &= D^{\varrho_1}_{t|T}(\varpi_1^l(x)\varpi_2(t)), \\ \psi &= D^{\varrho_2}_{t|T}(\tilde{\varpi}) \\ &= D^{\varrho_2}_{t|T}(\varpi_1^l(x)\varpi_2(t)). \end{aligned} \tag{11.63}$$

The functions ϖ_1 and ϖ_2 are given by

$$\varpi_1(x) = \Phi\left(\frac{|x|}{K}\right) \tag{11.64}$$

and

$$\varpi_2(t) = \left(1-\frac{t}{T}\right)_+^l, \tag{11.65}$$

and $\Phi(s)$ is a smooth nonincreasing function such that

$$\Phi(s) = \begin{cases} 1, & 0 \le s \le 1, \\ 0, & 2 \le s, \end{cases} \tag{11.66}$$

with $1 \le l$ and K is constant will be chosen later.

We denote by Ω_K the support of ϖ_1 such that

$$\Omega_K = \{x \in R^n; |x|^2 \le 2K\}.$$

Now, let us use the definition of the weak solution

$$\begin{aligned}
&\Gamma(\varrho_1)\int_0^T\int_{\mathbb{R}^n} J_{0|t}^{\varrho_1}(|v|^p)D_{t|T}^{\varrho_1}(\tilde{\varpi})dxdt + \int_{\mathbb{R}^n} u_1(x)D_{t|T}^{\varrho_1}(\tilde{\varpi}(t=0,x))dx\\
&\quad - \int_{\mathbb{R}^n} u_0(x)\partial_t D_{t|T}^{\varrho_1}(\tilde{\varpi}(t=0,x))dx + b_1(0)\int_{\mathbb{R}^n} u_0(x)D_{t|T}^{\varrho_1}(\tilde{\varpi}(t=0,x))dx\\
&= \int_0^T\int_{\mathbb{R}^n} u\partial_t^2 D_{t|T}^{\varrho_1}(\tilde{\varpi})dxdt - \int_0^T\int_{\mathbb{R}^n} u\Delta D_{t|T}^{\varrho_1}(\tilde{\varpi})dxdt\\
&\quad - \int_0^T\int_{\mathbb{R}^n} ub_1(t)\partial_t D_{t|T}^{\varrho_1}(\tilde{\varpi})dxdt - \int_0^T\int_{\mathbb{R}^n} ub_1'(t)D_{t|T}^{\varrho_1}(\tilde{\varpi})dxdt,
\end{aligned}$$

and

$$\begin{aligned}
&\Gamma(\varrho_2)\int_0^T\int_{\mathbb{R}^n} J_{0|t}^{\varrho_2}(|u|^p)D_{t|T}^{\varrho_2}(\tilde{\varpi})dxdt + \int_{\mathbb{R}^n} v_1(x)D_{t|T}^{\varrho_2}(\tilde{\varpi}(t=0,x))dx\\
&\quad - \int_{\mathbb{R}^n} v_0(x)\partial_t D_{t|T}^{\varrho_2}(\tilde{\varpi}(t=0,x))dx + b_2(0)\int_{\mathbb{R}^n} v_0(x)D_{t|T}^{\varrho_2}(\tilde{\varpi}(t=0,x))dx\\
&= \int_0^T\int_{\mathbb{R}^n} v\partial_t^2 D_{t|T}^{\varrho_2}(\tilde{\varpi})dxdt - \int_0^T\int_{\mathbb{R}^n} v\Delta D_{t|T}^{\varrho_2}(\tilde{\varpi})dxdt\\
&\quad - \int_0^T\int_{\mathbb{R}^n} vb_2(t)\partial_t D_{t|T}^{\varrho_2}(\tilde{\varpi})dxdt - \int_0^T\int_{\mathbb{R}^n} vb_2'(t)D_{t|T}^{\varrho_2}(\tilde{\varpi})dxdt.
\end{aligned}$$

Then, using the definition (11.63), one has

$$\begin{aligned}
&\Gamma(\varrho_1)\int_0^T\int_{\mathbb{R}^n} J_{0|t}^{\varrho_1}(|v|^p)D_{t|T}^{\varrho_1}(\tilde{\varpi})dxdt + \int_{\mathbb{R}^n} u_1(x)\varpi_1^l(x)D_{t|T}^{\varrho_1}(\varpi_2(0))dx\\
&\quad - \int_{\mathbb{R}^n} u_0(x)\varpi_1^l(x)\partial_t D_{t|T}^{\varrho_1}(\varpi_2(0))dx + b_1(0)\int_{\mathbb{R}^n} u_0(x)\varpi_1^l(x)D_{t|T}^{\varrho_1}(\varpi_2(0))dx\\
&= \int_0^T\int_{\mathbb{R}^n} u\varpi_1^l(x)\partial_t^2 D_{t|T}^{\varrho_1}(\varpi_2(t))dxdt - \int_0^T\int_{\mathbb{R}^n} u\Delta\varpi_1^l(x)D_{t|T}^{\varrho_1}(\varpi_2(t))dxdt\\
&\quad - \int_0^T\int_{\mathbb{R}^n} ub_1(t)\varpi_1^l(x)\partial_t D_{t|T}^{\varrho_1}(\varpi_2(t))dxdt - \int_0^T\int_{\mathbb{R}^n} ub_1'(t)\varpi_1^l(x)D_{t|T}^{\varrho_1}(\varpi_2(t))dxdt,
\end{aligned}$$

and

$$\Gamma(\varrho_2)\int_0^T\int_{\mathbb{R}^n} J_{0|t}^{\varrho_2}(|u|^p)D_{t|T}^{\varrho_2}(\tilde{\varpi})dxdt + \int_{\mathbb{R}^n} v_1(x)\varpi_1^l(x)D_{t|T}^{\varrho_2}(\varpi_2(0))dx$$
$$- \int_{\mathbb{R}^n} v_0(x)\varpi_1^l(x)\partial_t D_{t|T}^{\varrho_2}(\varpi_2(0))dx + b_2(0)\int_{\mathbb{R}^n} v_0(x)\varpi_1^l(x)D_{t|T}^{\varrho_2}(\varpi_2(0))dx$$
$$= \int_0^T\int_{\mathbb{R}^n} v\varpi_1^l(x)\partial_t^2 D_{t|T}^{\varrho_2}(\varpi_2(t))dxdt - \int_0^T\int_{\mathbb{R}^n} v\Delta\varpi_1^l(x)D_{t|T}^{\varrho_2}(\varpi_2(t))dxdt$$
$$- \int_0^T\int_{\mathbb{R}^n} vb_2(t)\varpi_1^l(x)\partial_t D_{t|T}^{\varrho_2}(\varpi_2(t))dxdt - \int_0^T\int_{\mathbb{R}^n} vb_2'(t)\varpi_1^l(x)D_{t|T}^{\varrho_2}(\varpi_2(t))dxdt.$$

With Corollary 11.3 and (11.56), we get

$$\Gamma(\varrho_1)\int_0^T\int_{\mathbb{R}^n} J_{0|t}^{\varrho_1}(|v|^p)D_{t|T}^{\varrho_1}(\tilde{\varpi})dxdt + CT^{-\varrho_1}\int_{\mathbb{R}^n} u_1(x)\varpi_1^l(x)dx$$
$$+ CT^{-\varrho_1-1}\int_{\mathbb{R}^n} u_0(x)\varpi_1^l(x)dx + CT^{-\varrho_1}b_1(0)\int_{\mathbb{R}^n} u_0(x)\varpi_1^l(x)dx \tag{11.67}$$
$$= \int_0^T\int_{\mathbb{R}^n} u\varpi_1^l(x)D_{t|T}^{\varrho_1+2}(\varpi_2(t))dxdt - \int_0^T\int_{\mathbb{R}^n} u\Delta\varpi_1^l(x)D_{t|T}^{\varrho_1}(\varpi_2(t))dxdt$$
$$+ \int_0^T\int_{\mathbb{R}^n} ub_1(t)\varpi_1^l(x)D_{t|T}^{\varrho_1+1}(\varpi_2(t))dxdt - \int_0^T\int_{\mathbb{R}^n} ub_1'(t)\varpi_1^l(x)D_{t|T}^{\varrho_1}(\varpi_2(t))dxdt,$$

and

$$\Gamma(\varrho_2)\int_0^T\int_{\mathbb{R}^n} J_{0|t}^{\varrho_2}(|u|^p)D_{t|T}^{\varrho_2}(\tilde{\varpi})dxdt + CT^{-\varrho_2}\int_{\mathbb{R}^n} v_1(x)\varpi_1^l(x)dx$$
$$+ CT^{-\varrho_2-1}\int_{R\mathbb{R}^n} v_0(x)\varpi_1^l(x)dx + CT^{-\varrho_2}b_2(0)\int_{\mathbb{R}^n} v_0(x)\varpi_1^l(x)dx \tag{11.68}$$
$$= \int_0^T\int_{\mathbb{R}^n} v\varpi_1^l(x)D_{t|T}^{\varrho_2+2}(\varpi_2(t))dxdt - \int_0^T\int_{\mathbb{R}^n} v\Delta\varpi_1^l(x)D_{t|T}^{\varrho_2}(\varpi_2(t))dxdt$$
$$+ \int_0^T\int_{\mathbb{R}^n} vb_2(t)\varpi_1^l(x)D_{t|T}^{\varrho_2+1}(\varpi_2(t))dxdt - \int_0^T\int_{\mathbb{R}^n} vb_2'(t)\varpi_1^l(x)D_{t|T}^{\varrho_2}(\varpi_2(t))dxdt.$$

1. To deal with the left-hand side of (11.67) and (11.68) of the weak formulation of solutions, we start with the first term in each one using the formula of integration by parts from Proposition 11.6:

$$\begin{cases} \int_0^T \int_{\mathbb{R}^n} J_{0|t}^{\varrho_1}(|v|^p) D_{t|T}^{\varrho_1}(\tilde{\varpi}) dxdt = \int_0^T \int_{\mathbb{R}^n} D_{t|T}^{\varrho_1}(J_{0|t}^{\varrho_1}(|v|^p))\tilde{\varpi} dxdt, \\ \int_0^T \int_{\mathbb{R}^n} J_{0|t}^{\varrho_2}(|u|^p) D_{t|T}^{\varrho_2}(\tilde{\varpi}) dxdt = \int_0^T \int_{\mathbb{R}^n} D_{t|T}^{\varrho_2}(J_{0|t}^{\varrho_2}(|u|^p))\tilde{\varpi} dxdt. \end{cases} \tag{11.69}$$

Thanks to (11.55) in Proposition 11.7, we conclude that

$$\begin{cases} \int_0^T \int_{\mathbb{R}^n} J_{0|t}^{\varrho_1}(|v|^p) D_{t|T}^{\varrho_1}(\tilde{\varpi}) dxdt = \int_0^T \int_{\mathbb{R}^n} |v|^p \tilde{\varpi} dxdt, \\ \int_0^T \int_{\mathbb{R}^n} J_{0|t}^{\varrho_2}(|u|^p) D_{t|T}^{\varrho_2}(\tilde{\varpi}) dxdt = \int_0^T \int_{\mathbb{R}^n} |u|^p \tilde{\varpi} dxdt. \end{cases} \tag{11.70}$$

For the other terms, we use the fact that the support of ϖ_1 is included in

$$\Omega_{T^d} = \{x \in \mathbb{R}^n; |x|^2 \leq 2T^d\}.$$

In addition, the conditions (11.61) imply that

$$\begin{cases} \int_\Omega u_i(x)\varpi_1^l(x)dx > 0, \\ \int_\Omega v_i(x)\varpi_1^l(x)dx > 0, \end{cases} \quad i = 0, 1,$$

and $u_i, v_i \in L^1_{\text{loc}}(\mathbb{R}^n)$ for $i = 0$. Then, using the Lebesgue dominated convergence theorem, we obtain for $i = 0$ and $i = 1$ and ($\Omega = \Omega_{T^d}$),

$$\begin{cases} \lim_{T\to\infty} \int_\Omega u_i(x)\varpi_1^l(x)dx = \int_{\mathbb{R}^n} u_i(x)dx, \\ \lim_{T\to\infty} \int_\Omega v_i(x)\varpi_1^l(x)dx = \int_{\mathbb{R}^n} v_i(x)dx. \end{cases}$$

Hence,

$$\begin{cases} \int_\Omega u_i(x)\varpi_1^l(x)dx > 0, \\ \int_\Omega v_i(x)\varpi_1^l(x)dx > 0, \end{cases} \quad i = 0, 1.$$

Now, for T large enough, we can rewrite (11.67) and (11.68) as follows:

$$\begin{aligned} &\int_{Q_T} |v|^p \tilde{\varpi} dxdt \\ &\quad < \int_{Q_T} \left|u\varpi_1^l(x) D_{t|T}^{\varrho_1+2}(\varpi_2(t))\right| dxdt + \int_{Q_T} \left|u\Delta\varpi_1^l(x) D_{t|T}^{\varrho_1}(\varpi_2(t))\right| dxdt \\ &\quad + \int_{Q_T} \left|ub_1(t)\varpi_1^l(x) D_{t|T}^{\varrho_1+1}(\varpi_2(t))\right| dxdt + \int_{Q_T} \left|ub_1'(t)\varpi_1^l(x) D_{t|T}^{\varrho_1}(\varpi_2(t))\right| dxdt, \end{aligned} \tag{11.71}$$

and

$$\begin{aligned}
&\int_{Q_T} |u|^q \tilde{\varpi} dxdt \\
&\quad < \int_{Q_T} |v\varpi_1^l(x) D_{t|T}^{\varrho_2+2}(\varpi_2(t))| dxdt + \int_{Q_T} |v\Delta\varpi_1^l(x) D_{t|T}^{\varrho_2}(\varpi_2(t))| dxdt \\
&\quad + \int_{Q_T} |vb_2(t)\varpi_1^l(x) D_{t|T}^{\varrho_2+1}(\varpi_2(t))| dxdt + \int_{Q_T} |vb_2'(t)\varpi_1^l(x) D_{t|T}^{\varrho_2}(\varpi_2(t))| dxdt,
\end{aligned} \tag{11.72}$$

with

$$Q_T = [0,T] \times \mathbb{R}^n.$$

2. To treat the left-hand side of (11.71) and (11.72) of the weak formulation of solutions, we use Hölder's inequality with parameters p, p', q and q'. We start with the first terms:

$$\begin{aligned}
&\int_{Q_T} |u\varpi_1^l(x) D_{t|T}^{\varrho_1+2}(\varpi_2(t))| dxdt \\
&\quad = \int_{Q_T} |u|\tilde{\varpi}^{\frac{1}{q}} \tilde{\varpi}^{-\frac{1}{q}} \varpi_1^l(x) D_{t|T}^{\varrho_1+2}(\varpi_2(t)) dxdt \\
&\quad < \left[\int_{Q_T} |u|^q \tilde{\varpi} dxdt\right]^{\frac{1}{q}} \left[\int_{Q_T} \varpi_1^l(x) \varpi_2^{-\frac{1}{q-1}}(t) [D_{t|T}^{\varrho_1+2}(\varpi_2(t))]^{q'} dxdt\right]^{\frac{1}{q'}}.
\end{aligned}$$

In the same way, we have

$$\begin{aligned}
&\int_{Q_T} |v\varpi_1^l(x) D_{t|T}^{\varrho_2+2}(\varpi_2(t))| dxdt \int_{Q_T} |v|\tilde{\varpi}^{\frac{1}{p}} \tilde{\varpi}^{-\frac{1}{p}} \varpi_1^l(x) D_{t|T}^{\varrho_2+2}(\varpi_2(t)) dxdt \\
&\quad < \left[\int_{Q_T} |v|^p \tilde{\varpi} dxdt\right]^{\frac{1}{p}} \left[\int_{Q_T} \varpi_1^l(x) \varpi_2^{-\frac{1}{p-1}}(t) [D_{t|T}^{\varrho_2+2}(\varpi_2(t))]^{p'} dxdt\right]^{\frac{1}{p'}}.
\end{aligned}$$

For the second terms, we can use

$$\Delta\varpi_1^l = l\varpi_1^{l-1}\Delta\varpi_1 + l(l-1)\varpi_1^{l-2}|\nabla\varpi_1|^2,$$

with the fact that $0 \le \varpi_1 \le 1$ to get

$$|l\varpi_1^{l-1}\Delta\varpi_1 + l(l-1)\varpi_1^{l-2}|\nabla\varpi_1|^2| \le C\varpi_1^{l-2}(|\Delta\varpi_1| + |\nabla\varpi_1|^2).$$

Then

$$\int_{Q_T} |u\Delta\varpi_1^l(x) D_{t|T}^{\varrho_1}(\varpi_2(t))| dxdt < \int_{Q_T} |u|\varpi_1^{l-2}(|\Delta\varpi_1| + |\nabla\varpi_1|^2) D_{t|T}^{\varrho_1}(\varpi_2(t)) dxdt,$$

$$\int_{Q_T} |v\Delta\varpi_1^l(x) D_{t|T}^{\varrho_2}(\varpi_2(t))| dxdt < \int_{Q_T} |v|\varpi_1^{l-2}(|\Delta\varpi_1| + |\nabla\varpi_1|^2) D_{t|T}^{\varrho_2}(\varpi_2(t)) dxdt.$$

Using Hölder's inequality, we get

$$\begin{aligned}\int_{Q_T} |u\Delta\varpi_1^l(x) D_{t|T}^{\varrho_1}(\varpi_2(t))| dxdt &< \left[\int_{Q_T} |u|^q \tilde{\varpi} dxdt\right]^{\frac{1}{q}} \\ &\times \left[\int_{Q_T} \varpi_1^{l-2q'}(x)\varpi_2^{-\frac{1}{q-1}}(t)(|\Delta\varpi_1| + |\nabla\varpi_1|^2)^{q'} [D_{t|T}^{\varrho_1}(\varpi_2(t))]^{q'} dxdt\right]^{\frac{1}{q'}},\end{aligned}$$

and

$$\begin{aligned}\int_{Q_T} |v\Delta\varpi_1^l(x) D_{t|T}^{\varrho_2}(\varpi_2(t))| dxdt &< \left[\int_{Q_T} |v|^p \tilde{\varpi} dxdt\right]^{\frac{1}{p}} \\ &\times \left[\int_{Q_T} \varpi_1^{l-2p'}(x)\varpi_2^{-\frac{1}{p-1}}(t)(|\Delta\varpi_1| + |\nabla\varpi_1|^2)^{p'} [D_{t|T}^{\varrho_2}(\varpi_2(t))]^{p'} dxdt\right]^{\frac{1}{p'}}.\end{aligned}$$

For the third terms, we use the same way and we get

$$\begin{aligned}\int_{Q_T} |ub_1(t)\varpi_1^l(x) D_{t|T}^{\varrho_1+1}(\varpi_2(t))| dxdt &= \int_{Q_T} |u|\tilde{\varpi}^{\frac{1}{q}}\tilde{\varpi}^{-\frac{1}{q}} b_1(t)\varpi_1^l(x) D_{t|T}^{\varrho_1+1}(\varpi_2(t)) dxdt \\ &< \left[\int_{Q_T} |u|^q \tilde{\varpi} dxdt\right]^{\frac{1}{q}}\end{aligned}$$

$$\times\left[\int_{Q_T} \varpi_1^l(x)\varpi_2^{-\frac{1}{q-1}}(t)|b_1(t)|^{q'}\left[D_{t|T}^{\varrho_1+1}(\varpi_2(t))\right]^{q'}dxdt\right]^{\frac{1}{q'}},$$

and

$$\begin{aligned}&\int_{Q_T}|vb_2(t)\varpi_1^l(x)D_{t|T}^{\varrho_2+1}(\varpi_2(t))|dxdt\\&=\int_{Q_T}|v|\tilde{\varpi}^{\frac{1}{p}}\tilde{\varpi}^{-\frac{1}{p}}b_2(t)\varpi_1^l(x)D_{t|T}^{\varrho_2+1}(\varpi_2(t))dxdt\\&<\left[\int_{Q_T}|v|^p\tilde{\varpi}dxdt\right]^{\frac{1}{p}}\\&\quad\times\left[\int_{Q_T}\varpi_1^l(x)\varpi_2^{-\frac{1}{p-1}}(t)|b_2(t)|^{p'}\left[D_{t|T}^{\varrho_2+1}(\varpi_2(t))\right]^{p'}dxdt\right]^{\frac{1}{p'}}.\end{aligned}$$

For the fourth terms, using Hölder's inequality again, we conclude that

$$\begin{aligned}&\int_{Q_T}|ub_1'(t)\varpi_1^l(x)D_{t|T}^{\varrho_1}(\varpi_2(t))|dxdt\\&=\int_{Q_T}|u|\tilde{\varpi}^{\frac{1}{q}}\tilde{\varpi}^{-\frac{1}{q}}b_1'(t)\varpi_1^l(x)D_{t|T}^{\varrho_1}(\varpi_2(t))dxdt\\&<\left[\int_{Q_T}|u|^q\tilde{\varpi}dxdt\right]^{\frac{1}{q}}\\&\quad\times\left[\int_{Q_T}\varpi_1^l(x)\varpi_2^{-\frac{1}{q-1}}(t)|b_1'(t)|^{q'}\left[D_{t|T}^{\varrho_1}(\varpi_2(t))\right]^{q'}dxdt\right]^{\frac{1}{q'}},\end{aligned}$$

and

$$\begin{aligned}&\int_{Q_T}|vb_2'(t)\varpi_1^l(x)D_{t|T}^{\varrho_2}(\varpi_2(t))|dxdt\\&=\int_{Q_T}|v|\tilde{\varpi}^{\frac{1}{p}}\tilde{\varpi}^{-\frac{1}{p}}b_2'(t)\varpi_1^l(x)D_{t|T}^{\varrho_2}(\varpi_2(t))dxdt\\&<\left[\int_{Q_T}|v|^p\tilde{\varpi}dxdt\right]^{\frac{1}{p}}\\&\quad\times\left[\int_{Q_T}\varpi_1^l(x)\varpi_2^{-\frac{1}{p-1}}(t)|b_2'(t)|^{p'}\left[D_{t|T}^{\varrho_2}(\varpi_2(t))\right]^{p'}dxdt\right]^{\frac{1}{p'}}.\end{aligned}$$

Now, we include all the results that we get in (11.71) and (11.72), and we obtain

$$\begin{cases} \int_{Q_T} |v|^p \tilde{\varpi} dxdt \le [\int_{Q_T} |u|^q \tilde{\varpi} dxdt]^{\frac{1}{q}} \times \mathrm{M}, \\ \int_{Q_T} |u|^q \tilde{\varpi} dxdt \le [\int_{Q_T} |v|^p \tilde{\varpi} dxdt]^{\frac{1}{p}} \times \mathrm{N} \end{cases}$$

with

$$\begin{aligned} \mathrm{M} = & \Big[\int_{Q_T} \varpi_1^l(x) \varpi_2^{-\frac{1}{q-1}}(t) [D_{t|T}^{\varrho_1+2}(\varpi_2(t))]^{q'} dxdt\Big]^{\frac{1}{q'}} \\ & + \Big[\int_{Q_T} \varpi_1^l(x) \varpi_2^{-\frac{1}{q-1}}(t) |b_1'(t)|^{q'} [D_{t|T}^{\varrho_1}(\varpi_2(t))]^{q'} dxdt\Big]^{\frac{1}{q'}} \\ & + \Big[\int_{Q_T} \varpi_1^l(x) \varpi_2^{-\frac{1}{q-1}}(t) |b_1(t)|^{q'} [D_{t|T}^{\varrho_1+1}(\varpi_2(t))]^{q'} dxdt\Big]^{\frac{1}{q'}} \\ & + \Big[\int_{Q_T} \varpi_1^{l-2q'}(x) \varpi_2^{-\frac{1}{q-1}}(t) (|\Delta \varpi_1| + |\nabla \varpi_1|^2)^{q'} [D_{t|T}^{\varrho_1}(\varpi_2(t))]^{q'} dxdt\Big]^{\frac{1}{q'}}, \end{aligned} \tag{11.73}$$

and

$$\begin{aligned} \mathrm{N} = & \Big[\int_{Q_T} \varpi_1^l(x) \varpi_2^{-\frac{1}{p-1}}(t) [D_{t|T}^{\varrho_2+2}(\varpi_2(t))]^{p'} dxdt\Big]^{\frac{1}{p'}} \\ & + \Big[\int_{Q_T} \varpi_1^l(x) \varpi_2^{-\frac{1}{p-1}}(t) |b_2'(t)|^{p'} [D_{t|T}^{\varrho_2}(\varpi_2(t))]^{p'} dxdt\Big]^{\frac{1}{p'}} \\ & + \Big[\int_{Q_T} \varpi_1^l(x) \varpi_2^{-\frac{1}{p-1}}(t) |b_2(t)|^{p'} [D_{t|T}^{\varrho_2+1}(\varpi_2(t))]^{p'} dxdt\Big]^{\frac{1}{p'}} \\ & + \Big[\int_{Q_T} \varpi_1^{l-2p'}(x) \varpi_2^{-\frac{1}{p-1}}(t) (|\Delta \varpi_1| + |\nabla \varpi_1|^2)^{p'} [D_{t|T}^{\varrho_2}(\varpi_2(t))]^{p'} dxdt\Big]^{\frac{1}{p'}}, \end{aligned} \tag{11.74}$$

with $p' = \frac{p}{p-1}$ and $q' = \frac{q}{q-1}$. Combining (11.73) and (11.74), we get

$$\begin{cases} [\int_{Q_T} |v|^p \tilde{\varpi} dxdt]^{1-\frac{1}{qp}} < \mathrm{N}^{\frac{1}{q}} \mathrm{M}, \\ [\int_{Q_T} |u|^q \tilde{\varpi} dxdt]^{1-\frac{1}{qp}} \le \mathrm{M}^{\frac{1}{p}} \mathrm{N}. \end{cases} \tag{11.75}$$

3. Now, we will deal with the right-hand side and find δ and δ' such that

$$\begin{cases} [\int_{Q_T} |v|^p \tilde{\varpi} dxdt]^{1-\frac{1}{qp}} < T^{-\delta}, \\ [\int_{Q_T} |u|^q \tilde{\varpi} dxdt]^{1-\frac{1}{qp}} < T^{-\delta'}. \end{cases} \tag{11.76}$$

To simplify, let us define

$$\begin{aligned} I_1^{i,k} &= \int_{Q_T} \varpi_1^l(x)\varpi_2^{-\frac{1}{k-1}}(t)[D_{t|T}^{\varrho_i+2}(\varpi_2(t))]^{k'} dxdt, \\ I_2^{i,k} &= \int_{Q_T} \varpi_1^l(x)\varpi_2^{-\frac{1}{k-1}}(t)|b_i(t)|^{k'} [D_{t|T}^{\varrho_i+1}(\varpi_2(t))]^{k'} dxdt, \\ I_3^{i,k} &= \int_{Q_T} \varpi_1^l(x)\varpi_2^{-\frac{1}{k-1}}(t)|b_i'(t)|^{k'} [D_{t|T}^{\varrho_i}(\varpi_2(t))]^{k'} dxdt, \\ I_4^{i,k} &= \int_{Q_T} \varpi_1^{l-2k'}(x)\varpi_2^{-\frac{1}{k-1}}(t)(|\Delta\varpi_1| + |\nabla\varpi_1|^2)^{k'} [D_{t|T}^{\varrho_i}(\varpi_2(t))]^{k'} dxdt, \end{aligned}$$

for $i = 1, 2, 3, 4$ and $k' = \frac{k}{k-1}$ with $k \in \{p, q\}$. Then we can rewrite M and N as follows:

$$\begin{aligned} \mathrm{M} &= [I_1^{1,q}]^{\frac{1}{q'}} + [I_2^{1,q}]^{\frac{1}{q'}} + [I_3^{1,q}]^{\frac{1}{q'}} + [I_4^{1,q}]^{\frac{1}{q'}}, \\ \mathrm{N} &= [I_1^{2,p}]^{\frac{1}{p'}} + [I_2^{2,p}]^{\frac{1}{p'}} + [I_3^{2,p}]^{\frac{1}{p'}} + [I_4^{2,p}]^{\frac{1}{p'}}. \end{aligned}$$

We shall estimate the integrals $I_j^{i,k}$ for $j = 1, 2, 3, 4$. For that, we choose the scaled variables

$$x = T^{\frac{d}{2}} y \quad \text{and} \quad t = T\tau. \tag{11.77}$$

We recall that

$$\operatorname{supp} \varpi_1 = \Omega_{T^d}.$$

By Fubini's theorem, we obtain

$$\begin{aligned} I_1^{i,k} &= \int_{Q_T} \varpi_1^l(x)\varpi_2^{-\frac{1}{k-1}}(t)[D_{t|T}^{\varrho_i+2}(\varpi_2(t))]^{k'} dxdt \\ &= \Big(\int_{\Omega_{T^d}} \varpi_1^l(x)dx\Big)\Big(\int_0^T \varpi_2^{-\frac{1}{k-1}}(t)[D_{t|T}^{\varrho_i+2}(\varpi_2(t))]^{k'} dt\Big) \\ &= I_{11}^{i,k} I_{12}^{i,k}. \end{aligned} \tag{11.78}$$

First, we have

$$I_{11}^{i,k} = \int_{\Omega_{T^d}} \varpi_1^l(x)dx = T^{\frac{nd}{2}} \int_{|y|\le 2} \Phi^l(y^2)dy = CT^{\frac{nd}{2}}, \tag{11.79}$$

with $C > 0$. Next, we will use Corollary 11.2 with $l \le 1$ to ensure that $l-(\varrho_i+2)k' > 0$. These results assume that the next integral exists

$$\begin{aligned}
I_{12}^{i,k} &= \int_0^T \varpi_2^{-\frac{1}{k-1}}(t)[D_{t|T}^{\varrho_i+2}(\varpi_2(t))]^{k'} dt \\
&= \int_0^1 (1-\tau)^{-\frac{l}{k-1}}[T^{-(\varrho_i+2)}(1-\tau)^{l-(\varrho_i+2)}]^{k'} d\tau \\
&= T^{1-(\varrho_i+2)k'} \int_0^1 (1-\tau)^{l-(\varrho_i+2)k'} d\tau \\
&= CT^{1-(\varrho_i+2)k'}.
\end{aligned} \tag{11.80}$$

Putting (11.79) and (11.80) in (11.78), we arrive at

$$I_1^{i,k} = T^{1-(\varrho_i+2)k'+\frac{nd}{2}}. \tag{11.81}$$

We estimate $I_2^{i,k}$ and $I_3^{i,k}$ as follows:

$$\begin{aligned}
I_2^{i,k} &= \int_{Q_T} \varpi_1^l(x)\varpi_2^{-\frac{1}{k-1}}(t)|b_i(t)|^{k'}[D_{t|T}^{\varrho_i+1}(\varpi_2(t))]^{k'} dxdt \\
&= \Big(\int_{\Omega_{T^d}} \varpi_1^l(x)dx\Big)\Big(\int_0^T \varpi_2^{-\frac{1}{k-1}}(t)|b_i(t)|^{k'}[D_{t|T}^{\varrho_i+1}(\varpi_2(t))]^{k'} dt\Big) \\
&= I_{21}^{i,k} I_{22}^{i,k}
\end{aligned} \tag{11.82}$$

and

$$\begin{aligned}
I_3^{i,k} &= \int_{Q_T} \varpi_1^l(x)\varpi_2^{-\frac{1}{k-1}}(t)|b_i'(t)|^{k'}[D_{t|T}^{\varrho_i}(\varpi_2(t))]^{k'} dxdt \\
&= \Big(\int_{\Omega_{T^d}} \varpi_1^l(x)dx\Big)\Big(\int_0^T \varpi_2^{-\frac{1}{k-1}}(t)|b_i'(t)|^{k'}[D_{t|T}^{\varrho_i}(\varpi_2(t))]^{k'} dt\Big) \\
&= I_{31}^{i,k} I_{32}^{i,k}.
\end{aligned} \tag{11.83}$$

Noting that

$$I_{11}^{i,k} = I_{21}^{i,k} = I_{31}^{i,k}.$$

Then we have

$$b_i(t) = (1+t)^{r_i},$$

for $r_i \in (0,1)$ with $i = 1$ and $i = 2$. It is clear that b_i is strictly increasing. Then we have for all $t \in [0,1]$,

$$b_i(t) \le b_i(T) = (1+T)^{r_i}. \tag{11.84}$$

Hence,

$$I_{22}^{i,k} < (1+T)^{k'r_i} T^{1-(\varrho_i+1)k'} < T^{1-(\varrho_i+1-r_i)k'}, \tag{11.85}$$

for T sufficiently large. Then we can get from (11.85) and (11.79),

$$I_2^{i,k} < T^{1+\frac{nd}{2}-(\varrho_i+1-r_i)k'}. \tag{11.86}$$

Now, we deal with $I_32^{i,k}$. Using the fact that

$$b_i'(t) \sim (1+t)^{r_i-1} \quad \text{for } r_i \in (0,1).$$

First, we have to see how to estimate the integral (put $\tau = \frac{t}{T}$),

$$T^{1-\varrho_i k'} \int_0^1 (1+\tau T)^{(r_i-1)k'} (1-\tau)^{-\frac{l}{k-1}} (1-\tau)^{(l-\varrho_i)k'} \, d\tau.$$

For this reason, we split the integral

$$\int_0^1 = \int_0^{T^{-\lambda}} + \int_{T^{-\lambda}}^1,$$

into two integrals. So, let us choose $\lambda \in (0,1)$. If l is large, then we may estimate the first integral (the integral is bounded) as follows:

$$\int_0^{T^{-\lambda}} (1+\tau T)^{(r_i-1)k'} (1-\tau)^{-\frac{l}{k-1}} (1-\tau)^{(l-\varrho_i)k'} \, d\tau \le CT^{-\lambda}.$$

In the second integral, we have

$$(1+\tau T)^{(r_i-1)k'} \le \left(1+T^{1-\lambda}\right)^{(r_i-1)k'}.$$

If l is large, then we can estimate

$$\begin{aligned}
&\int_{T^{-\lambda}}^{1} (1+\tau T)^{(r_i-1)k'}(1-\tau)^{-\frac{l}{k-1}}(1-\tau)^{(l-\varrho_i)k'}\,d\tau \\
&\qquad \le \left(1+T^{1-\lambda}\right)^{(r_i-1)k'} \int_{T^{-\lambda}}^{1} (1-\tau)^{-\frac{l}{k-1}}(1-\tau)^{(l-\varrho_i)k'}\,d\tau \\
&\qquad < \left(1+T^{1-\lambda}\right)^{(r_i-1)k'}.
\end{aligned}$$

The optimal choice for λ is given by the condition for T large,

$$T^{-\lambda} \sim T^{(1-\lambda)(r_i-1)k'},$$

That is,

$$\lambda = \frac{(r_i-1)k'}{(r_i-1)k'-1}.$$

Therefore, we may conclude that

$$I_3^{i,k} < T^{1+\frac{nd}{2}-\varrho_i k' - \frac{(r_i-1)k'}{(r_i-1)k'-1}}. \tag{11.87}$$

Finally, we deal with last one $I_4^{i,k}$. We have

$$\begin{aligned}
I_4^{i,k} &= \int_{Q_T} \varpi_1^{l-2k'}(x)\varpi_2^{-\frac{1}{k-1}}(t)\left(|\Delta\varpi_1| + |\nabla\varpi_1|^2\right)^{k'}\left[D_{t|T}^{\varrho_i}(\varpi_2(t))\right]^{k'}\,dxdt \\
&< \int_{Q_T} \varpi_1^{l-2k'}(x)\varpi_2^{-\frac{1}{k-1}}(t)\left(|\Delta\varpi_1|^{k'} + |\nabla\varpi_1|^{2k'}\right)\left[D_{t|T}^{\varrho_i}(\varpi_2(t))\right]^{k'}\,dxdt.
\end{aligned}$$

We put

$$\Pi(\varpi_1) = |\Delta\varpi_1|^{k'} + |\nabla\varpi_1|^{2k'}.$$

Then we have

$$\begin{aligned}
I_4^{i,k} &= \left(\int_{\Omega_{T^d}} \Pi(\varpi_1)\varpi_1^{l-2k'}\,dx\right)\left(\int_0^T \varpi_2^{-\frac{1}{k-1}}(t)\left[D_{t|T}^{\varrho_i}(\varpi_2(t))\right]^{k'}\,dt\right) \\
&= I_{41}^{i,k} I_{42}^{i,k}.
\end{aligned}$$

For $I_{41}^{i,k}$, we should see

$$I_{41}^{i,k} = \int\limits_{\Omega_{T^d}} \Pi(\varpi_1)\varpi_1^{l-2k'}\,dx < T^{\rho}.$$

We will now find ρ. Since ϖ_1 is bounded, we have that $\varpi_1^{l-2k'}$ is also bounded. We choose l such that $l - 2k' > 0$. Next, we have for all $j = 1, \ldots, n$ the relation

$$\partial_{x_i}\varpi_1 = \partial_{x_i}\Phi\left(\frac{|x|^2}{V}\right) = \frac{2}{V}\Phi'\left(\frac{|x|^2}{V}\right)x_j.$$

So,

$$|\nabla\varpi_1|^{2k'} = \left(\frac{2}{V}\right)^{2k'}\left|\Phi'\left(\frac{|x|^2}{V}\right)\right|^{2k'}|x|^{2k'}.$$

Using the change of variables (11.77) such that $V = T^d$ and $|\Phi'(s)| \le \frac{c}{1+s}$, we arrive at

$$\begin{aligned}\int\limits_{\mathbb{R}^n} |\nabla\varpi_1|^{2k'}\,dx &= \left(\frac{2}{T^d}\right)^{2k'} T^{\frac{nd}{2}} \int\limits_{|y|^2\le 2} |\Phi'(|y|^2)|^{2k'}\left(T^d|y|^2\right)^{k'}dy \\ &< T^{-dk'+\frac{nd}{2}} \int\limits_{|y|^2\le 2} \frac{|y|^{2k'}}{(1+|y|^2)^{2k'}}\,dy \le C_0 T^{-dk'+\frac{nd}{2}}.\end{aligned} \tag{11.88}$$

On the other hand, we have for $j = 1, \ldots, n$ the following relations:

$$\partial_{x_i}^2\varpi_1 = \frac{2}{V}\partial_{x_i}\left(\Phi'\left(\frac{|x|^2}{V}\right)x_j\right) = \frac{4}{V^2}\Phi''\left(\frac{|x|^2}{V}\right)x_j^2 + \frac{2}{V}\Phi'\left(\frac{|x|^2}{V}\right).$$

So,

$$|\Delta\varpi_1(x)|^{k'} = \left|\frac{4}{V^2}\Phi''\left(\frac{|x|^2}{V}\right)|x|^2 + \frac{2n}{V}\Phi'\left(\frac{|x|^2}{V}\right)\right|^{k'}.$$

Then

$$\int\limits_{\mathbb{R}^n} |\Delta\varpi_1(x)|^{k'}\,dx = \int\limits_{\mathbb{R}^n}\left|\frac{4}{V^2}\Phi''\left(\frac{|x|^2}{V}\right)|x|^2 + \frac{2n}{V}\Phi'\left(\frac{|x|^2}{V}\right)\right|^{k'}dx. \tag{11.89}$$

Now, we are going to prove the estimate

$$\int\limits_{\mathbb{R}^n} |\Delta\varpi_1(x)|^{k'}\,dx \le C_0 T^{-dk'+\frac{nd}{2}}. \tag{11.90}$$

Applying Minkowski's inequality and using (11.89), we get the following:

$$
\begin{aligned}
&\Big(\int_{\mathbb{R}^n} |\Delta\varpi_1(x)|^{k'}\,dx\Big)^{\frac{1}{k'}}\\
&\quad < \Big(\int_{\mathbb{R}^n} \Big|\frac{4}{V^2}\Phi''\Big(\frac{|x|^2}{V}\Big)|x|^2\Big|^{k'}\,dx\Big)^{\frac{1}{k'}}\\
&\quad\quad + \Big(\int_{\mathbb{R}^n} \Big|\frac{2n}{V}\Phi'\Big(\frac{|x|^2}{V}\Big)\Big|^{k'}\,dx\Big)^{\frac{1}{k'}}\\
&\quad < \Big(4^{k'}V^{\frac{n}{2}-2k'} \int_{|y|^2\le 2} |\Phi''(|y|^2)(V|y|^2)|^{k'}\,dy\Big)^{\frac{1}{k'}}\\
&\quad\quad + \Big((2n)^{k'}V^{\frac{n}{2}-2k'} \int_{|y|^2\le 2} |\Phi'(|y|^2)|^{k'}\,dy\Big)^{\frac{1}{k'}}\\
&\quad < (C_1 4^{k'}V^{\frac{n}{2}-2k'})^{\frac{1}{k'}} + (C_2(2n)^{k'}V^{\frac{n}{2}-2k'})^{\frac{1}{k'}}\\
&\quad < V^{\frac{n}{2k'}-1},
\end{aligned}
$$

where

$$
C_1 = \int_{|y|^2\le 2} |\Phi''(|y|^2)(K|y|^2)|^{k'}\,dy,
$$

and

$$
C_2 = \int_{|y|^2\le 2} |\Phi'(|y|^2)|^{k'}\,dy.
$$

We can find (11.90) setting $V = T^d$ and $k' = \frac{k}{k-1}$. Finally, from (11.90) and (11.88), we get

$$
I_{41}^{i,k} < T^{-dk'+\frac{nd}{2}}. \tag{11.91}
$$

In addition, we have the relation

$$
\begin{aligned}
I_{42}^{i,k} &= \int_0^T \varpi_2^{-\frac{1}{k-1}}(t)[D_{t|T}^{\varrho_i}(\varpi_2(t))]^{k'}\,dt\\
&= \int_0^1 (1-\tau)^{-\frac{l}{k-1}}[T^{-\varrho_i}(1-\tau)^{l-\varrho_i}]^{k'}\,T d\tau
\end{aligned} \tag{11.92}
$$

$$= T^{-\varrho_i k'+1}\int_0^1 (1-\tau)^{l-\varrho_i k'}\,d\tau = CT^{-\varrho_i k'+1}.$$

Now, from (11.92) and (11.91), we may conclude

$$I_4^{i,k} = \int_{Q_T} \varpi_1^{l-2k'}(x)\varpi_2^{-\frac{1}{k-1}}(t)\Pi(\varpi_1)\big[D_{t|T}^{\varrho_i}(\varpi_2(t))\big]^{k'}\,dxdt < T^{-(\varrho_i+d)k'+\frac{nd}{2}+1}. \tag{11.93}$$

From (11.81), (11.86), (11.87) and (11.93), we may write

$$\begin{aligned} \mathrm{M} &< \big[T^{-(\varrho_1+2)q'+\frac{nd}{2}+1}\big]^{\frac{1}{q'}} + \big[T^{-(\varrho_1+1-r_1)q'+\frac{nd}{2}+1}\big]^{\frac{1}{q'}} \\ &\quad + \big[T^{-(\varrho_1+d)q'+\frac{nd}{2}+1}\big]^{\frac{1}{q'}} + \big[T^{-\varrho_1 q'-\frac{(r_1-1)q'}{(r_1-1)q'-1}+\frac{nd}{2}+1}\big]^{\frac{1}{q'}}, \end{aligned} \tag{11.94}$$

and

$$\begin{aligned} \mathrm{N} &< \big[T^{-(\varrho_2+2)p'+\frac{nd}{2}+1}\big]^{\frac{1}{p'}} + \big[T^{-(\varrho_2+1-r_2)p'+\frac{nd}{2}+1}\big]^{\frac{1}{p'}} \\ &\quad + \big[T^{-(\varrho_2+d)p'+\frac{nd}{2}+1}\big]^{\frac{1}{p'}} + \big[T^{-\varrho_2 p'-\frac{(r_2-1)p'}{(r_2-1)p'-1}+\frac{nd}{2}+1}\big]^{\frac{1}{p'}}. \end{aligned} \tag{11.95}$$

Then we want to see that all the exponents of T on the right-hand side are negative. It is enough to guarantee

$$\begin{cases} -(\varrho_1+d)q' + \frac{nd}{2} + 1 < 0, \\ -\varrho_1 q' - \frac{(r_1-1)q'}{(r_1-1)q'-1} + \frac{nd}{2} + 1 < 0, \quad d > 0, \\ -(\varrho_2+d)p' + \frac{nd}{2} + 1 < 0, \\ -\varrho_2 p' - \frac{(r_2-1)p'}{(r_2-1)p'-1} + \frac{nd}{2} + 1 < 0, \quad d > 0. \end{cases} \tag{11.96}$$

From the first inequality, in each line, we get (we put $\zeta = \frac{nd}{2}$ and $k' = \frac{k}{k-1}$ with $k \in \{p,q\}$),

$$q' > \frac{\zeta+1}{\varrho_1 + \frac{2\zeta}{n}} \quad \text{and} \quad p' > \frac{\zeta+1}{\varrho_2 + \frac{2\zeta}{n}}.$$

Moreover, we have from the second inequality, in each line,

$$q' > \sqrt{\frac{\zeta+1}{\varrho_1(1-r_1)} + \left(\frac{\varrho_1-(1-r_1)\zeta}{2\varrho_1(1-r_1)}\right)^2} - \frac{\varrho_1-(1-r_1)\zeta}{2\varrho_1(1-r_1)},$$

$$p' > \sqrt{\frac{\zeta+1}{\varrho_2(1-r_2)} + \left(\frac{\varrho_2-(1-r_2)\zeta}{2\varrho_2(1-r_2)}\right)^2} - \frac{\varrho_2-(1-r_2)\zeta}{2\varrho_2(1-r_2)}.$$

Finally,

$$\begin{cases} q' > \max\{\frac{\zeta+1}{\varrho_1+\frac{2\zeta}{n}}; \sqrt{\frac{\zeta+1}{\varrho_1(1-r_1)} + (\frac{\varrho_1-(1-r_1)\zeta}{2\varrho_1(1-r_1)})^2} - \frac{\varrho_1-(1-r_1)\zeta}{2\varrho_1(1-r_1)}\}, \\ p' > \max\{\frac{\zeta+1}{\varrho_2+\frac{2\zeta}{n}}; \sqrt{\frac{\zeta+1}{\varrho_2(1-r_2)} + (\frac{\varrho_2-(1-r_2)\zeta}{2\varrho_2(1-r_2)})^2} - \frac{\varrho_2-(1-r_2)\zeta}{2\varrho_2(1-r_2)}\}. \end{cases}$$

We arrive to this condition

$$\begin{cases} \frac{q}{q-1} > \inf_{d>0} \max\{\frac{\frac{nd}{2}+1}{\varrho_1+d}; \sqrt{\frac{\frac{nd}{2}+1}{\varrho_1(1-r_1)} + (\frac{\varrho_1-(1-r_1)\frac{nd}{2}}{2\varrho_1(1-r_1)})^2} - \frac{\varrho_1-(1-r_1)\frac{nd}{2}}{2\varrho_1(1-r_1)}\}, \\ \frac{p}{p-1} > \inf_{d>0} \max\{\frac{\frac{nd}{2}+1}{\varrho_2+d}; \sqrt{\frac{\frac{nd}{2}+1}{\varrho_2(1-r_2)} + (\frac{\varrho_2-(1-r_2)\frac{nd}{2}}{2\varrho_2(1-r_2)})^2} - \frac{\varrho_2-(1-r_2)\frac{nd}{2}}{2\varrho_2(1-r_2)}\}. \end{cases} \tag{11.97}$$

These are conditions (11.61), (11.59) in the Theorem 11.4 for $\varrho_i = 1 - \gamma_i$ with $i = 1$ and $i = 2$. Then, from (11.94), (11.95) and (11.97), it follows that there exist:

$$\delta_1 = \delta_1(q, n, \varrho_1, r_1) > 0,$$

and

$$\delta_2 = \delta_2(p, n, \varrho_2, r_2) > 0,$$

such that

$$\begin{cases} \mathrm{M} \le T^{-\delta_1}, \\ \mathrm{N} \le T^{-\delta_2}. \end{cases} \tag{11.98}$$

From (11.75) and (11.98), we get

$$\begin{cases} [\int_{Q_T} |v|^p \tilde{\varpi} dxdt]^{1-\frac{1}{qp}} < T^{-(\frac{\delta_2}{q}+\delta_1)}, \\ [\int_{Q_T} |u|^q \tilde{\varpi} dxdt]^{1-\frac{1}{qp}} < T^{-(\frac{\delta_1}{p}+\delta_2)}. \end{cases} \tag{11.99}$$

Then $\delta = \frac{\delta_2}{q} + \delta_1$ and $\delta' = \frac{\delta_1}{p} + \delta_2$ in (11.76). We take the limit when $T \to +\infty$ in the right-hand side (11.99). Then the left-hand side will converge to zero with the fact that $pq > 1$ with

$$\lim_{T\to+\infty} \int_{Q_T} |v|^p \tilde{\varpi} dxdt = \int_{\mathbb{R}_+} \int_{\mathbb{R}^n} |v|^p \tilde{\varpi} dxdt,$$

$$\lim_{T\to+\infty} \int_{Q_T} |u|^q \tilde{\varpi} dxdt = \int_{\mathbb{R}_+} \int_{\mathbb{R}^n} |u|^q \tilde{\varpi} dxdt.$$

Finally, we may conclude that $(u, v) \equiv (0, 0)$. Then it is a contradiction with conditions (11.57) and (11.58).

1. For $r_1 = r_2 = 0$, we have $b_1 \equiv b_2 \equiv 0$. We will repeat the same way to estimate but in this time we arrive at $I_3^{i,k} = 0$. So, we get

$$\begin{cases} \mathrm{M} < [T^{-(\varrho_1+2)q'+\frac{nd}{2}+1}]^{\frac{1}{q'}} + [T^{-(\varrho_1+1)q'+\frac{nd}{2}+1}]^{\frac{1}{q'}} + [T^{-(\varrho_1+d)q'+\frac{nd}{2}+1}]^{\frac{1}{q'}}, \\ \mathrm{N} < [T^{-(\varrho_2+2)p'+\frac{nd}{2}+1}]^{\frac{1}{p'}} + [T^{-(\varrho_2+1)p'+\frac{nd}{2}+1}]^{\frac{1}{p'}} + [T^{-(\varrho_2+d)p'+\frac{nd}{2}+1}]^{\frac{1}{p'}}. \end{cases}$$

Then, with the same way, we may see the condition

$$\begin{cases} \frac{q}{q-1} > \inf_{d>0} \max\{\frac{\frac{nd}{2}+1}{\varrho_1+1}; \frac{\frac{nd}{2}+1}{\varrho_1+d}\} = \frac{\frac{n}{2}+1}{\varrho_1+1}, \\ \frac{p}{p-1} > \inf_{d>0} \max\{\frac{\frac{nd}{2}+1}{\varrho_2+1}; \frac{\frac{nd}{2}+1}{\varrho_2+d}\} = \frac{\frac{n}{2}+1}{\varrho_2+1}. \end{cases}$$

These are the conditions (11.62) and (11.60) with $\varrho_i = 1 - \gamma_i$ for $i = 1$ and $i = 2$. If we take the same way as the previous subsection, we arrive to $(u, v) \equiv (0, 0)$. Then it is a contradiction with conditions (11.57) and (11.58).

2. For $r_i \in (-1, 0)$, the functions

$$b_i = b_i(t) = (1 + t)^{r_i},$$

are decreasing. So, the strategy we will use is to estimate in an effective way the integrals $I_{(j+2)2}^{i,k}$ for $j = 0, 1$ such that

$$I_{(2+j)2}^{i,k} = \int_0^T |b_i^{(j)}(t)|^{k'} \varpi_2^{-\frac{1}{k-1}}(t) |D_{t|T}^{\varrho_i+1-j} \varpi_2(t)|^{k'} \, dt,$$

(see (11.82) and (11.83)) then we look for some way to estimate the integral (put $\tau = \frac{t}{T}$),

$$T^{1-(\varrho_i+1-j)k'} \int_0^1 (1 + \tau T)^{-(r_i+j)k'} (1 - \tau)^{-\frac{1}{k-1}} (1 - \tau)^{(l-\varrho_i-1+j)k'} \, d\tau.$$

For that, we split the integral in two integrals as follows:

$$\int_0^1 = \int_0^{T^{-\lambda_j}} + \int_{T^{-\lambda_j}}^1 .$$

We choose $\lambda_j \in (0, 1)$. If l is large, then we may estimate the integral (the integral is bounded) as follows:

$$\int_0^{T^{-\lambda_j}} (1+\tau T)^{-(r_i+j)k'}(1-\tau)^{-\frac{1}{k-1}}(1-\tau)^{(l-\varrho_i-1+j)k'}\,d\tau < CT^{-\lambda_j}, \quad j=0,1.$$

We use the second integral

$$(1+\tau T)^{-(r_i+j)k'} \le \left(1+T^{1-\lambda_j}\right)^{-(r_i+j)k'}, \quad j=0,1.$$

If l is large, then, for $j=0,1$, we may estimate

$$\begin{aligned}
&\int_{T^{-\lambda_j}}^{1} (1+\tau T)^{-(r_i+j)k'}(1-\tau)^{-\frac{1}{k-1}}(1-\tau)^{(l-\varrho_i-1+j)k'}\,d\tau \\
&\quad \le \left(1+T^{1-\lambda_j}\right)^{-(r_i+j)k'} \int_{T^{-\lambda_j}}^{1} (1-\tau)^{-\frac{1}{k-1}}(1-\tau)^{(l-\varrho_i-1)k'}\,d\tau \\
&\quad < \left(1+T^{1-\lambda_j}\right)^{-(r_i+j)k'}.
\end{aligned}$$

The optimal choice for λ_j is given by the condition (for large T)

$$T^{-\lambda_j} \sim \left(1+T^{1-\lambda_j}\right)^{-(r_i+j)k'}.$$

That is,

$$\lambda_j = (1-\lambda_j)(r_i+j)k'.$$

We find at the end

$$\lambda_j = \frac{k(r_i+j)}{k-1+k(r_i+j)}.$$

In addition, if we take that result with (11.79), we get the following estimate:

$$I_2^{i,k} + I_3^{i,k} < T^{-(\varrho_i+1)k' + \frac{kr_i}{k-1+k(r_i+j)} + \frac{nd}{2} + 1} + T^{-(\varrho_i)k' - \frac{k(1-r_i)}{k-1+k(1-r_i)} + \frac{nd}{2} + 1}.$$

Next, we use the estimates (11.81) and (11.93) to conclude that

$$\begin{aligned}
\mathrm{M} &< \left[T^{-(\varrho_1+2)q' + \frac{nd}{2} + 1}\right]^{\frac{1}{q'}} + \left[T^{-(\varrho_1+1)q' + \frac{qr_1}{q-1-qr_1} + \frac{nd}{2} + 1}\right]^{\frac{1}{q'}} \\
&\quad + \left[T^{-(\varrho_1+d)q' + \frac{nd}{2} + 1}\right]^{\frac{1}{q'}} + \left[T^{-\varrho_1 q' - \frac{(1-r_1)q}{q-1+(1-r_1)q} + \frac{nd}{2} + 1}\right]^{\frac{1}{q'}},
\end{aligned} \tag{11.100}$$

and

$$
\begin{aligned}
N < {} & \left[T^{-(\varrho_2+2)p'+\frac{nd}{2}+1}\right]^{\frac{1}{p'}} + \left[T^{-(\varrho_2+1)p'++\frac{pr_2}{p-1-pr_2}++\frac{nd}{2}+1}\right]^{\frac{1}{p'}} \\
& + \left[T^{-(\varrho_2+d)p'+\frac{nd}{2}+1}\right]^{\frac{1}{p'}} + \left[T^{-\varrho_2 p'-\frac{(r_2-1)p'}{(r_2-1)p'-1}+\frac{nd}{2}+1}\right]^{\frac{1}{p'}}.
\end{aligned} \tag{11.101}
$$

Now, we want to see all the exponents of T on the right-hand side are negative. It is enough to guarantee

$$
\begin{cases}
-(\varrho_1+d)q' + \frac{nd}{2} + 1 < 0, \\
-\varrho_1 q' - \frac{(r_1-1)q'}{(r_1-1)q'-1} + \frac{nd}{2} + 1 < 0, \quad d > 0, \\
-(\varrho_2+d)p' + \frac{nd}{2} + 1 < 0, \\
-\varrho_2 p' - \frac{(r_2-1)p'}{(r_2-1)p'-1} + \frac{nd}{2} + 1 < 0, \quad d > 0.
\end{cases} \tag{11.102}
$$

It is the same condition (11.96). Then it leads us to the same result and the proof is now completed. □

Bibliography

[1] V. I. Arnold. Mathematical Methods of Classical Mechanics. Springer-Verlag, New York, 1989.

[2] A. A. Kilbas, O. Marichev and S. Samko. Fractional Integrals and Derivatives (Theory and Applications). Gordon and Breach Science Publishers, Yverdon, Switzerland, 1993.

[3] H. D. A. Levine and G. Todorova. *Blow up of solutions of the Cauchy problem for a wave equation with nonlinear damping and source terms and positive initial energy*. Proc. Am. Math. Soc. 129 (3) (2000), 793–805.

[4] J. L. Lions. Quelques méthodes de résolution des problemes aux limites non lineaires. Dunod, Paris, 1969.

[5] M. Reed and B. Simon. Methods of Mathematical Physics III: Scattering Theory. Academic Press, New York, 1979.

https://doi.org/10.1515/9783111633787-012

Index

https://doi.org/10.1515/9783111633787-013

De Gruyter Series in Nonlinear Analysis and Applications

Volume 44
Yong Zhou, Jia Wei He
Nonautonomous Fractional Evolution Equations, 2024
ISBN 978-3-11-139057-4, e-ISBN (PDF) 978-3-11-139124-3, e-ISBN (EPUB) 978-3-11-139218-9

Volume 43
Ranis Ibragimov
Lie Group Analysis of Differential Equations. Invariant Solutions of Nonlinear Physical Phenomena, 2024
ISBN 978-3-11-138699-7, e-ISBN (PDF) 978-3-11-138749-9, e-ISBN (EPUB) 978-3-11-138780-2

Volume 44
Yong Zhou, Jia Wei He
Nonautonomous Fractional Evolution Equations, 2024
ISBN 978-3-11-139057-4, e-ISBN (PDF) 978-3-11-139124-3, e-ISBN (EPUB) 978-3-11-139218-9

Volume 43
Ranis Ibragimov
Lie Group Analysis of Differential Equations. Invariant Solutions of Nonlinear Physical Phenomena, 2024
ISBN 978-3-11-138699-7, e-ISBN (PDF) 978-3-11-138749-9, e-ISBN (EPUB) 978-3-11-138780-2

Volume 42
Catherine Bandle, Alfred Wagner
Shape Optimization. Variations of Domains and Applications, 2023
ISBN 978-3-11-102526-1, e-ISBN (PDF) 978-3-11-102543-8, e-ISBN (EPUB) 978-3-11-102581-0

Volume 41
Jesús Garcia-Falset, Khalid Latrach
Nonlinear Functional Analysis and Applications, 2023
ISBN 978-3-11-103096-8, e-ISBN (PDF) 978-3-11-103181-1, e-ISBN (EPUB) 978-3-11-103208-5

Volume 40
Dung Le
Cross Diffusion Systems. Dynamics, Coexistence and Persistence, 2022
ISBN 978-3-11-079498-4, e-ISBN (PDF) 978-3-11-079513-4, e-ISBN (EPUB) 978-3-11-079517-2

Kolibri 360 GmbH
Ferdinand-Jühlke-Straße 7
99095 Erfurt, Deutschland
produktsicherheit@kolibri360.de